# Alternative Transportation

# FUELS

## Utilisation in Combustion Engines

# Alternative Transportation
# FUELS

## Utilisation in Combustion Engines

## M. K. Gajendra Babu
## K. A. Subramanian

**CRC Press**
Taylor & Francis Group
Boca Raton   London   New York

CRC Press is an imprint of the
Taylor & Francis Group, an **informa** business

CRC Press
Taylor & Francis Group
6000 Broken Sound Parkway NW, Suite 300
Boca Raton, FL 33487-2742

First issued in paperback 2017

© 2013 by Taylor & Francis Group, LLC
CRC Press is an imprint of Taylor & Francis Group, an Informa business

No claim to original U.S. Government works

Version Date: 20130305

ISBN 13: 978-1-138-07666-2 (pbk)
ISBN 13: 978-1-4398-7281-9 (hbk)

---

**Library of Congress Cataloging-in-Publication Data**

---

Babu, M. K. Gajendra.
　　Alternative transportation fuels : utilisation in combustion engines / M.K. Gajendra Babu, K.A. Subramanian.
　　　pages cm
　　Summary: "Co-authored by a renowned researcher in energy studies, this book discusses the production, distribution, and applications of alternative fuels with respect to different modes of transportation (land, sea, and aviation). Much of the emphasis is on applications of these fuels in internal combustion engines. The authors provide an in-depth analysis of engine combustion, including injection, spray, combustion, performance, and emissions. The book also highlights greenhouse gases (GHGs), in view of climate change and global warming"-- Provided by publisher.
　　Includes bibliographical references and index.
　　ISBN 978-1-4398-7281-9 (hardback)
　　1. Motor fuels. 2. Fuel switching. 3. Spark ignition engines--Alternative fuels. 4. Diesel motor--Alternative fuels. I. Subramanian, K. A., 1971- II. Title.

TP343.B25 2013
662.6--dc23

2013007333

---

**Visit the Taylor & Francis Web site at**
**http://www.taylorandfrancis.com**

**and the CRC Press Web site at**
**http://www.crcpress.com**

# Contents

# *Preface*

During the past couple of decades, there has been a large expansion in the transport sector, which resulted in a significant increase in the consumption of petroleum-based fuels such as gasoline and diesel. This situation is likely to pave the way for the depletion of fossil fuel–based reserves and to deteriorate the quality of the environment. Hence, there exists a definite need to stem this problem to the maximum possible extent by exploring the feasibility of using alternative fuels that could pave the way for the sustained operation of the transport sector. In this direction, this book exposes the reader to the assessment of the potential avenues that could be contemplated for using different alternative fuels in the transport sector.

Chapter 1 briefly highlights several modes of transport and their effect on the environment, while Chapters 2 and 3 discuss conventional and alternative fuels for land transport. Fuels for the aviation sector are covered in Chapter 4. Experimental investigations relating to the utilisation of alternative fuels in internal combustion engines are reported in Chapter 5. Fuel quality characterisation and a modelling of alternative-fuelled engines are briefly highlighted in Chapters 6 and 7, respectively. Chapter 8 briefly describes alternative-powered vehicles. Potential alternative fuels for rail, marine and aviation applications are presented in Chapters 9, 10 and 11, respectively. Chapter 12 highlights potential global warming and climate change on account of utilising conventional and alternative fuels. Some of the material in this book is based on the authors' own experience at different laboratories around the globe.

We are indeed grateful to the College of Engineering, Guindy, Chennai; Indian Institute of Technology, Madras; Indian Institute of Technology, Delhi; Indian Institute of Petroleum, Dehradun; University of Tokyo; University of Melbourne; University of Manchester Institute of Science and Technology; and Hosei University for providing the necessary facilities for us to undertake some of the research activities indicated in this book.

We wish to acknowledge the support extended by our teachers and former colleagues who motivated us in our early stages. Notable among them are Professors B.S. Murthy, T. Asanuma, T. Obokata, H.C. Watson, D. Winterbone, Satoshi Okajima, A. Ramesh, P.A. Janakiraman, and T.R. Jagadeesan and Drs. P.A. Laksminarayanan and K. Kumar.

The contributions made by our former research scholars Drs. D.S. Khatri, Alok Kumar, P.G. Tewari, K. Subba Reddi, and Ragupathy during their PhD programs and our MTech scholars Silesh, Kavitha Kuppula, Supriya Sukla, Jaspreet Hira, and Ameet Srivastave have been used in some chapters. We are grateful for the sponsorship provided by the Department of Science and Technology under the Funds for Improvement in Science and

Technology (FIST) as some of the results generated under this scheme are used in the book.

We thank Vinay C. Mathad, project associate, for his hard work in bringing the book to a proper shape. His contribution to the book is invaluable as he devoted a considerable amount of time to draw figures, format manuscripts and develop numerical problems.

We acknowledge our research scholars Subhash Lahane, Venkateshwaralu, B.L. Salvi, Ashok Kumar, Sunmeet Singh and MTech scholars Charan, Vaibav Vasuntre, Ram Kumar Bhardan, Apporva Milind Moon and Navin Shukla for their contribution in searching the relevant literature for the book.

Our sincere thanks to the Centre for Energy Studies, IIT, Delhi, for encouraging us to write this book. We thank the management of RMK Engineering College, Chennai, for permitting us to use their facilities during the preparation of this book.

Finally, we wish to thank our family members for extending their support during the preparation of this book.

**M.K. Gajendra Babu**
**K.A. Subramanian**

# *Authors*

**M.K. Gajendra Babu** is a senior professor at RMK Engineering College, Chennai after retiring from the Indian Institute of Technology (IIT), Delhi, where he had served as professor and head of the Centre for Energy Studies. He was also a Henry Ford Chair Professor at IIT Madras and a visiting faculty at the University of Tokyo, University of Melbourne, University of Manchester, and University of Hosei, Japan.

Dr. Babu has been working in the field of computer simulation, alternative fuels, instrumentation, and emission controls for internal combustion engines for the past 44 years.

Dr. Babu has published about 250 research papers in several national and international journals and conferences. He is a Fellow of the Society of Automotive Engineers (SAE) International. He has been awarded the Indian Automobile Engineer of the Year award, the Indian Society of Technical Education's Anna University Outstanding Academic Award, the Indian Society of Environment's honorary fellowships from A. P. J. Abdul Kalam and the SAE India Foundation's Automotive Education Award for his outstanding contribution to automotive education in India.

**K.A. Subramanian** is an associate professor in the Centre for Energy Studies, Indian Institute of Technology (IIT), Delhi. He is a former scientist in the Indian Institute of Petroleum.

Dr. Subramanian's research area includes utilisation of alternative fuels (biodiesel, compressed natural gas, hydrogen, etc.) in internal combustion engines, the development of the homogeneous charge compression ignition (HCCI) concept engine, greenhouse control in transport engines, sustainable power generation using a hybrid energy system, computer simulation, and computational fluid dynamics (CFD). He is involved in several R&D projects, including the development of a biodiesel–CNG-based dual-fuel diesel engine, the utilisation of enriched biogas in automotive vehicles, and hydrogen utilisation in a multicylinder spark ignition engine. A patent application has been filed in his name.

Dr. Subramanian has jointly supervised three doctoral scholars and supervised about 10 PhD scholars. He was nominated to participate in the project Study Mission on Energy Efficiency, sponsored by the Asian Productivity Organization, Japan in 2009.

# Abbreviations

| | |
|---|---|
| AC | Alternate current |
| AFVs | Alternate fuel vehicles |
| AK | Antiknock |
| $Al_2O_3$ | Alumina |
| ASTM | American Standards Testing Material |
| ATR | Auto thermal reforming |
| B0 (or) D100 | Diesel |
| B20 | 20% Biodiesel + 80% diesel |
| B40 | 40% Biodiesel + 60% diesel |
| B60 | 60% Biodiesel + 40% diesel |
| B80 | 80% Biodiesel + 20% diesel |
| BD | Biodiesel |
| BIS | Bureau of Indian Standards |
| BMEP | Brake mean effective pressure |
| BP | Brake power |
| BSEC | Brake specific energy consumption |
| BSFC | Brake specific fuel consumption |
| BTL | Biomass to liquid |
| Btoe | Billion tonnes of oil equivalent |
| $C_2H_5OH$ | Ethanol |
| $C_3H_6O_3$ | Dimethyl carbonate |
| $C_3H_7OH$ | Propanol |
| $C_4H_9OH$ | Butanol |
| CAFE | Corporate average fuel economy |
| CAGR | Compound annual growth rate |
| CDM | Clean development mechanism |
| CFC | Chloro fluoro carbons |
| CFR | Cooperative Fuel Research |
| $CH_3OH$ | Methanol |
| $CH_4$ | Methane |
| CI | Compression ignition |
| CI engine | Compression ignition engine |
| CN | Cetane number |
| CNG | Compressed natural gas |
| Co | Cobalt |
| CO | Carbon monoxide |
| $CO_2$ | Carbon dioxide |
| CR | Compression ratio |
| CRC | Coordinating Research Council |
| CRDI | Common rail direct injection |

| | |
|---|---|
| CTL | Coal to liquid |
| Cu | Copper |
| CVS | Constant volume sample |
| °C | Degree Celsius |
| DC | Direct current |
| DC | Discounted cost |
| DEE | Diethyl ether |
| DICI | Direct injected compression ignition engine |
| DISI | Direct injection spark ignition engine |
| DME | Dimethyl ether or Di methyl ether |
| E10 | Ethanol 10% |
| E15 | 15% Ethanol + 85% gasoline |
| E30 | 30% Ethanol + 70% gasoline |
| E70 | 70% Ethanol + 30% gasoline |
| E85 | Ethanol 85% |
| E95 | Ethanol 95% |
| EAMA | European Automobile Manufacturer's Association |
| EGR | Exhaust gas recirculation |
| EPEEE | European Program on Emissions, Fuels and Engine Technologies |
| EU | European Union |
| EU27 | European Union of 27 member states |
| EV | Electrical vehicles |
| FBP | Final boiling point |
| FC | Fuel cell |
| FCC | Fluidised catalytic converter |
| FCV | Fuel cell vehicles |
| Fe | Iron |
| FFA | Free fatty acids |
| Fp | Finished product |
| Fpe | Finished products to the end user |
| FRJ | Fermentation renewable jet |
| FSU | Former Soviet Union |
| FT | Fischer–Tropsch |
| F–T Diesel | Fisher–Tropsch diesel |
| FTS | Fischer–Tropsch synthesis |
| g/km | Gram per kilometre |
| g/kWh | Gram per kilowatt hour |
| G-8 | Group of eight |
| GDI | Gasoline direct ignition |
| GDP | Gross domestic product |
| GHGs | Green house gases |
| GTL | Gas to liquid |
| GWP | Global warming potential |
| $H_2$ | Hydrogen |
| $H_2O$ | Water |

| | |
|---|---|
| HC | Hydrocarbons |
| HCV | Higher commercial vehicles |
| HDS | Hydro desulphurisation |
| HDT | Hydro treating |
| HEV | Hybrid electric vehicle |
| Hp | Horse power |
| HRJ | Hydro-treated renewable jet |
| HVAC | Heating ventilation and air conditioning |
| IATA | International Airline Industry Association |
| IBP | Initial boiling point |
| IC | Internal combustion |
| ICAO | International Civil Aviation Organization |
| ICE | Internal combustion engine |
| IFO | Intermediate fuel oil |
| IMEP | Indicated mean effective pressure |
| IPCC | Intergovernmental Panel on Climate Change |
| IT | Injection timing |
| JASO | Japanese Automobile Standard Organization |
| K | Kelvin |
| kg | Kilogram |
| km | Kilometre |
| km/h | Kilometre per hour |
| KOH | Potassium hydroxide |
| kW | Kilowatt |
| Lb-ft | Pounds-foot |
| LCV | Light commercial vehicles |
| $LH_2$ | Liquefied hydrogen |
| LNG | Liquefied natural gas |
| LOME | Line seed oil methyl ester |
| LPG | Liquefied petroleum gas |
| $m^3$ | Cubic metre |
| M15 | 15% Methanol + 85% gasoline |
| M85 | 85% Methanol + 15% gasoline |
| MAP | Manifold absolute pressure |
| MEA | Mono-ethanol amine |
| MJ | Mega Joule |
| mm | Millimetre |
| MMT | Million metric tons |
| MON | Motor octane number |
| Mpg | Miles per gallon |
| Mph | Miles per hour |
| MT | Million ton |
| MTBE | Methyl tetra butyl ether |
| MTOE | Million ton oil equivalent |
| MUV | Multi-utility vehicle |

| | |
|---|---|
| $N_2$ | Nitrogen |
| $N_2O$ | Nitrous oxide |
| NaOH | Sodium hydroxide |
| NASA | National Aeronautical Space Association |
| NEDC | New European Driving Cycle |
| NGV | Natural gas vehicles |
| $NH_3$ | Ammonia |
| $NO_x$ | Oxides of nitrogen |
| $O_3$ | Ozone |
| OC | Operating fuel cost |
| OECD | Organization for Economic Co-operation and Development |
| OEM | Original Engine Manufacturer |
| ON | Octane number |
| PAH | Poly aromatic hydrocarbon |
| PAHC | Polycyclic aromatic hydrocarbon |
| PISI | Port injected spark ignition engine |
| PJ | Pico Joule |
| PM | Particulate matter |
| POX | Partial oxidation |
| ppm | Parts per million |
| psi | Pounds square inch |
| PV | Photo voltaic |
| R&D | Research & Development |
| RFO | Refined fuel oil |
| Rm | Raw material |
| RON | Research octane number |
| RPK | revenue passenger kilometre |
| rpm | Revolutions per minute |
| s | Seconds |
| SCR | Selective catalytic reduction |
| SEC | Specific energy consumption |
| SI | Spark ignition |
| SI engine | Spark ignition engine |
| $SiO_2$ | Silicon oxide |
| SIT | Self-ignition temperature |
| SMD | Sauter mean diameter |
| SMR | Steam methane reforming |
| $SO_x$ | Sulphur oxides |
| SPK | Synthetic paraffinic kerosene |
| SRM | Steam reforming method |
| SUV | Sports utility vehicles |
| TAN | Total acid number |
| TBN | Total base number |
| TC | Transportation cost |
| $T_{EG}$ | Exhaust temperature |

| | |
|---|---|
| TFp | Transportation cost of finished product from downstream of industries to retailer |
| TFpe | Transportation cost of finished product from retailer to end user |
| TRm | Transportation cost of raw material |
| TWe | Transportation cost of waste effluent |
| UBHC | Unburnt hydrocarbon |
| UC | Utility cost |
| UHC/UBHC | Unburnt hydrocarbons |
| UKCCC | United Kingdom Committee on Climate Change |
| US/USA | United States/United States of America |
| We | Waste effluent |
| Wh/kg | Watt-hour/kilogram |
| Wh/L | Watt-hour/litre |
| WOT | wide open throttle |
| WTW | Well to wheel |
| wt/wt | weight by weight |
| XTL | Anything to liquid (synthesis fuel) |
| ZEV | Zero emission vehicles |
| ZnO | Zinc oxide |
| $ZnO_2$ | Zinc dioxide |

# 1

# Introduction: Land, Sea and Air Transportation

## 1.1 Transportation

Transportation can be defined as the movement of people, livestock and all types of goods from one place to another through the use of self-power, manpower, motor power or combinations of any two or all of the above. Nature is a good example of movement as planets rotate in their orbit around the sun. Six billion people in the world have to journey from their origin/home/ house to other places for their education, office work, industry and other general purposes. The materials normally transported include commodities such as food items and industrial goods from the origin of production to the customer's destination for the survival of human life.

A raw material has to be moved from its source to end-users as depicted in Figure 1.1. Raw materials ($Rm_1$, $Rm_2$, $Rm_3$, ..., $Rm_n$) have to be transported from their place of origin to upstream of an industry. The finished products ($Fp_1$, $Fp_2$, $Fp_3$, ..., $Fp_n$) have to be transported from downstream of an industry to the retailer's end and then to the end-users ($Fpe_1$, $Fpe_2$, $Fpe_3$, ..., $Fpe_n$). The waste effluent ($We_1$, $We_2$, $We_3$, ..., $We_n$) from the industry needs to be disposed of to a safe place in view of its environmental concern. Thus, the transportation cycle of consumer goods is completed in this manner for most of the industries such as cement, paper, sugar and petrochemical. Total transportation cost could be written as a summation of all transportation costs such as raw materials, finished product to retailer, retailer to end-users and waste effluent disposal, as shown in Equation 1.1

$$\text{Total transport cost} \sum_{i}^{n} TRm_i = \sum_{i}^{n} TFp_i \sum_{i}^{n} TFpe_i \sum_{i}^{n} TWe_i \qquad (1.1)$$

where $i = 1,2,3,4,..., n$.

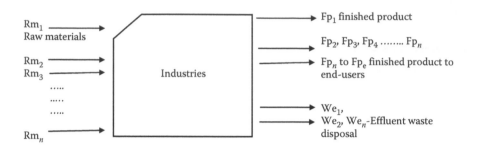

**FIGURE 1.1**
Schematic diagram of transportation of raw materials to end-users.

## 1.2 Modes of Different Transportation

A line diagram showing different modes of transportation is given in Figure 1.2. Two-, three- and four-wheelers are used for personal transportation for travelling short distances, whereas air mode transportation is used for longer distances. In case of mass transportation, heavy goods are transported using land mode transportation by internal combustion engine-powered vehicles and locomotives. Air mode is used for light goods and it is the fastest service that is preferred by all sectors. However, it is an expensive mode of transportation that is used in situations depending on the emergency or time-bound activities. Even though sea mode transportation is the cheapest as compared to other modes, it is only possible for ocean-bounded countries.

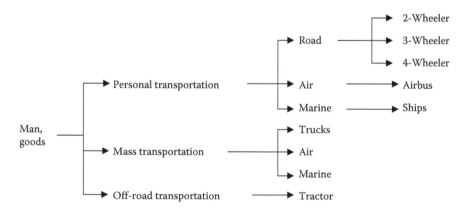

**FIGURE 1.2**
Line diagram of different modes of transportation.

## 1.3 Indigenous Production Levels of Crude Oil from Different Countries

The indigenous production of crude oil, shale oil, oil sands and NGLs (the liquid content of natural gas from where this is recovered separately) from different countries is given in Table 1.1. A comparison of the production of crude oil from different countries is shown in Figure 1.3.

## 1.4 Import and Export Levels in Different Countries

The production and import scenarios of the United States, China and India are shown in Figure 1.4a, b and c. The United States imports 68% of its crude oil requirement (Figure 1.4a), whereas India imports 79% of its crude oil requirement from other countries, as shown in Figure 1.4c. Trade movements of crude oil from different countries are shown in Table 1.2. Details of India's crude oil production and import from other countries are shown in Figure 1.5.

- Europe's energy deficit remains roughly at today's levels for oil and coal but increases by 65% for natural gas (Figure 1.6). This is matched by gas production growth in the former Soviet Union (FSU) [4].
- Among energy-importing regions, North America is an exception, with growth in biofuel supplies and unconventional oil and gas turning today's energy deficit (mainly oil) into a small surplus by 2030.
- In aggregate, today's energy importers will need to import 40% more in 2030 than they do today, with deficits in Europe and Asia Pacific met by supply growth in the Middle East, the FSU, Africa and South and Central America.
- China's energy deficit increases by 0.8 Btoe (billion tonnes of oil equivalent, spread across all fuels) while India's import requirement grows by 0.4 Btoe (mainly oil and coal). The rest of Asia Pacific remains a big oil importer at similar levels to today.
- Asian energy requirements are partially met by increased Middle East and African production, but the rebalancing of global energy trade as a result of the improved net position in the Americas is also a key factor [4].
- Import dependency, measured as the share of demand met by net imports, increases for most major energy importers except the United States (Figure 1.7).

**TABLE 1.1**

Production of Crude Oil, Shale Oil, Oil Sands and NGLs (Excludes Liquid Fuels from Other Sources Such as Biomass and Coal Derivatives)

| Production (Million Tonnes) | 2000 | 2001 | 2002 | 2003 | 2004 | 2005 | 2006 | 2007 | 2008 | 2009 | 2010 |
|---|---|---|---|---|---|---|---|---|---|---|---|
| United States | 352.6 | 349.2 | 346.8 | 338.4 | 329.2 | 313.3 | 310.2 | 309.8 | 304.9 | 328.6 | 339.1 |
| Canada | 126.9 | 126.1 | 135.0 | 142.6 | 147.6 | 144.9 | 153.4 | 158.3 | 156.8 | 156.1 | 162.8 |
| Mexico | 171.2 | 176.6 | 178.4 | 188.8 | 190.7 | 187.1 | 183.1 | 172.7 | 157.7 | 147.5 | 146.3 |
| Argentina | 40.4 | 41.5 | 40.9 | 40.2 | 37.8 | 36.2 | 35.8 | 34.9 | 34.1 | 33.8 | 32.5 |
| Brazil | 63.2 | 66.3 | 74.4 | 77.0 | 76.5 | 84.6 | 89.2 | 90.4 | 93.9 | 100.4 | 105.7 |
| Colombia | 35.3 | 31.0 | 29.7 | 27.9 | 27.3 | 27.3 | 27.5 | 27.6 | 30.5 | 34.1 | 39.9 |
| Ecuador | 20.9 | 21.2 | 20.4 | 21.7 | 27.3 | 27.6 | 27.7 | 26.5 | 26.2 | 25.2 | 25.2 |
| Peru | 4.9 | 4.8 | 4.8 | 4.5 | 4.4 | 5.0 | 5.1 | 5.1 | 5.3 | 6.4 | 6.9 |
| Trinidad and Tobago | 6.8 | 6.5 | 7.5 | 7.9 | 7.3 | 8.3 | 8.3 | 7.2 | 6.9 | 6.8 | 6.5 |
| Venezuela | 167.3 | 161.6 | 148.8 | 131.4 | 150.0 | 151.0 | 144.2 | 133.9 | 131.5 | 124.8 | 126.6 |
| Other South and Central America | 6.6 | 6.9 | 7.8 | 7.8 | 7.3 | 7.2 | 7.0 | 7.1 | 7.0 | 6.7 | 6.6 |
| Azerbaijan | 14.1 | 15.0 | 15.4 | 15.5 | 15.6 | 22.4 | 32.5 | 42.8 | 44.7 | 50.6 | 50.9 |
| Denmark | 17.7 | 17.0 | 18.1 | 17.9 | 19.1 | 18.4 | 16.7 | 15.2 | 14.0 | 12.9 | 12.2 |
| Italy | 4.6 | 4.1 | 5.5 | 5.6 | 5.5 | 6.1 | 5.8 | 5.9 | 5.2 | 4.6 | 5.1 |
| Kazakhstan | 35.3 | 40.1 | 48.2 | 52.4 | 60.6 | 62.6 | 66.1 | 68.4 | 72.0 | 78.2 | 81.6 |
| Norway | 160.2 | 162.0 | 157.3 | 153.0 | 149.9 | 138.2 | 128.7 | 118.6 | 114.2 | 108.8 | 98.6 |
| Romania | 6.3 | 6.2 | 6.1 | 5.9 | 5.7 | 5.4 | 5.0 | 4.7 | 4.7 | 4.5 | 4.3 |
| Russian Federation | 323.3 | 348.1 | 379.6 | 421.4 | 458.8 | 470.0 | 480.5 | 491.3 | 488.5 | 494.2 | 505.1 |
| Turkmenistan | 7.2 | 8.0 | 9.0 | 10.0 | 9.6 | 9.5 | 9.2 | 9.8 | 10.3 | 10.4 | 10.7 |
| United Kingdom | 126.2 | 116.7 | 115.9 | 106.1 | 95.4 | 84.7 | 76.6 | 76.8 | 71.7 | 68.2 | 63.0 |
| Uzbekistan | 7.5 | 7.2 | 7.2 | 7.1 | 6.6 | 5.4 | 5.4 | 4.9 | 4.8 | 4.5 | 3.7 |
| Other Europe and Eurasia | 22.4 | 22.2 | 23.6 | 24.0 | 23.5 | 22.0 | 21.7 | 21.6 | 20.6 | 19.6 | 18.2 |
| Iran | 191.3 | 191.4 | 180.9 | 203.7 | 207.8 | 206.3 | 208.2 | 209.7 | 209.9 | 201.5 | 203.2 |

| | | | | | | | | | | | |
|---|---|---|---|---|---|---|---|---|---|---|---|
| Iraq | 128.8 | 123.9 | 104.0 | 66.1 | 100.0 | 90.0 | 98.1 | 105.2 | 119.5 | 119.8 | 120.4 |
| Kuwait | 109.1 | 105.8 | 98.2 | 114.8 | 122.3 | 129.3 | 132.7 | 129.9 | 137.2 | 121.7 | 122.5 |
| Oman | 46.4 | 46.1 | 43.4 | 39.6 | 38.1 | 37.4 | 35.7 | 34.5 | 35.9 | 38.7 | 41.0 |
| Qatar | 36.1 | 35.7 | 35.2 | 40.8 | 46.0 | 47.3 | 50.9 | 53.6 | 60.8 | 57.9 | 65.7 |
| Saudi Arabia | 456.3 | 440.6 | 425.3 | 485.1 | 506.0 | 526.8 | 514.3 | 494.2 | 515.3 | 464.7 | 467.8 |
| Syria | 27.3 | 28.9 | 27.2 | 26.2 | 24.7 | 22.4 | 21.6 | 20.6 | 19.8 | 18.6 | 19.1 |
| United Arab Emirates | 122.1 | 118.0 | 110.2 | 124.5 | 131.7 | 137.3 | 145.5 | 140.7 | 142.9 | 126.3 | 130.8 |
| Yemen | 21.3 | 21.5 | 21.5 | 21.1 | 19.9 | 19.6 | 17.9 | 16.3 | 14.4 | 13.5 | 12.5 |
| Other Middle East | 2.2 | 2.2 | 2.2 | 2.2 | 2.2 | 1.6 | 1.4 | 1.6 | 1.5 | 1.7 | 1.7 |
| Algeria | 66.8 | 65.8 | 70.9 | 79.0 | 83.6 | 86.4 | 86.2 | 86.5 | 85.6 | 77.9 | 77.7 |
| Angola | 36.9 | 36.6 | 44.6 | 42.8 | 54.5 | 69.0 | 69.6 | 82.5 | 92.2 | 87.4 | 90.7 |
| Chad | — | — | — | 1.2 | 8.8 | 9.1 | 8.0 | 7.5 | 6.7 | 6.2 | 6.4 |
| Brazzaville | 13.1 | 12.1 | 12.3 | 11.2 | 11.6 | 12.6 | 14.3 | 11.7 | 12.4 | 13.9 | 15.1 |
| Egypt | 38.8 | 37.3 | 37.0 | 36.8 | 35.4 | 33.9 | 33.7 | 34.1 | 34.6 | 35.3 | 35.0 |
| Equatorial Guinea | 4.5 | 8.8 | 11.4 | 13.2 | 17.4 | 17.7 | 16.9 | 17.3 | 17.2 | 15.2 | 13.6 |
| Gabon | 16.4 | 15.0 | 14.7 | 12.0 | 11.8 | 11.7 | 11.7 | 11.5 | 11.8 | 11.5 | 12.2 |
| Libya | 69.5 | 67.1 | 64.6 | 69.8 | 76.5 | 81.9 | 84.9 | 85.0 | 85.3 | 77.1 | 77.5 |
| Nigeria | 105.4 | 110.8 | 102.3 | 109.3 | 119.0 | 122.1 | 117.8 | 112.1 | 103.0 | 99.1 | 115.2 |
| Sudan | 8.6 | 10.7 | 11.9 | 13.1 | 14.9 | 15.0 | 16.3 | 23.1 | 23.7 | 23.6 | 23.9 |
| Tunisia | 3.7 | 3.4 | 3.5 | 3.2 | 3.4 | 3.4 | 3.3 | 4.6 | 4.2 | 4.0 | 3.8 |
| Other Africa | 7.2 | 6.6 | 6.7 | 6.8 | 8.1 | 7.7 | 7.6 | 8.3 | 8.1 | 7.7 | 7.1 |
| Australia | 35.3 | 31.8 | 31.5 | 26.6 | 24.8 | 24.5 | 23.2 | 23.5 | 23.7 | 21.9 | 23.8 |
| Brunei | 9.4 | 9.9 | 10.2 | 10.5 | 10.3 | 10.1 | 10.8 | 9.5 | 8.5 | 8.2 | 8.4 |
| China | 162.6 | 164.8 | 166.9 | 169.6 | 174.1 | 181.4 | 184.8 | 186.3 | 190.4 | 189.5 | 203.0 |
| India | 34.2 | 34.1 | 35.2 | 35.4 | 36.3 | 34.6 | 35.8 | 36.1 | 36.1 | 35.4 | 38.9 |
| Indonesia | 71.5 | 67.9 | 63.0 | 57.3 | 55.2 | 53.1 | 48.9 | 47.5 | 49.0 | 47.9 | 47.8 |
| Malaysia | 33.7 | 32.9 | 34.5 | 35.6 | 36.5 | 34.4 | 33.5 | 34.2 | 34.6 | 33.1 | 32.1 |
| Thailand | 7.0 | 7.5 | 8.2 | 9.6 | 9.1 | 10.8 | 118 | 12.5 | 13.3 | 13.7 | 13.8 |
| Vietnam | 16.2 | 17.1 | 17.3 | 17.7 | 20.8 | 19.4 | 17.8 | 16.4 | 15.4 | 16.8 | 18.0 |
| Other Asia Pacific | 9.4 | 9.1 | 9.0 | 9.1 | 10.5 | 12.5 | 13.2 | 13.9 | 14.7 | 14.3 | 13.6 |
| Total world | 3611.8 | 3601.6 | 3584.2 | 3701.1 | 3877.0 | 3906.6 | 3916.2 | 3904.3 | 3933.7 | 3831.0 | 3913.7 |

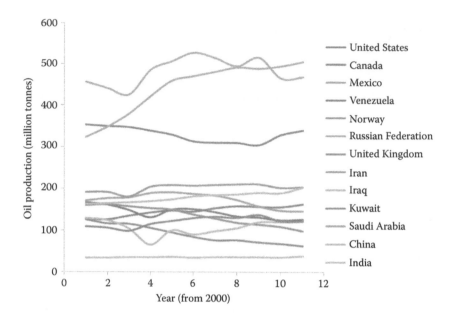

**FIGURE 1.3**
Comparison of production of crude oil from different countries.

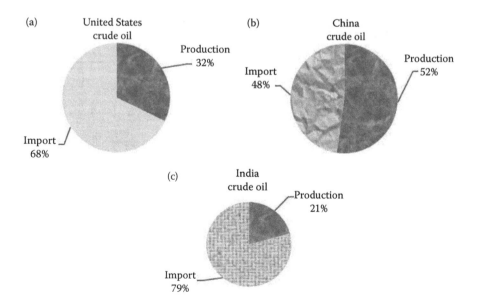

**FIGURE 1.4**
Production and import scenarios of crude oil in the (a) USA, (b) China and (c) India. (Adapted from IEA World Energy Outlook, www.iea.org, 2009; www.petroleum.nic.)

**TABLE 1.2**

Trade Movements of Crude Oil Worldwide

| Year | 2000 | 2001 | 2002 | 2003 | 2004 | 2005 | 2006 | 2007 | 2008 | 2009 | 2010 |
|---|---|---|---|---|---|---|---|---|---|---|---|
| *Imports (Thousand Barrels Daily)* | | | | | | | | | | | |
| United States | 11,092 | 11,618 | 11,357 | 12,254 | 12,898 | 13,525 | 13,612 | 13,632 | 12,872 | 11,453 | 11,689 |
| Europe | 11,070 | 11,531 | 11,895 | 11,993 | 12,538 | 13,261 | 13,461 | 13,953 | 13,751 | 12,486 | 12,094 |
| Japan | 5329 | 5202 | 5070 | 5314 | 5203 | 5225 | 5201 | 5032 | 4925 | 4263 | 4567 |
| Rest of the world | 15,880 | 16,436 | 16,291 | 17,191 | 18,651 | 19,172 | 20,287 | 22,937 | 23,078 | 24,132 | 25,160 |
| Total world | 43,371 | 44,787 | 44,613 | 46,752 | 49,290 | 51,182 | 52,561 | 55,554 | 54,626 | 52,333 | 53,510 |
| *Exports (Thousand Barrels Daily)* | | | | | | | | | | | |
| United States | 890 | 910 | 904 | 921 | 991 | 1129 | 1317 | 1439 | 1967 | 1947 | 2154 |
| Canada | 1703 | 1804 | 1959 | 2096 | 2148 | 2201 | 2330 | 2457 | 2498 | 2518 | 2599 |
| Mexico | 1814 | 1882 | 1966 | 2115 | 2070 | 2065 | 2102 | 1975 | 1609 | 1449 | 1539 |
| South and Central America | 3079 | 3143 | 2965 | 2942 | 3233 | 3528 | 3681 | 3570 | 3616 | 3748 | 3568 |
| Europe | 1967 | 1947 | 2234 | 2066 | 1993 | 2149 | 2173 | 2273 | 2023 | 2034 | 1888 |
| Former Soviet Union | 4273 | 4679 | 5370 | 6003 | 6440 | 7076 | 7155 | 8334 | 8184 | 7972 | 8544 |
| Middle East | 18,944 | 19,098 | 18,062 | 18,943 | 19,630 | 19,821 | 20,204 | 19,680 | 20,128 | 18,409 | 18,883 |
| North Africa | 2732 | 2724 | 2620 | 2715 | 2917 | 3070 | 3225 | 3336 | 3260 | 2938 | 2871 |
| West Africa | 3293 | 3182 | 3134 | 3612 | 4048 | 4358 | 4704 | 4830 | 4587 | 4364 | 4601 |
| Asia Pacific | 3736 | 3914 | 3848 | 3978 | 4189 | 4243 | 4312 | 6004 | 5392 | 5631 | 6226 |
| Rest of the world | 940 | 1506 | 1551 | 1361 | 1631 | 1542 | 1359 | 1656 | 1363 | 1323 | 637 |
| Total world | 43,371 | 44,789 | 44,613 | 46,752 | 49,290 | 51,182 | 52,561 | 55,554 | 54,626 | 52,333 | 53,510 |

*Source:* Adapted from BP Statistical Review of World Energy, http://www.bp.com/statistical review, June 2011.

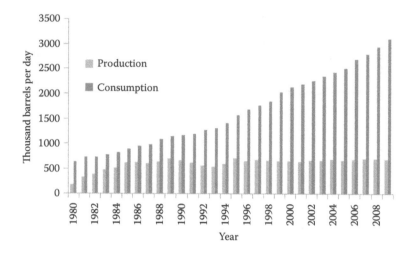

**FIGURE 1.5**
India's crude oil production and import from other countries. (Adapted from www.petroleum.nic.)

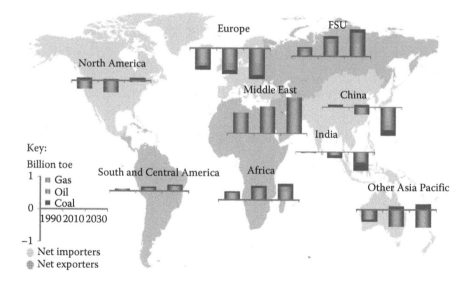

**FIGURE 1.6**
Net imports and exports of gas, oil and coal—world view. (Adapted from BP Energy Outlook 2030, London, January 2012.)

- The import share of oil demand and the volume of oil imports in the United States will fall below the 1990s levels, largely due to the rising production of domestic shale oil and ethanol, displacing crude imports. The United States also becomes a net exporter of natural gas.

- In China, imports of oil and natural gas rise sharply as the growth in demand outpaces domestic supply. Oil continues to dominate

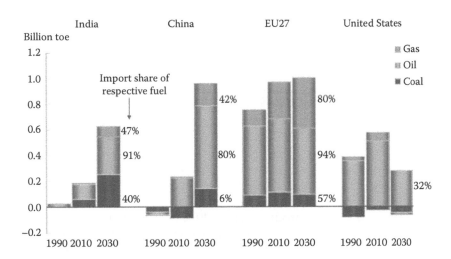

**FIGURE 1.7**
Import dependency rises in Asia and Europe. (Adapted from BP Energy Outlook 2030, London, January 2012.)

China's energy imports, although gas imports increase by a factor of 16. China also becomes a major importer of coal.

- India will increasingly have to rely on imports of all three—oil, coal and natural gas—to supply its growing energy needs.
- European net imports (and imports as a share of consumption) rise significantly due to the declining domestic oil and gas production and rising gas consumption. Virtually all of the growth in net imports is from natural gas [4].

## 1.5 Refining Capacities of Petrol and Diesel Worldwide

People have used naturally available crude oil for thousands of years. The ancient Chinese and Egyptians, for example, burned oil to produce light. Before the 1850s, Americans often used whale oil for light. When whale oil became scarce, people began looking for other oil sources. In some places, oil seeped naturally to the surface of ponds and streams. People skimmed this oil and made it into kerosene. Kerosene was commonly used to light America's homes before the arrival of the electric light bulb.

As demand for kerosene grew, a group of businessmen hired Edwin Drake to drill for oil in Titusville, PA. After much hard work and slow progress, he discovered oil in 1859. Drake's well was 21.18 metres deep, very shallow as compared to today's wells. Drake refined the oil from his well into kerosene for lighting. Gasoline and other products made during refining were simply thrown away because people had no use for them. In 1892, the horseless carriage, or automobile, solved this problem since it required gasoline. By 1920, there were nine million motor vehicles in the United States and gas stations were opening everywhere.

Although research has improved the odds since Edwin Drake's days, petroleum exploration today is still a risky business. Geologists study underground rock formations to find areas that might yield oil. Even with advanced methods, only 23% of exploratory wells found oil in 2009. Developmental wells fared slightly better as 38% of them found oil.

When the potential for oil production is found onshore, a petroleum company brings in a 15–30 m drilling rig and raises a derrick that houses the drilling tools. Today's oil wells average 1600 m deep and may sink below 6000 m. The average well produces about 10 barrels of oil a day.

Oil's first stop after being pumped from a well is an oil refinery. A refinery is a plant where crude oil is processed. Sometimes, refineries are located near oil wells, but usually the crude oil has to be delivered to the refinery by ship, barge, pipeline, truck or train. After the crude oil has reached the refinery, large cylinders store the oil until it is ready to be processed. Tank farms are sites with many storage tanks.

An oil refinery cleans and separates the crude oil into various fuels and by-products. The most important one is gasoline. Some other petroleum products are diesel fuel, heating oil and jet fuel.

Refineries use many different methods to make these products. One method is a heating process called distillation. Since oil products have different boiling points, the end products can be distilled, or separated. For example, asphalts have a higher boiling point than gasoline, allowing the two to be separated.

Refineries have another job. They remove contaminants from the oil. A refinery removes sulphur from gasoline, for example, to increase its efficiency and to reduce air pollution. Nine per cent of the energy in the crude oil is used to operate the refineries. The various products that are produced from one barrel of oil (1 barrel of oil = 159.11 L) are shown in Figure 1.8. The refining capacity of different countries is shown in Table 1.3.

## 1.6 Energy Consumption: World View

Crude oil is the world's largest total primary energy consumed as shown in Table 1.4. The crude oil-derived diesel and gasoline fuels are used as fuels in internal combustion engine-powered vehicles. However, some countries like Russia mostly use natural gas due to its abundant availability. It is clearly seen that crude oil usage influences the economic development of a country. If the crude oil cost fluctuates, there is an unstable economic development of a nation.

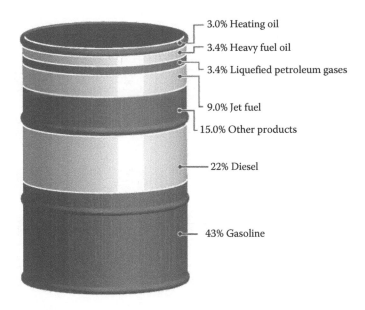

**FIGURE 1.8**
Products produced from a barrel of oil.

It can be observed from Figure 1.9 that economic growth is always associated with growth in energy consumption and associated emission. The growth rates of primary energy consumption, GDP and $CO_2$ emission were 4.9%, 4.7% and 4.7%, respectively. Passenger transportation by car is the highest for all countries as compared to other modes as shown in Table 1.5. The second largest transportation is by air for EU27 and the United States. The freight transportation for EU27, the United States, Japan, China and Russia is shown in Table 1.5. The rail transportation for freight is the highest in the United States, whereas sea transport is the highest in China. It is dependent on the geological structure and the country's political policy.

Based on the above discussion for different countries, it could be concluded that the passenger car, air and bus play a vital role for passenger transportation, whereas the rail and sea modes play a pivotal role for freight transportation.

It can be seen from the above discussion that the world economic development is primarily based on crude oil. However, the oil resources gradually deplete year after year as the demand increases steeply. The reserve-to-production ratio of fossil fuels such as oil, natural gas and coal is shown in Figure 1.10. If this ratio of oil reduces below a minimum value, it results in severe worldwide energy crisis. If the demand continues at the same rate and no new oilfield is explored, the oil resources may get depleted in about 50 years. Otherwise, it may not be possible to meet the required demand. As the crude oil price fluctuates, it affects the economic development of a country directly. The average crude oil price for the year 2008 peaked about 100 $/barrel as compared to the past couple of decades as shown in Figure 1.11. In countries that have a higher

**TABLE 1.3**

Refinery Capacities from Various Countries

| Capacity (Thousand Barrels Daily) | 2000 | 2001 | 2002 | 2003 | 2004 | 2005 | 2006 | 2007 | 2008 | 2009 | 2010 |
|---|---|---|---|---|---|---|---|---|---|---|---|
| United States | 16,595 | 16,785 | 16,757 | 16,894 | 17,125 | 17,339 | 17,443 | 17,594 | 17,672 | 17,688 | 17,594 |
| Canada | 1861 | 1917 | 1923 | 1959 | 1915 | 1896 | 1914 | 1907 | 1951 | 1976 | 1914 |
| Mexico | 1481 | 1481 | 1463 | 1463 | 1463 | 1463 | 1463 | 1463 | 1463 | 1463 | 1463 |
| Argentina | 626 | 619 | 619 | 620 | 623 | 627 | 623 | 634 | 634 | 635 | 638 |
| Brazil | 1849 | 1849 | 1854 | 1915 | 1915 | 1916 | 1916 | 1935 | 2045 | 2095 | 2095 |
| Netherlands Antilles | 320 | 320 | 320 | 320 | 320 | 320 | 320 | 320 | 320 | 320 | 320 |
| Venezuela | 1269 | 1269 | 1269 | 1269 | 1284 | 1291 | 1294 | 1303 | 1303 | 1303 | 1303 |
| Other South and Central America | 2207 | 2189 | 2234 | 2229 | 2235 | 2251 | 2260 | 2310 | 2356 | 2335 | 2351 |
| Belgium | 770 | 785 | 803 | 805 | 782 | 778 | 774 | 745 | 745 | 823 | 823 |
| France | 1984 | 1961 | 1987 | 1967 | 1982 | 1978 | 1959 | 1962 | 1971 | 1873 | 1703 |
| Germany | 2262 | 2274 | 2286 | 2304 | 2320 | 2322 | 2390 | 2390 | 2366 | 2362 | 2091 |
| Greece | 403 | 412 | 412 | 412 | 412 | 418 | 425 | 425 | 425 | 425 | 440 |
| Italy | 2485 | 2485 | 2485 | 2485 | 2497 | 2515 | 2526 | 2497 | 2396 | 2396 | 2396 |
| Netherlands | 1277 | 1278 | 1282 | 1282 | 1284 | 1274 | 1274 | 1236 | 1280 | 1280 | 1274 |
| Norway | 318 | 307 | 310 | 310 | 310 | 310 | 310 | 310 | 310 | 310 | 310 |
| Russian Federation | 5655 | 5628 | 5590 | 5454 | 5457 | 5522 | 5599 | 5596 | 5549 | 5527 | 5555 |
| Spain | 1330 | 1330 | 1330 | 1347 | 1372 | 1377 | 1377 | 1377 | 1377 | 1377 | 1427 |
| Sweden | 422 | 422 | 422 | 422 | 422 | 422 | 422 | 422 | 422 | 422 | 422 |
| Turkey | 713 | 713 | 713 | 713 | 693 | 613 | 613 | 613 | 613 | 613 | 613 |

| | | | | | | | | | | | |
|---|---|---|---|---|---|---|---|---|---|---|---|
| United Kingdom | 1778 | 1769 | 1785 | 1813 | 1848 | 1819 | 1836 | 1819 | 1827 | 1757 | 1757 |
| Other Europe and Eurasia | 6002 | 5912 | 5754 | 5691 | 5687 | 5650 | 5537 | 5573 | 5559 | 5596 | 5705 |
| Iran | 1597 | 1597 | 1597 | 1607 | 1642 | 1642 | 1727 | 1772 | 1805 | 1860 | 1860 |
| Iraq | 740 | 740 | 740 | 740 | 740 | 743 | 748 | 755 | 744 | 763 | 856 |
| Kuwait | 740 | 759 | 809 | 909 | 931 | 931 | 931 | 931 | 931 | 931 | 931 |
| Saudi Arabia | 1806 | 1806 | 1810 | 1890 | 2075 | 2100 | 2100 | 2100 | 2100 | 2100 | 2100 |
| United Arab Emirates | 440 | 674 | 711 | 645 | 620 | 620 | 620 | 625 | 673 | 673 | 673 |
| Other Middle East | 1168 | 1170 | 1248 | 1248 | 1248 | 1248 | 1283 | 1339 | 1345 | 1491 | 1491 |
| Total Africa | 2897 | 3164 | 3228 | 3177 | 3116 | 3224 | 3049 | 3037 | 3171 | 3022 | 3292 |
| Australia | 828 | 815 | 829 | 756 | 763 | 711 | 694 | 733 | 734 | 734 | 740 |
| China | 5407 | 5643 | 5933 | 6295 | 6603 | 7165 | 7865 | 8399 | 8722 | 9479 | 10,121 |
| India | 2219 | 2261 | 2303 | 2293 | 2558 | 2558 | 2872 | 2983 | 2992 | 3574 | 3703 |
| Indonesia | 1127 | 1127 | 1092 | 1057 | 1057 | 1057 | 1133 | 1157 | 1068 | 1106 | 1158 |
| Japan | 5010 | 4705 | 4721 | 4683 | 4567 | 4529 | 4542 | 4598 | 4650 | 4621 | 4463 |
| Singapore | 1255 | 1255 | 1255 | 1255 | 1255 | 1255 | 1255 | 1255 | 1385 | 1385 | 1385 |
| South Korea | 2598 | 2598 | 2598 | 2598 | 2598 | 2598 | 2633 | 2671 | 2712 | 2712 | 2712 |
| Taiwan | 732 | 874 | 1159 | 1159 | 1159 | 1159 | 1140 | 1197 | 1197 | 1197 | 1197 |
| Thailand | 899 | 1064 | 1068 | 1068 | 1068 | 1078 | 1125 | 1125 | 1175 | 1240 | 1253 |
| Other Asia Pacific | 1403 | 1512 | 1487 | 1416 | 1410 | 1428 | 1435 | 1443 | 1459 | 1605 | 1662 |
| Total world | 82,473 | 83,469 | 84,183 | 84,468 | 85,355 | 86,147 | 87,427 | 88,552 | 89,446 | 91,068 | 91,791 |

**TABLE 1.4**

World Energy Consumption by Different Fuel Type (Year: 2009)

| S. No. | Region | Oil (MT) | Natural Gas (MTOE) | Coal (MTOE) | Nuclear Energy (MTOE) | Hydroelectricity (MTOE) | Total (MTOE) |
|---|---|---|---|---|---|---|---|
| 1 | Total North America | 1025.5 | 736.6 | 531.3 | 212.7 | 158.3 | 2664.4 |
| 2 | Total South and Central America | 256 | 121.2 | 22.5 | 4.7 | 158.4 | 562.9 |
| 3 | Total Europe and Eurasia | 913.9 | 952.8 | 456.4 | 265 | 182 | 2770 |
| 4 | Total Middle East | 336.3 | 311 | 9.2 | — | 2.4 | 659 |
| 5 | Total Africa | 144.2 | 84.6 | 107.3 | 2.7 | 22 | 360.8 |
| 6 | Total Asia Pacific | 1206.2 | 446.9 | 2151.6 | 125.3 | 217.1 | 4147.2 |
|  | Total world | 3882.1 | 2653.1 | 3278.3 | 610.5 | 740.3 | 11,164.3 |

*Source:* Adapted from BP Statistical Review of World Energy, http://www.bp.com/statistical review S, June 2010.

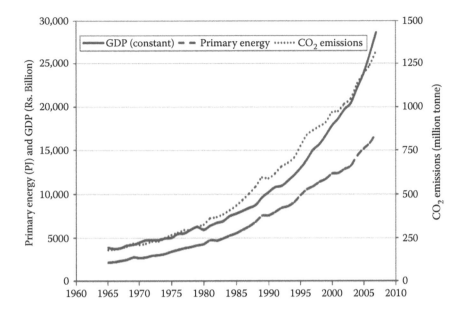

**FIGURE 1.9**
Trends in GDP, primary energy consumption and $CO_2$ emission in India. 1 U.S. dollar = INR 54.020 (on September 2012). (Adapted from D. R. Balachandra and N. H. Ravindranath, *Journal of Policy*, 38, 6428–6438, 2010.)

GDP growth rate, the high dependency on crude oil import from other countries leads to unstabilised economic growth. The solution to this problem lies in the exploration of alternative fuels for sustainable transportation.

## 1.7 Transportation Sector: Current Scenario

The primary modes of transportation are land, air and water. The land mode plays a pivotal role in personal, mass and off-road transportation as already depicted in Figure 1.2. The personal transport over a short distance is normally done by two-, three- and four-wheelers.

*Man—personal transportation:* Two-wheeler (motorbike, mopeds and scooters), three-wheeler (auto), four-wheeler (car, MUV), bus (six-wheeler)

*Mass transportation:* Three- (LCV), four- (LCV and HCV), six- and more wheelers (HCV)

*Off-road transportation:* Tractors and farm equipments such as tillers and grinders

**TABLE 1.5**

Passenger and Freight Transport from Different Countries

| S. No. | Mode of Transport | Passenger Transport (Billon Passenger Kilometre) | | | | |
|---|---|---|---|---|---|---|
|  |  | EU27 | USA | Japan | China | Russia |
| 1 | Passenger car | 4725 | 7201.8[a] | 769.1[b] | 1263.6[c] | — |
| 2 | Bus + trolley bus + coach | 546.7 | 243 | 89.9 | — | 124.8 |
| 3 | Railway | 409.2 | 37.1 | 404.6 | 777.9 | 175.9 |
| 4 | Tram + metro | 89.0 | 21.1 |  |  | 51.6 |
| 5 | Waterborne | 40.9 | 0.6 | 5.5[d] | 7.5 | 0.9 |
| 6 | Air (domestic/intra-EU-27) | 561 | 977.8 | 81 | 288.3 | 122.6 |
|  |  | Freight Transport (Billon Tonnes Kilometre) | | | | |
| 1 | Road | 1877.7 | 1922.9 | 346.4 | 1135.5 | 216.3 |
| 2 | Rail | 442.7 | 2656.6 | 22.3 | 2379.7 | 2116.2 |
| 3 | Inland waterways | 145.3 | 472.3 | — | 1559.9 | 64 |
| 4 | Oil pipeline | 124.1 | 814.2 | — | 186.6[e] | 2464.0 |
| 5 | Sea (domestic/intra-EU-27) | 1498.0 | 333.0 | 187.5 | 4868.6 | 85.0 |

*Source:*  Adapted from Eurostat, Japan Statistics Bureau, US Bureau of Transportation Statistics,
Goskom STAT (Russia), National Bureau of Statistics of China, International Transport
Forum. EU and transport in figures, Statistical pocket book 2010, European Union 2010.
[a]  United States: including light trucks/vans.
[b]  Japan: including light motor vehicles and taxis.
[c]  China: including buses and coaches.
[d]  Japan: included in railway passenger kilometre (pkm).
[e]  China: oil and gas pipelines.

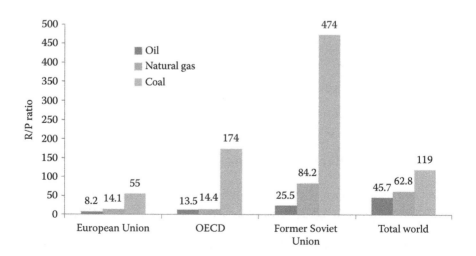

**FIGURE 1.10**
Fossil fuel reserves to production (R/P) ratio. (Adapted from Repowering Transport—A Cross-
Industry Report, World Economic Forum, Geneva, February 2011.)

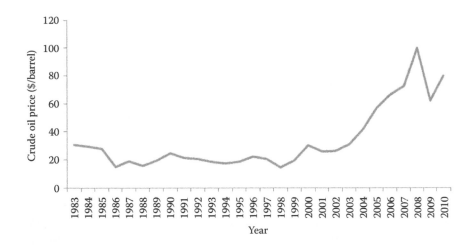

**FIGURE 1.11**
Crude oil price from the year 1983 to 2010. (Adapted from IEA World Energy Outlook, www. iea.org, 2009.)

The statistical data of vehicles in India and around the globe are given in Table 1.6. The engine power of a passenger car normally lies in the range of 15–300 kW.

### 1.7.1 Mass Transportation: Diesel Buses and Trucks

The medium distance is generally covered by public transportation such as buses. The mass transportation for goods and raw material is also undertaken by trucks, lorries and light commercial vehicles. Goods from industries to retailers are transported by trucks. The local distribution is undertaken by a three- or four-wheeled LCV.

### 1.7.2 High-Power Rail Transportation

The larger-distance transportation is usually done by using rail transportation. The advantages of rail transportation are lower travel cost and fast services due to a free traffic. However, it cannot provide services like road transport.

### 1.7.3 Aviation Sector: Gas Turbines

Air transportation is the second largest sector for passenger transportation in several developed countries. Gas turbines are used in airbuses. The growth in this sector has been significant due to an enhanced increase in economic growth. The fuels used for the airbus are aviation kerosene and jet fuels. $NO_x$ and $CO_2$ are major pollutants in the aviation sector. In addition to this, noise pollution is another irritant to the public.

**TABLE 1.6**

Statistical Data of Vehicles (2008)

| Mode of Transport | Range of Engine Power (in kW) | Type of Engine | Number of Vehicles in India | Number of Vehicles in the World |
|---|---|---|---|---|
| *Road Transport* | | | | |
| Mopeds, scooters and motorcycles | 0.75–7 | SI, 2/4 stroke | 71,025,312 | 117,264,312 |
| Small passenger cars | 15–75 | SI, 4 stroke | 12,546,841 | 165,641,327 |
| Large passenger cars | 75–200 | SI, 4 stroke | 1,120,369 | 196,565,887 |
| Light commercial | 35–150 | SI/CI, 4 stroke, | 3,368,923 | 268,794,513 |
| Heavy commercial | 120–300 | CI, 4 stroke | 2,125,676 | 255,615,515 |
| Agricultural | 3–150 | SI/CI, 2/4 stroke | 2,748,686 | 25,680,124 |
| *Rail Transport* | | | | |
| Railway locomotives | 400–3000 | CI, 2/4 stroke | 5022 | 26,415 |
| *Air Transport* | | | | |
| Helicopters | 45–1500 | SI, 4 stroke | 298 | 56,200 |
| Aeroplanes | 45–2700 | SI, 4 stroke | 1151 | 418,899 |
| *Marine Transport* | | | | |
| Ships | 3500–22,000 | CI, 2/4 stroke | 735 | 43,457 |

*Source:* Adapted from www.data.worldbank.org.

### 1.7.4 Global Vehicle Fleet

The global vehicle fleet (commercial vehicles and passenger cars) will grow rapidly—by 60% from around 1 billion today to 1.6 billion by 2030 (Figure 1.13). Most of the growth will be in the developing world with some mature markets at saturation levels. More than three quarters of the total fleet growth will occur in the non-OECD countries, where the vehicle population will rise from 340 million to 840 million over the next 20 years—a 2½-fold inscrease (Figure 1.12). From 2010 to 2030, the vehicle density per 1000 population will grow from approximately 50 to 140 in China (5.7% p.a.) and from 20 to 65 in India (6.7% p.a.) as shown in Figure 1.13. China is expected to follow a slower path to vehicle ownership than is seen historically in other countries. This reflects the impact of current and assumed future policies, designed to limit oil import dependency and congestion, including rising fuel taxation, widespread mass transportation options and relatively uneven income distribution [4].

- Transport fuel in 2030 is expected to be dominated by oil (87%) and biofuels (7%). Other fuels gain share, such as natural gas and electricity (4% and 1%, respectively, in 2030), are constrained by limited policy support combined with a general lack of infrastructure in all but a handful of markets.

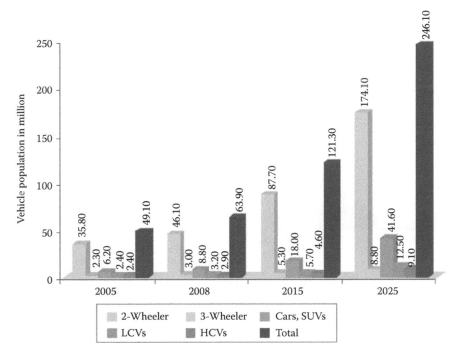

**FIGURE 1.12**
Projections of vehicle fleet. (Adapted from Asian Development Bank (ADB), Energy efficiency and climate change considerations for on-road transport in Asia, Asian Development Bank, Philippines, 2006.)

- Despite the projected 60% increase in vehicles over the next 20 years, energy consumption in total transport is forecast to grow only 26% (1.2% p.a.—down from 1.9% p.a. between 1990 and 2010).

- The growth rate of energy (Figure 1.13) used for transport declines due to accelerating improvements in fuel economy and the impact of high oil prices on driving behaviour. Vehicle saturation in the OECD countries and a likely increase in taxation (or subsidy reduction) and the development of mass transportation in the non-OECD countries are other factors.

- Vehicle fuel economy improvements are driven by a tightening policy ($CO_2$ emission limits in Europe and corporate average fuel economy (CAFE) standards in the United States), which is enabled by improving technology. Prices also play a role, since high fuel costs provide an additional incentive to improve vehicle efficiency [4].

- Assuming no changes to vehicle usage, efficiency and the use of alternatives, oil demand in road transport would increase by a massive 23 Mb/d over the next 20 years, more than the total projected

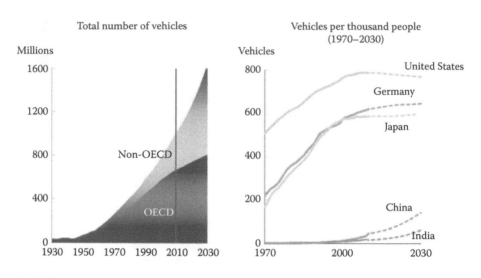

**FIGURE 1.13**
Growth of global vehicle fleet. (Adapted from BP Energy Outlook 2030, London, January 2012.)

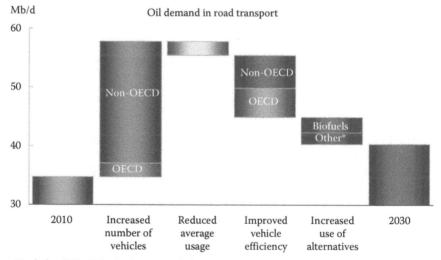

* Includes GTL, CTL, CNG, LNG and electricity.

**FIGURE 1.14**
Oil demand in road transport.

oil demand growth (16 Mb/d), mostly due to more vehicles in the non-OECD countries. Instead, we project oil demand growth for road transport to be 6 Mb/d. Vehicle fuel economy is forecast to improve by 1.1% p.a. in both the OECD and non-OECD countries as shown in Figure 1.14. These efficiency gains are equivalent to 11 Mb/d by

2030—saving approximately half of the incremental oil demand that would otherwise be required under the above 'no change' case.

- Average miles driven per vehicle is expected to fall (saving 2.4 Mb/d) as high fuel prices (partly due to rising taxes or reduced subsidies), congestion and mass transit outweigh the impact of rising incomes.

- Biofuels make up more than half of the incremental demand for alternative fuels in transport. The use of electric vehicles and compressed natural gas/liquefied natural gas (CNG/LNG) is growing, but there are still barriers delaying the scale-up. Alternative fuels, therefore, are not expected to have a material impact until 2030 [4].

## 1.8 Fossil Fuel Consumption in the Transport Sector

Transportation and fossil fuels are currently inextricably linked; more than 60% of the 84 million barrels of oil consumed every day powers the world's cars, trucks, planes and other modes of transportation [2,5] and more than 96% of current energy supply to the transport sector is from liquid fossil fuels. Many studies of energy usage in the transport sector show a significant growth in demand in the years ahead. The ramifications of this dependency are becoming more transparent every year. The increasing concentration of conventional oil production in fewer geographies and the increasing cost of new liquid fuels have combined to generate significant unease at the global and national levels regarding the security of supply (Figure 1.15a and b). Indeed, many countries rely on imported crude oil and refined oil products to fuel their transport sectors, and that dependence will only become more severe in the coming decade. U.S. crude oil imports are projected by the International Energy Agency (IEA) to account for 80% of the total consumed by 2020, up from 65% in 2008; China's reliance on imports will increase to 68% from 49% over this same time period.

Recent increases in the price of oil and persistent volatility in energy prices have exacerbated these concerns, and most forecasts point to even higher energy costs in the years to come (Figure 1.16).

The global transport sector consumes about 2200 million tonnes of oil equivalent (MTOE) of energy each year. Of this, more than 96% comes from oil, comprising over 60% of the world's total oil production (Figure 1.17). Road transport accounts for the majority of this energy consumption, with light-duty vehicles (LDVs) accounting for about 52% of the total, while buses and trucks combined represent a 21% share. While air and marine transport each account for roughly 10% of global transport energy consumption, aviation is by far the fastest-growing sector, with a forecast increase in revenue-tonne-kilometres of ~5.1% per year to 2030. The rail sector accounts for roughly 3% of total transport-related energy consumption.

(a)

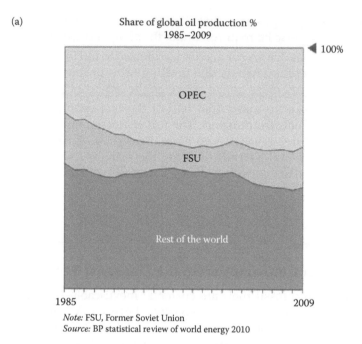

Note: FSU, Former Soviet Union
Source: BP statistical review of world energy 2010

(b)

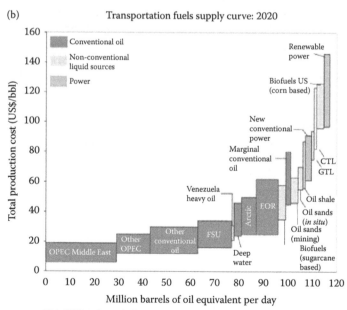

Note: EOR, enhanced oil recovery
Source: Bcoz and company analysis

**FIGURE 1.15**
(a) Share of global oil production. (b) Transportation fuel supply curve. (Adapted from IEA World Energy Outlook, www.iea.org, 2009.)

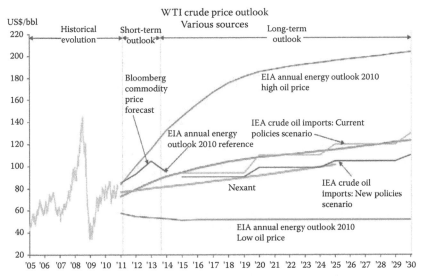

**FIGURE 1.16**
Long-term upward pressure on crude prices. (Adapted from Repowering transport—A cross-industry report, World Economic Forum, Geneva, February 2011.)

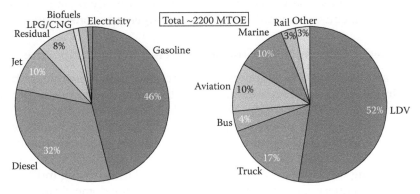

**FIGURE 1.17**
Global transport energy consumption. IEA/SMP, IMO, IATA, Carbon neutral skies team analysis, Repowering transport team analysis. (Adapted from Repowering transport—A cross-industry report, World Economic Forum, Geneva, February 2011.)

If energy consumption grows as per the forecasts made by the IEA and many others—about 1.7% per year until 2030—the world will consume about 40% more energy than it uses today, with little change to the mix of energy sources and both biofuels use and energy efficiency investments continuing at their current pace (Figure 1.18). By just about any measure—oil supplies,

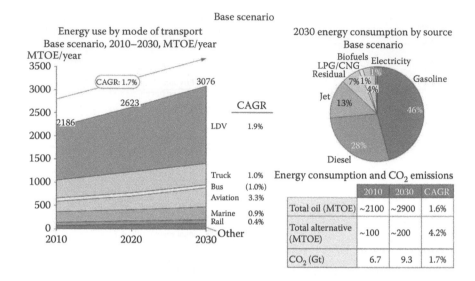

**FIGURE 1.18**
Establishing a baseline for 2030 energy demand. (Adapted from Repowering transport—A cross-industry report, World Economic Forum, Geneva, February 2011.)

energy security, $CO_2$ or other emissions—such growth in energy demand threatens the sustainability of the sector. In this baseline case, energy security risks—already perceived to be significant—would increase substantially, and since the energy source pattern would broadly remain the same, $CO_2$ and local emissions would rise steeply with energy consumption.

### 1.8.1 Energy Consumption in Transport Sector: Indian Perspective

Nearly all motorised vehicles necessitate the combustion of petroleum-based fuels. In India, transport accounted for nearly half of the petroleum product consumption in 2004–2005 [11]. The growth in transport demands directly weighs on the country's need for oil. India's oil dependency has increased over time and stood at 76% of total crude oil refinery requirements in 2005 (MOSPI, 2006). In 1990, the crude oil dependency was only 39%. This reflects the increasing need for petroleum products to feed the growing Indian vehicle market. Refinery capacity covers all the needs of the domestic market and exports a very small quantity.

Energy consumption in the transport sector is evenly distributed between freight and passenger transportation as shown in Figure 1.19. Road transport is the most used means of transport, followed by air and then rail. Finally, a very small quantity of energy is used for waterways transport.

In 2004, diesel and motor gasoline represented 90% of the final energy consumed in the transport sector, while jet kerosene represented 8% and electricity 2% [12]. Diesel is the most used form of energy, with a share of 66%, followed by motor gasoline with 24%. Statistics of energy consumption

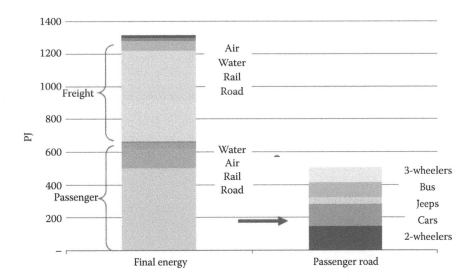

**FIGURE 1.19**
Transport energy consumption per mode in 2004. (Adapted from Bureau of Energy Efficiency (BEE), www.bee-india.nic.in, 2007.)

over time from the Ministry of Oil and Gas show a steady increase of motor gasoline; however, statistics for diesel consumption show uneven trends. In 1996, a serious break in the series occurred when diesel consumption in the transport sector plunged by 26%. In reality, no major activity disturbance or technology breakthrough can explain such a decline over a 1-year period. It is believed that a major restructuring in statistics accounting explains this trend; however, no official document or note was found to justify this argument. Hence, it is assumed that more recent statistics on diesel consumption for the transport sector reflect the real consumption. Figure 1.20 shows data for motor gasoline and diesel consumption based on the bottom-up model. Figure 1.20 also shows trends from fuel consumption data by different vehicles collected from the national statistics from MOSPI and the IEA.

Diesel consumption shows a different trend as compared to motor gasoline (Figure 1.20a and b). From 1999, instead of continuing its escalating trend, diesel consumption showed a fall for a couple of years and started increasing again after 2003. This is in contradiction with sales data on truck and diesel cars.

Figure 1.21 shows that the fossil fuel energy consumption (% of total) in India was 73.05 in 2009, according to a World Bank report published in 2010. The fossil fuel energy consumption (% of total) in India was reported at 71.03 in 2008, according to the World Bank [12]. Fossil fuel comprises coal, oil, petroleum and natural gas products.

Figure 1.22 shows the road sector diesel fuel consumption (kt of oil equivalent) in India, which was reported at 25,553.00 in 2008, according to the World Bank. Diesel is a heavy oil used as a fuel in diesel engines.

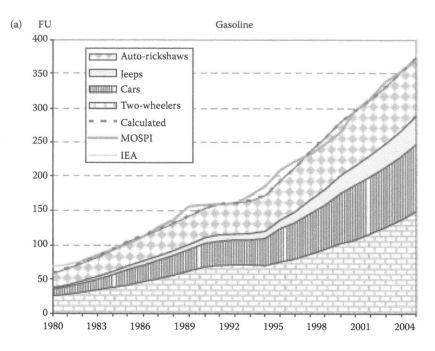

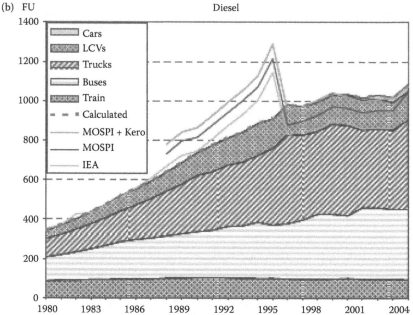

**FIGURE 1.20**
(a) Gasoline transport consumption. (b) Diesel transport consumption. (Adapted from International Energy Agency, Energy Balances of Non-OECD Countries 2004–2005, IEA/OECD, Paris, France, 2007a.)

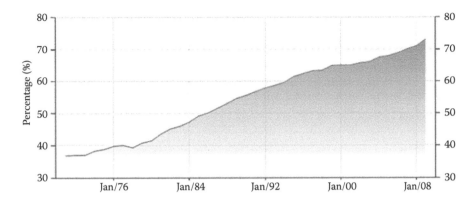

**FIGURE 1.21**
Fossil fuel energy consumption (% of total) in India. (Adapted from M. Karpoor, Vision 2020 Transport, prepared for the Planning Commission, 2002.)

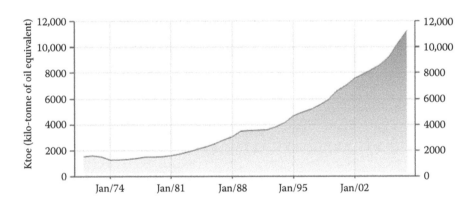

**FIGURE 1.22**
Road sector diesel consumption in India. (Adapted from M. Karpoor, Vision 2020 Transport, prepared for the Planning Commission, 2002.)

Figure 1.23 shows the road sector gasoline fuel consumption (kt of oil equivalent) in India, which was reported at 11,257.00 in 2008, according to the World Bank. Gasoline is a light hydrocarbon oil used in internal combustion engines such as motor vehicles, excluding aircraft.

The Asian Institute of Transport Development (AITD) [14] conducted a comparative assessment of rail and road transport in India from the perspectives of social and environmental sustainability. The study observed that the energy consumption on different inter-city rail sections in the case of freight traffic varied between 10.28% and 25.01% of the energy consumed by road transport in parallel stretches of state and national highways. In the case of passenger transport, the energy consumption on rail varied between 78.77% and 94.91% of the energy consumed by road transport (Figures 1.24 and 1.25).

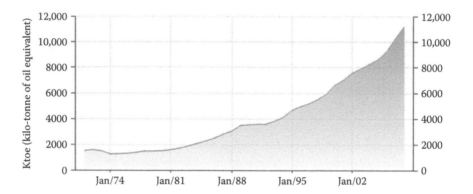

**FIGURE 1.23**
Road sector petrol consumption. (Adapted from M. Karpoor, Vision 2020 Transport, prepared for the Planning Commission, 2002.)

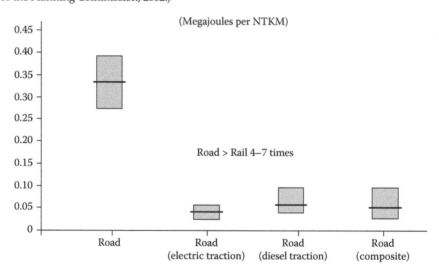

**FIGURE 1.24**
Energy consumption: inter-city freight road and rail. (Adapted from Asian Institute of Transport Development (AITD), Environmental and Social Sustainability of Transport: Comparative Study of Rail and Road, Asian Institute of Transport Development, New Delhi, 2002.)

Furthermore, the land requirement for road corridor has been observed to be two and a half times that of the rail corridor.

## 1.9 GHG Emissions from the Transportation Sector

The carbon dioxide ($CO_2$) concentration in the atmosphere, one of the main greenhouse gas (GHG) emissions, has drastically increased from 280 ppm

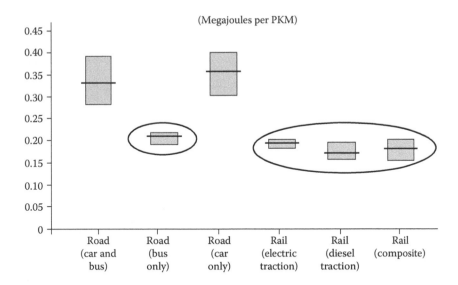

**FIGURE 1.25**
Energy consumption: inter-city passenger road and rail. (Adapted from Asian Institute of Transport Development (AITD), Environmental and Social Sustainability of Transport: Comparative Study of Rail and Road, Asian Institute of Transport Development, New Delhi, 2002.)

at the start of the industrial revolution in the year 1800 to 380 ppm in 2011. The threshold level of $CO_2$ emission is predicted to be 550 ppm; if it exceeds this level, it may lead to severe problems such as melting of the polar ice caps, a sea level rise of up to 1 m by the year 2100, an increased frequency of extreme climate events, permanent flooding of coastal cities, disruption of the ecosystem and extinction of species [15]. To avoid these wide-ranging consequences, international experts believe that it is absolutely essential to limit the global temperature rise within 1.5–2.5°C. In this direction, the G-8 nation summit recently decided to take some initiatives to at least halve the global $CO_2$ emissions by 2050. GHG's affecting the climate change has gained great momentum due to the Kyoto protocol. The ultimate objective of the Kyoto protocol is to stabilise the atmospheric concentration of GHGs and it commits industrialised nations to reduce their emissions of baskets of GHGs by around 5% between 2008 and 2012 as compared to 1990 levels [15]. It may be noted that India is a signatory to the multilateral treaty of the Intergovernmental Panel on Climate Change (IPCC). A transport vehicle gets more attention as it contributes about 20–25% of the $CO_2$ released to the atmosphere, and this share tends to increase [7].

Transportation is an important societal need for better human life and economic development. Considerable importance is being given for combating the regulated emissions such as CO, HC, $NO_x$ and particulate matter (PM) from transport vehicles. As the GHGs emission is not under the regulated criterion, much emphasis is not historically given to the GHG emissions from the vehicles.

**TABLE 1.7**

Transportation GHG Emissions by Mode, 2000 (US EPA)

| S. No. | Mode of Transportation | GHG Emission Contribution (2000) (%) |
|--------|------------------------|--------------------------------------|
| 1 | Passenger cars | 36 |
| 2 | Light trucks | 19 |
| 3 | Heavy trucks | 16 |
| 4 | Buses | 1 |
| 5 | Aircraft | 10 |
| 6 | Marine | 5 |
| 7 | Rail | 2 |
| 8 | Other (two- and three-wheelers) | 11 |

*Source:* Adapted from K. A. Subramanian et al., Control of GHG emissions from transport vehicles: Issues and challenges, SAE Paper No. 2008-28-0056, 2008.

GHG emissions from transport vehicles are mainly $CO_2$, $N_2O$ and $CH_4$. The transportation GHG emission contribution made by different types of vehicles is given in Table 1.7. It is clearly seen from the table that automobiles contribute about 70% GHG emission in the total transportation sectors. The development of new technology is so far mainly targeted to reduce regulated emissions enforced by the country's legislation, as the customer's expectation is primarily focussed on fuel economy, comfort and safety. There is an urgent need to reduce GHG emission from all sectors, including the transport sector, to avoid global warming consequences. In this direction, the European Union (EU) has taken a number of initiatives to control $CO_2$ emission from transport vehicles.

The GHG emission formation mechanism, control measures, issues and challenges from transport vehicles are reviewed and discussed in the following sections.

### 1.9.1 Mechanism of GHG Pollutant Formation in Internal Combustion Engines

The main GHG pollutants from internal combustion engines are $CO_2$, $N_2O$ and $CH_4$, and the pollutant formation mechanism is briefly described below.

### 1.9.2 $CO_2$ Emission

$CO_2$ is not considered as an obnoxious pollutant from the combustion point of view as a complete combustion is expected to yield higher $CO_2$ emissions. On the other hand, $CO_2$ is an important pollutant from the environmental point of view. The fuel containing carbon and hydrogen during combustion with oxygen is converted into $CO_2$ and water vapour as products. The carbon content of some fuels by weight are 84–87% in gasoline, 85–88% in

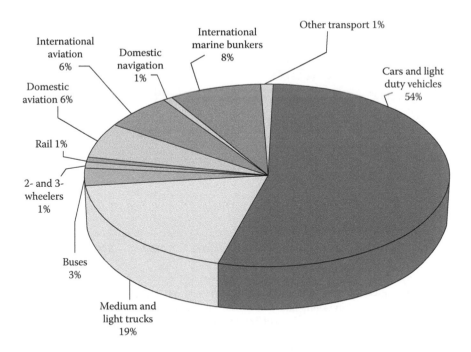

**FIGURE 1.26**
Modal shares of transport $CO_2$ emissions (2005). (Adapted from International Energy Agency (IEA), $CO_2$ Emissions from Fuel Combustion: 1971–2005, IEA, Paris, 2007c.)

diesel, 75% in CNG, 82% in propane, 37.5% in methanol, 52.2% in ethanol and 0% in hydrogen [2]. The development of new technologies are mainly targeted for reducing regulated emissions such as CO, HC and particulate emissions resulting in higher $CO_2$ emission. In general, conventional fuels are mostly carbon-based fuels that would generally lead to carbon-based emissions such as $CO_2$, CO, HC and PM.

Figure 1.26 provides the transport sector $CO_2$ emissions across modes for 2005. This shows that road transport is the largest contributor to emissions, and within this the key contributors are on-road vehicles such as cars and LDVs, that is, four-wheeled vehicles (including sports utility vehicles, small passenger vans with up to eight seats), and trucks. Between 1990 and 2005, $CO_2$ emission from the world transport sector rose by 37%. During the same period, road transport emissions increased by 29% in the industrialised countries and 61% in other countries, which includes primarily the developing countries [17]. The Intergovernmental Panel on Climate Change (IPCC) in its fourth assessment report (AR4) advises that to avoid the worst impact of climate change, global $CO_2$ emissions must be cut by at least 50% (Figure 1.27). Transport has a very significant role to play in realising that goal.

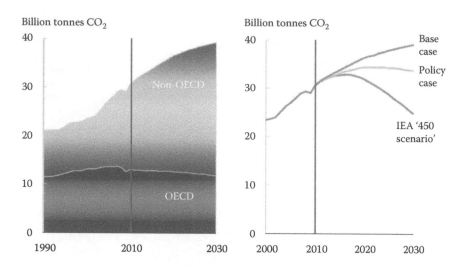

**FIGURE 1.27**

Global $CO_2$ emissions from energy use. (Adapted from Asian Institute of Transport Development (AITD), Environmental and Social Sustainability of Transport: Comparative Study of Rail and Road, Asian Institute of Transport Development, New Delhi, 2002.)

- It is desirable to continue tightening the policies to address climate change. Implementation of carbon abatement policies in the OECD would help in reducing emission in 2030. However, it would only be about 10% as compared to 2010.

- Non-OECD countries do make significant progress in reducing the carbon intensity of their economies, but this is outweighed by carbon increases due to rapid economic growth. The net result is a projected increase in global emissions of 28% by 2030.

- This leaves the world well above the required emission path to stabilise the concentration of GHGs at the level recommended by scientists (around 450 ppm).

- Our 'policy case' assumes a step change in the political commitment to action on carbon emissions. Even in this case, the path to reach 450 ppm remains elusive. However, a declining emissions path by 2030 is achievable, given the political will to shoulder the cost [4].

### 1.9.3 N₂O Emission

$N_2O$ emission occurs during combustion at low (<950°C) temperatures and it is affected by fuel type, operating conditions and excess air level, oxygen in fuel, and catalytic activity. Exhaust gas recirculation (EGR) is one of the promising technologies to enhance the control of $NO_x$ emission by means of reducing the combustion gas temperature, which results in a high $N_2O$ emission. Most of the research in the world is focussing on reducing $NO_x$

emission by reducing the gas temperature, but it may negatively result in high $N_2O$ emission. The global warming potential (GWP) of $N_2O$ is about 320 times higher than the $CO_2$ equivalent.

### 1.9.4 $CH_4$ Emission

$CH_4$ is typically formed due to incomplete combustion. This emission may be due to partial combustion, quenching and unburned hydrocarbon. This level may be higher in case of a CNG-fuelled vehicle.

Typical values of $CO_2$, $CH_4$ and $N_2O$ emissions from vehicles using different fuels are given in Table 1.8. It can be seen from the table that a CNG-fuelled vehicle gives the lowest $CO_2$ emission as compared to a petrol-, diesel- and LPG-fuelled vehicle. It may be attributed to the effect of CNG having the lowest carbon-to-hydrogen ratio among all fossil fuels. Further, it may be observed that the GWP of $N_2O$ and $CH_4$ emission is about 320 and 63 times more than the $CO_2$ equivalent, respectively. Some countries take measures to promote fuel-efficient vehicles around the world as given in Table 1.9. The countries are mainly aiming to reduce $CO_2$ emission by improving the fuel economy.

### 1.9.5 Roadmap and Strategy for $CO_2$ Emission Reduction by Other Countries

Some countries are setting the regulation for controlling $CO_2$ emission from vehicles. The yearwise targeted $CO_2$ emission of various countries and their strategies are given in Table 1.10.

Vehicle speed plays a vital role in fuel economy. Overspeed and a lower speed of vehicle would lead to fuel penalty. An internal combustion engine normally gives a higher fuel economy at medium speed. In Germany and the United States, the maximum speed will be normally up to 120 km/h. However, in countries like India, the vehicle speed is generally limited to 20–30 km/h in urban areas, which may lead to heavy fuel penalty as most of the time the engine operates at a lower speed and idling conditions.

**TABLE 1.8**

Direct Greenhouse Gas Emissions from Passenger Cars on Petrol, Diesel, LPG and CNG under Real-World Driving Conditions

| Fuel | $CO_2$ (g/km) | $CH_4$ (g/km) | $N_2O$ (g/km) | GHG Emission (g of $CO_2$ Equivalent/km) |
|---|---|---|---|---|
| Petrol | 208.1 | 0.009 | 0.003 | 209.2 |
| Diesel | 180.5 | 0.004 | 0.007 | 182.7 |
| LPG | 189.3 | 0.007 | 0.003 | 190.4 |
| CNG | 168.6 | 0.0074 | 0.001 | 170.6 |

*Source:* Adapted from P. Hendriksen et al., Evaluation of the environmental performance of modern passenger cars running on petrol, diesel, automotive LPG and CNG, TNO-report 03.OR.VM.055.1/PHE, TNO Automotive, December 2003.

**TABLE 1.9**

Measures to Promote Fuel-Efficient Vehicles around the World

| Country/Region | Fuel Efficiency Approach | Measures |
|---|---|---|
| United States, Japan, Canada, Australia, China, Taiwan, South Korea | Fuel economy standards | Numeric standard in mpg, km/L or L/100 km |
| European Union, California | GHG emission standards | G/km |
| European Union, Japan | High fuel taxes | Fuel taxes at least 50% greater than crude oil base price |
| European Union, Japan | Fiscal incentives | Tax relief based on engine size, efficiency and $CO_2$ emission |
| California | Technology mandates and targets | Sales requirement for ZEVs |
| United States | Economic penalties | Gas guzzler tax |
| Several U.S. states | Traffic control measures | Hybrid allowed in HOV lanes, ban on SUVs |
| India | Fuel economy standards | km/L |

*Source:* Adapted from K. A. Subramanian et al., Control of GHG emissions from transport vehicles: Issues and challenges, SAE Paper No. 2008-28-0056, 2008.

The Kyoto protocol objective is to reduce the $CO_2$ emission by about 5–8% in the target time frame of 2008–2012. For example, the EU has set a target of $CO_2$ emission reduction to about 120 g/km by the year 2012. As most of the vehicles may be in a production stage, which will be sold by the year 2012, it may be a big challenge to meet the $CO_2$ emission target. The industry could continue to cut $CO_2$ from cars through improved vehicular technology, but specific legal requirements cannot come into force before 2015 [16]. The lead time for $CO_2$ emission norms implementation needs to be formulated with the association of automobile manufacturers of the concerned country.

New vehicle models emitting $CO_2$ emission of 120 g/km launched by the year 2000 were facing a lack of market acceptance. For examples, battery electric-powered vehicle such as Audi Duo, Fiat Auto's Seicento Elettra, Ford's electric Think City vehicle, General Motor's Astra and Corsa Eco, Mercedes A 160, PSA's Electric Vehicles, Renault's Electric Vehicles, Volkswagen's Lupo and Golf CityStromer received poor acceptance from customers [16].

As the automotive industry is currently facing problems in meeting the stringent norms of regulated emissions, the additional burden of $CO_2$ emission reduction may lead to the slowdown of the overall economy.

The EU policy decision is focussed towards the goal of 120 g $CO_2$/km by 2012 through improved vehicular technology. This may cost around Euro 3600 on average per vehicle. This is up to 10 times more expensive than similar or even more effective measures such as an increased use of biofuels and adopting an economic driving style. Reducing $CO_2$ emissions through improved vehicular technology is up to 10 times more expensive than other traffic-related measures [16].

**TABLE 1.10**

Implemented/Targeting $CO_2$ Emission Standards

| Country Name | Target $CO_2$ by Year | Strategy |
|---|---|---|
| European Union (EU) | 130 $gCO_2$/km by 2012 | Infrastructure measures, driver behaviour, alternative fuels, etc., test method: EU NEDC |
| California; other states such as New York, New Jersey, Massachusetts, Connecticut, Maine, Rhode Island, Vermont, Oregon | 201 $gCO_2$/km by 2009<br>141 $gCO_2$/km by 2013<br>127 $gCO_2$/km by 2016 | Fuel economy: g/mile, test method: US CAFE |
| Japan | 138 $gCO_2$/km by 2015 | Integrated approach: 48% by vehicle technology, 52% by infrastructure adjustments such as dynamic traffic lights, lower road-rolling resistance, tax incentives for purchase of low-emission vehicles and others, test method: Japan 10–15 |
| United States | — | Fuel, test method: U.S. CAFE |
| China | — | Fuel, weight-based, test method: EU NEDC |
| Australia | — | Fuel, engine size, test method: EU NEDC |
| Taiwan, South Korea | — | Fuel, engine size, test method: U.S. CAFE |
| France | 140 $gCO_2$/km for new vehicle | 20% mandatory purchase of LPG, NGV or electric vehicle |
| Belgium | 145 $gCO_2$/km for diesel and 160 $gCO_2$/km for petrol cars | — |
| United Kindom | 10% of fleet cars by 2006 | Alternatively fuelled vehicles |

*Source:* Adapted from K. A. Subramanian et al., Control of GHG emissions from transport vehicles: Issues and challenges, SAE Paper No. 2008-28-0056, 2008.

## 1.10 Environmental Concerns

### 1.10.1 Near Term: Local Air Pollution

Road transportation is primarily based on internal combustion engine-fitted vehicles. Air transportation is based on gas turbine-fitted aircrafts and internal combustion engine-based helicopters. Marine transportation is based on internal combustion engine-fitted marine vehicles (ships). Rail transportation is based on diesel engines or electricity-driven locomotives. Road transportation contributes to local air pollution. Other transport sectors do not contribute directly to local air pollution but they contribute significantly to overall air pollution.

Local air pollution affects people's health as the tailpipe emissions come into direct contact with the people who are in the proximity of vehicles. Furthermore, these emissions do affect people in residential areas, hospitals and office buildings.

### 1.10.2 Toxic Air Pollutants

Automotive engines emit toxic emissions such as CO, HC, $NO_x$ and smoke/PM. The present emission norms are more stringent as compared to the norms of the previous decade. A steep rise in vehicular population did result in the loss of the advantages of emission norms. In other words, the total emission load will be almost the same as given by Equation 1.2.

$$\text{Total emission load of each pollutant (tonne)} = (\Sigma \text{ vehicle class } * \\ \text{emission factor (g/km)} * \text{distance travelled (km)}) * 10^{-6} \qquad (1.2)$$

It can be seen from Equation 1.2 that as the number of vehicles increases, the total emission load will be the same or even higher.

#### Example 1.1

A city implemented the new emission norms of 1.0 g/km for CO in the year 2010, whereas the old norm was 1.5 g/km for the year 2000. The number of vehicles increased from 1 million to 1.5 million in the decade. Assume that the vehicles are of the same class and that the vehicles travelled 100 km per day for both norms.

Vehicle class = $10^6$ (old), $1.5 \times 10^6$ (new)
Emission factor = 1.5 g/km (old), 1.0 g/km (new)
Distance travelled = 100 km/day

Emission load calculation using Equation 1.2 for the old emission norm:

Total old emission load per day = $1*10^6*1.5*100*10^{-6}$ = 1.5 tonnes of CO per day

Emission load calculation using Equation 1.2 for the new emission norm:

Total new emission load per day = $1.5*10^6*1.0*100*10^{-6}$ = 1.5 tonnes of CO per day

It is clearly seen from the simple example that the total emission load will be the same or higher with the stringent emission norms due to the steep increase in vehicular population. If the economy grows, the number of vehicles will also increase. Thus, the local air pollution poses bigger challenges, which need to be mitigated without affecting the economy.

### 1.10.3 Environmental: Long-Term, Climate Change/Global Warming Effect

The GHGs have a long-term threat to the environment. Internal combustion engine-powered vehicles emit GHGs emissions such as $CO_2$, $N_2O$ and $CH_4$. If 1 kg of fuel burns in a powered heat engine, it theoretically emits about 2.7 kg of $CO_2$.

If 1 kg of any hydrocarbon fuel burns, it theoretically emits about 2.7 kg of $CO_2$. The $CO_2$ emission will be 30,143.6 MMT for 11,164.3 MMT of oil equivalent of primary energy consumption. $CO_2$ emission is one of the GHG emissions and it is likely to create a long-term threat to the environment in terms of global warming and climate change.

## 1.11 Environmental Standards

Regulated emissions such as CO, HC, $NO_x$ and PM are implemented in vehicle fleet in many countries. The United States follows TIER emission norms, whereas the EU follows Euro norms (EURO 1, 2, 3, 4, 5 and 6) and India follows Bharat Stage norms (I, II, III and IV). However, unregulated emissions such as $CO_2$, $N_2O$, $CH_4$, aldehyde, ketones and $SO_x$ are not given enough emphasis. These emissions are equally important to formulate the emission standards and implementations in vehicle fleets. It may be noted that $CO_2$, $N_2O$ and $CH_4$ are known as GHGs, which are primarily responsible for global warming and climate change.

## 1.12 Sustainability Issues

### 1.12.1 Vehicle Attributes

The present and future automobiles should meet the following aspects: Vehicle price, fuel cost, performance, fuel economy, range of distance, comfort, lifetime and safety standards

### 1.12.2 Fuel Attributes

- Economic issues: low delivered price
- Performance issues: octane and cetane numbers, high energy density, easy and safe storage, other safety issues
- Environmental issues: carbon content, volatility and impurities (sulphur and hazardous air pollutants)

- Sustainability issues: will the fuel be available indefinitely or is it exhaustible?
- Costs: lifetime cost includes vehicle cost, fuel cost, maintenance, repair, licensing, insurance and end of life

Table 1.11 shows the life cycle inventory for different alternative fuels. For example, the localised emission of battery-operated vehicles is zero. However, the global net GHGs emission is higher than that of other fuels, resulting in unsustainable development. This analysis can be done using well-to-tank and tank-to-wheel analysis. Well-to-tank analysis indicates upstream sectors such as fuel production, transportation, storage and distribution, whereas tank-to-wheel analysis shows downstream sectors such as its utilisation in heat engines (internal combustion engine and gas turbines).

It is clearly seen from the above facts that alternative fuel is the best option available to tackle the fossil fuel depletion crisis, high demand–supply gap, energy self-sufficiency, environmental problems and sustainable economy development. The following chapter deals with the production of conventional fuels from resources, transportation, storage and distribution and its utilisation in internal combustion-powered vehicles and gas turbines.

Figure 1.28 shows the energy outlook for 2011, which is as discussed below:

- Nuclear prospects have been revised down after the Fukushima incident and the resulting policy changes in Japan and Europe.
- Biofuels growth (while still very robust) has been reduced due to more modest expectations of penetration of next-generation fuels.

**TABLE 1.11**

Comparison of Life Cycle Inventory Studies: Well-to-Tank Efficiencies and GHG Emissions

| S. No. | Fuel/Source | Efficiency (%) Range | GHGs (gCO$_2$ Equivalent/ MJ Fuel) Range |
|--------|-------------|----------------------|------------------------------------------|
| 1 | Gasoline | 80–87 | 15–26 |
| 2 | Diesel | 83–90 | 12–18 |
| 3 | FT-diesel | 52–59 | — |
| 4 | CNG | 83–91 | 10–27 |
| 5 | Methanol | 57–66 | 22–41 |
| 6 | Ethanol (corn): total energy | 45–67 | −19 to +90 |
| 7 | Ethanol (corn): fossil energy | 60–67 | — |
| 8 | Ethanol (lignocell): fossil | 26–56 | −85 to +14 |
| 9 | Ethanol (lignocell): fossil | 83–96 | — |
| 10 | Hydrogen (various sources) | 23–76 | 76–332 |
| 11 | Electricity (various sources) | 29–48 | 127–198 |

*Source:* Adapted from K. A. Subramanian et al., Control of GHG emissions from transport vehicles: Issues and challenges, SAE Paper No. 2008-28-0056, 2008.

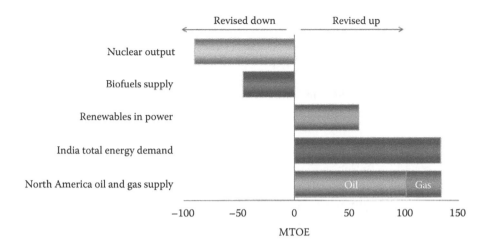

**FIGURE 1.28**
Changes in 2030 levels versus the 2011 energy outlook.

- Renewable power generation has been revised to a higher side due to improved prospects for cost reductions. They also play a role in replacing the lost nuclear output in Japan and Europe.

- Indian energy consumption has been revised upwards on a reassessment of the country's economic development path.

- North American oil and natural gas supply outlooks have been revised higher due to evolving expectations for shale gas plays.

- Population and income growth are the two most powerful driving forces behind the demand for energy. Since 1900, the world population has more than quadrupled, whereas real income has grown by a factor of 25 and primary energy consumption by a factor of 22.5 (Figure 1.29) [20].

- The next 20 years are likely to see continued global integration and a rapid growth of low- and medium-income economies. Population growth is trending down, but income growth is showing an upward trend.

- Over the last 20 years, the world population has increased by 1.6 billion people, and it is projected to rise by 1.4 billion over the next 20 years. The world's real income has risen by 87% over the past 20 years and it is likely to rise by 100% over the next 20 years.

- At the global level, the most fundamental relationship in energy economics remains robust—more people with more income means that the production and consumption of energy will rise.

- Today, the amount of energy used to produce a unit of GDP ('energy intensity') is declining steadily in most countries (Figure 1.30).

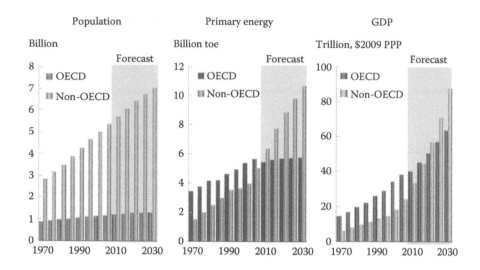

**FIGURE 1.29**
The world view of population, energy and GDP. (Adapted from BP Energy Outlook 2030 (60 Years BP Statistical Review), London, January 2011.)

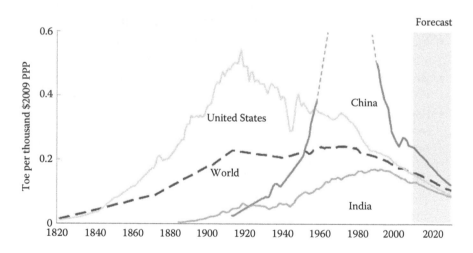

**FIGURE 1.30**
Energy usage per unit of GDP. (Adapted from BP Energy Outlook 2030 (60 Years BP Statistical Review), London, January 2011.)

- Historically, a common pattern emerges in energy intensity
  - Increases: As countries industrialise, the share of the energy-intensive industry in GDP rises relative to other sectors.
  - Peaks: Usually coincides with a peak in the share of the industrial sector in GDP. Also, the nature of the industry changes (from

heavy and energy intensive to lighter and high value added) and the industry becomes more energy efficient.

- Converges across countries: Driven by energy trade, the use of common technologies and similarities in consumption patterns.

- As one would expect, regional peak levels decline over time (as energy efficiency improves) and are higher in countries with abundant energy resources. Global competition and open markets drive convergence. Before proceeding to the study of alternative fuels, it becomes imperative to know about the conventional fuels, especially their production and utilisation, which is discussed in the next chapter.

## References

1. BP Statistical Review of World Energy, http://www.bp.com/statistical review, June 2011.
2. IEA World Energy Outlook, www.iea.org, 2009.
3. www.petroluem.nic.
4. BP Energy Outlook 2030, London, January 2012.
5. BP Statistical Review of World Energy, http://www.bp.com/statistical review S, June 2010.
6. D. R. Balachandra and N. H. Ravindranath, Energy efficiency in India: Assessing the policy regimes and their impacts, *Journal of Policy*, 38, 6428–6438, 2010.
7. Report "Energy and Transport in Figs 2010", Part 3: Transport, European Union, European Commission, Directorate General for Energy and Transport in co-operation with Eurostat, 2010.
8. www.data.worldbank.org.
9. Asian Development Bank (ADB), Energy efficiency and climate change considerations for on-road transport in Asia, Asian Development Bank, Philippines, 2006.
10. Repowering transport—A cross-industry report, World Economic Forum, Geneva, February 2011.
11. Bureau of Energy Efficiency (BEE), www.bee-india.nic.in, 2007.
12. International Energy Agency, Energy balances of non-OECD countries 2004–2005, IEA/OECD, Paris, France, 2007.
13. M. Karpoor, Vision 2020 Transport, prepared for the Planning Commission, 2002.
14. Asian Institute of Transport Development (AITD), Environmental and social sustainability of transport: Comparative study of rail and road, Asian Institute of Transport Development, New Delhi, 2002.
15. www.ipcc.ch.
16. K. A. Subramanian et al., Control of GHG emissions from transport vehicles: Issues and challenges, SAE Paper No. 2008-28-0056, 2008.
17. World Energy Outlook, International Energy Agency (IEA), Paris, 2007.

18. International Energy Agency (IEA), $CO_2$ emissions from fuel combustion: 1971–2005, IEA, Paris, 2007.
19. P. Hendriksen et al., Evaluation of the environmental performance of modern passenger cars running on petrol, diesel, automotive LPG and CNG, TNO-report 03.OR.VM.055.1/PHE, TNO Automotive, December 2003.
20. BP Energy Outlook 2030 (60 years BP Statistical Review), London, January 2011.

# 2

# Conventional Fuels for Land Transportation

## 2.1 Conventional Fuels for Spark Ignition Engines/Vehicles

Conventional fuels such as gasoline or diesel are produced in oil refineries from crude oil whose composition is given in Table 2.1.

As the crude oil cannot be used directly in the transportation sector, it is refined in an oil refinery to obtain gasoline or diesel. A typical refining process is given in Figure 2.1. Crude oil contains different hydrocarbon chains, molecular structures and boiling point ranges. The lighter fraction of crude oil, such as refinery fuel and liquefied petroleum gas (LPG) that has a lower boiling point, is separated using the atmospheric distillation process. The fuel quality of a relatively higher-boiling-point-range distilled product is upgraded by different conversion processes such as isomerisation, polymerisation and reforming to obtain gasoline fuel. Jet aviation fuel is produced from the distilled crude oil fraction after it undergoes the hydro-treating (HDT) and hydro-desulphurisation (HDS) process. The heavier fraction, which has a relatively higher boiling point, is distilled by using the vacuum distillation process. Diesel fuel is obtained by chemical treatments such as HDT, HDS, hydro-cracking, fluid catalytic cracking (FCC), alkylation and a reforming process. The tangible products such as residual fuel oil (RFO), lubricating oil, wax and tar are produced from the residual oil. The common fractions that can be obtained from crude oil are presented in Table 2.2.

Petroleum refining operations can be broadly categorised into five basic areas:

1. *Distillation*: It is the separation of crude oil in atmospheric and vacuum distillation towers into groups of hydrocarbon compounds of different boiling point ranges as described below.

   a. *Atmospheric distillation:* Distillation is carried out first under atmospheric pressure as shown in Figure 2.2. The maximum temperature up to which crude oil can be heated without cracking is about 400°C. This sets the maximum amount of distillate products.

   b. *Vacuum distillation:* The residue from this primary distillation (atmospheric) is distilled under a vacuum. Steam is used to

**TABLE 2.1**

Average Crude Oil Composition

| S. No. | Composition | Values (%) |
| --- | --- | --- |
| 1 | Carbon | 84 |
| 2 | Hydrogen | 14 |
| 3 | Sulphur, including hydrogen sulphide, sulphides, disulphides and elemental sulphur | 1–3 |
| 4 | Nitrogen (basic compound with amine groups) | <1 |
| 5 | Oxygen (found in organic compounds such as carbon dioxide, phenols, ketones and carboxylic acids) | <1 |
| 6 | Metals (nickel, iron, vanadium, copper, arsenic) | <1 |
| 7 | Salts (sodium chloride, magnesium chloride and calcium chloride) | <1 |

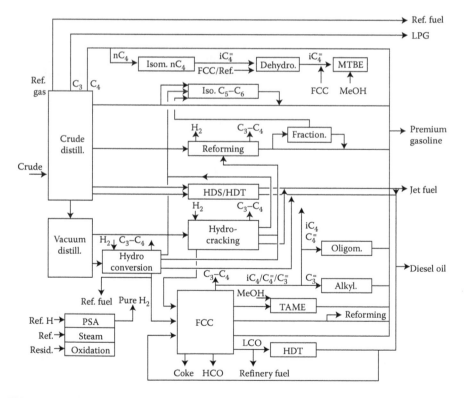

**FIGURE 2.1**

Long-term refinery configuration (2010–2020). (Adapted from D. Decroocq, Major Scientific and Technical Challenges about Development of New Processes in Refining and Petrochemistry Oil and Gas Science and Technology—Rev. IFP 52(5), September–October 1997.)

**TABLE 2.2**

Common Fractions from Crude Petroleum

| S. No. | Fraction | Approximate Boiling Range (°C) | Products |
|---|---|---|---|
| 1 | Fuel gas | −162 to −42 | Methane, ethane |
| 2 | Propane | −42 | LPG |
| 3 | Butane | −12 to −0.5 | LPG, gasoline |
| 4 | Light naphtha | −1 to −150 | Straight-run gasoline |
| 5 | Heavy naphtha | 150–205 | Solvents |
| 6 | Kerosene | 205–260 | Jet fuels |
| 7 | Stove oil | 205–290 | Illuminant fuel |
| 8 | Light gas oil | 205–315 | Fuel oil, diesel fuel |
| 9 | Heavy gas oil | 315–430 | Fuel after catalytic cracking |
| 10 | Vacuum gas oil | 430–540 | Lubricating oil |
| 11 | Residue | 595 and above | Coke, bitumen |

facilitate the vapourisation of the high-boiling-point constituents. A maximum temperature of up to 600°C is possible under vacuum distillation.

2. *Conversion*: It includes processes to change the size and structure of hydrocarbons, such as decomposition (dividing) by the thermal and

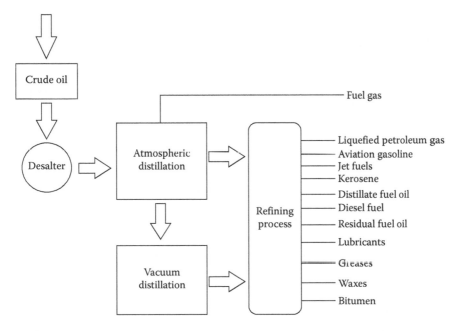

**FIGURE 2.2**
Atmospheric distillation.

catalytic cracking process, unification (combining) through the alkylation and polymerisation process, and alteration (rearranging) with the isomerisation and catalytic reforming process. Isomerisation is a chemical change that involves a rearrangement of atoms and bonds within a molecule, without changing the molecular formulae.

3. *Treatment*: The finished product still needs additional processes to improve its quality by the removal of impurities and undesirable contaminants. It may involve chemical and physical separation such as dissolving, absorbing, desalting and hydro-desulphurising.

4. *Formulating and blending*: It is the process of mixing and combining hydrocarbon fractions, additives and other components to produce finished products with specific performance properties.

5. *Other refining operations*: These include storage and handling, waste and waste water treatment and sulphur recovery.

## 2.2 Production of Gasoline/Diesel Fuels

Locating an oilfield is the first obstacle to be overcome. Today, petroleum engineers use instruments such as gravimeters and magnetometers in the search for petroleum. Generally, the first stage in the extraction of crude oil is to drill a well into the underground reservoir. Often, many wells (called multilateral wells) are drilled into the same reservoir to ensure that the extraction rate will be economically viable. Also, some wells (secondary wells) may be used to pump water, steam, acids or various gas mixtures into the reservoir to raise or maintain the reservoir pressure, and so maintain an economically viable extraction rate.

### 2.2.1 Primary Oil Recovery

If the underground pressure in the oil reservoir is sufficient, then the oil will be forced to the surface under this pressure. Gaseous fuels or natural gas are usually present, which also supply the needed underground pressure. In this situation, it is sufficient to place a complex arrangement of valves on the wellhead to connect the well to a pipeline network for storage and processing. This is called primary oil recovery. Usually, only about 20% of the oil in a reservoir can be extracted this way.

### 2.2.2 Secondary Oil Recovery

Over the lifetime of the well, the pressure will fall, and at some point, there will be insufficient underground pressure to force the oil to the surface. If economically viable, and it often is, the remaining oil in the well is extracted using secondary oil recovery methods. Secondary oil recovery uses various techniques to aid in recovering oil from depleted or low-pressure reservoirs. Sometimes, pumps, such as beam pumps and electrical submersible pumps, are used to bring the oil to the surface. Other secondary recovery techniques increase the reservoir's pressure by water injection, natural gas re-injection and gas lift, which inject air, carbon dioxide or some other gas into the reservoir. Together, primary and secondary recovery allows 25–35% of the reservoir's oil to be recovered.

### 2.2.3 Tertiary Oil Recovery

Tertiary oil recovery reduces the oil's viscosity to increase oil production. Tertiary recovery is started when secondary oil recovery techniques are no longer enough to sustain production, but only when the oil can still be extracted profitably. This depends on the cost of the extraction method and the current price of crude oil. When prices are high, previously unprofitable wells are brought back into production and when they are low, production is curtailed. Thermally enhanced oil recovery methods (TEOR) are tertiary recovery techniques that heat the oil and make it easier to extract.

- Steam injection is the most common form of TEOR. It is often done with a cogeneration plant. In this type of plant, a gas turbine is used to generate electricity and the waste heat is used to produce steam, which is then injected into the reservoir.
- *In situ* burning is another form of TEOR, but instead of steam, some of the oil is burnt to heat the surrounding oil.
- Occasionally, detergents are also used to decrease the oil's viscosity.

Tertiary recovery allows another 5–15% of the reservoir's oil to be recovered.

## 2.3 Refining Process to Produce Gasoline/Diesel

The petroleum industry can be divided into two broad groups: *upstream* producers (exploration, development and production of crude oil or natural gas) and *downstream* transporters (tanker, pipeline transport, refiners, retailers and consumers).

Raw oil or unprocessed crude oil is not very useful in the form as it comes out of the ground. It needs to be broken down into parts and refined before use in a solid material such as plastics and foams, or as petroleum fossil fuels as in the case of automobile and aeroplane engines. An oil refinery is an industrial process plant where crude oil is processed in three ways to be useful petroleum products:

1. *Separation*—separates crude oil into various fractions

   Oil can be used in so many ways because it contains hydrocarbons of varying molecular masses and lengths such as paraffins, aromatics, naphthenes (or cycloalkanes), alkenes, dienes and alkynes. Hydrocarbons are molecules of varying length and complexity made of hydrogen and carbon. The trick in the separation of different streams in the oil refinement process is the difference in boiling points between the hydrocarbons, which means they can be separated by distillation. Figure 2.3 shows the typical distillation scheme of an oil refinery.

2. *Conversion*—conversion to saleable products by skeletal alteration

   Once separated and after all contaminants and impurities have been removed, either the oil can be sold without any further processing or smaller molecules such as *iso*-butene and propylene or butylenes can be recombined to meet the specified octane number requirements by processes such as alkylation or, less commonly, dimerisation.

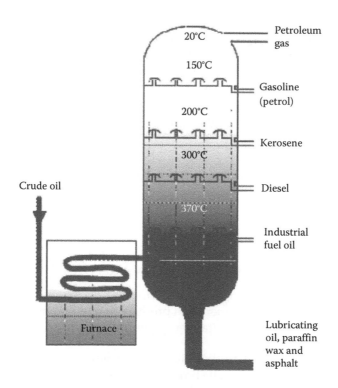

**FIGURE 2.3**
Schematic of the distillation of crude oil.

**TABLE 2.3**

Common Process Units in an Oil Refinery

| Unit Process | Function |
|---|---|
| Atmospheric distillation unit | Distils crude oil into fractions |
| Vacuum distillation unit | Further distils residual bottoms after atmospheric distillation |
| Hydro-treater unit | Desulphurises naphtha from atmospheric distillation, before sending to a catalytic reformer unit |
| Catalytic reformer unit | Reformate paraffins to aromatics, olefins and cyclic hydrocarbons that have a high octane number |
| Fluid catalytic cracking | Break down heavier fractions into lighter, more valuable products—by means of a catalytic system |
| Hydro-cracker unit | Break down heavier fractions into lighter, more valuable products—by means of a steam |
| Alkylation unit | Produces a high octane component by increasing branching or alkylation |
| Dimerisation unit | Smaller olefinic molecules of less octane number are converted to molecules of higher octane number by dimerisation of the smaller olefins |
| Isomerisation unit | Straight-chain normal alkenes of less octane number are isomerised to branched-chain alkenes of higher octane number |

Octane number requirements can also be improved by catalytic reforming, which strips the hydrogen out of the hydrocarbons to produce aromatics, which have higher octane ratings. Intermediate products such as gas oils can even be reprocessed to break heavy, long-chained oil into a lighter, short-chained one by various forms of cracking such as fluid catalytic cracking, thermal cracking and hydro-cracking. The final step in gasoline production is the blending of fuels with different octane ratings, vapour pressures and other properties to meet the product specification. The common process units and their function in an oil refinery are shown in Table 2.3.

3. *Finishing*—purification of the product streams

## 2.3.1 Details of Unit Processes

### 2.3.1.1 Hydro-Treater

A hydro-treater uses hydrogen to saturate aromatics and olefins as well as to remove undesirable compounds of elements such as sulphur and nitrogen.

The common major elements of a hydro-treater unit are a heater, a fixed-bed catalytic reactor and a hydrogen compressor. The catalyst promotes the reaction of the hydrogen with the sulphur compounds such as mercaptans to produce hydrogen sulphide, which is then usually bled off and treated with amine in an amine treater. The hydrogen is also saturated with hydrocarbon double bonds, which help to raise the stability of the fuel.

### 2.3.1.2 Catalytic Reforming

A catalytic reforming process converts a feed stream containing paraffins, olefins and naphthenes into aromatics to be used either as a motor fuel blending stock or as a source for specific aromatic compounds, namely benzene, toluene and xylene for use in petrochemicals production. The product stream of the reformer is generally referred to as a reformate. Reformate produced by this process has a high octane rating. Significant quantities of hydrogen are also produced as a by-product. Catalytic reforming is normally facilitated by a bifunctional catalyst that is capable of rearranging and breaking long-chain hydrocarbons as well as removing hydrogen from naphthenes to produce aromatics. This process is different from steam reforming, which is also a catalytic process that produces hydrogen as the main product.

### 2.3.1.3 Cracking

In an oil refinery, cracking processes allow the production of light products (such as LPG and gasoline) from heavier crude oil distillation fractions (such as gas oils) and residues. FCC produces a high yield of gasoline and LPG while hydro-cracking is a major source of jet fuel, gasoline components and LPG. Thermal cracking is currently used to upgrade very heavy fractions or to produce light fractions or distillates, burner fuel and/or petroleum coke. Two extremes of thermal cracking in terms of product range are represented by the high-temperature process called steam cracking or pyrolysis (750–900°C or more), which produces valuable ethylene and other feedstocks for the petrochemical industry, and the milder-temperature delayed coking (500°C), which can produce, under the right conditions, valuable needle coke, a highly crystalline petroleum coke used in the production of electrodes for the steel and aluminium industries.

Fuel oil is a fraction obtained from petroleum distillation. It is made of a long hydrocarbon chain such as paraffin (alkenes), naphthenic (cycloalkanes) and aromatics. The fuel oil carbon chain length is typically in the range of 20–70 atoms.

#### 2.3.1.3.1 Thermal Cracking

Thermal cracking is carried out without any catalyst at a temperature of 450–750°C and a pressure of 1–70 atmospheres. The important reactions are decomposition, dehydrogenation, isomerisation and polymerisation as shown in Equations 2.1 and 2.2. The paraffinic molecules are first decomposed into lower-molecular-weight hydrocarbons, usually paraffin and olefin.

$$CH_3-(CH_2)_8-CH_3 \rightarrow CH_3-(CH_2)_4-CH_3 + CH_3-CH_2-CH=CH_2$$

$$\textit{n-decane} \qquad\qquad \textit{n-hexane} \qquad\qquad \text{butene}$$

(2.1)

Subsequently, the olefins can undergo isomerisation, dehydrogenation and polymerisation reactions

$$2(CH_3-CH_2-CH=CH_2) \rightarrow \underset{\underset{\underset{CH_3}{|}}{\underset{|}{CH_3}}{CH_3-C-CH_2-C=CH_2}} \quad (2.2)$$

butene

*iso*-octene

### 2.3.1.3.2 Catalytic Cracking

Catalytic cracking is more advantageous than thermal cracking due to the following reasons:

- The catalyst accelerates the reaction rate.
- A new type of reaction is possible with the catalyst for more yield.
- There is uncontrolled over-cracking and thermal degradation with thermal cracking.
- Thermal degradation and over-cracking lead to a higher volumetric flow rate of gas.
- The main effect of a catalyst may be to direct the cracking of paraffin towards the centre of the molecule.
- The use of a catalyst minimises the formation of coke during cracking.
- Crackability is higher for heavier fractions.
- Paraffinic feed is more crackable than aromatic.

Coke particles settling on the catalyst surface lead to deactivation. Regeneration of the catalyst is necessary for economic viability as well as to speed up the process. The catalyst is recirculated to the reactor and regenerator, which is called FCC.

### 2.3.1.3.3 Hydro-Cracking

Hydro-cracking is a two-stage process combining catalytic cracking and hydrogenation. Heavier feedstocks are cracked in the presence of hydrogen to produce more desirable products. This process employs high pressure, high temperature, a catalyst and hydrogen. Hydro-cracking is used for feedstocks that are difficult to process by either catalytic cracking or reforming. Feedstock is usually characterised by a high polycyclic aromatic content and/or high concentrations of the two principal catalyst poisons, the sulphur and nitrogen compounds.

The hydro-cracking process largely depends on the nature of the feedstock and the relative rates of the two competing reactions, hydrogenation and cracking. Heavy aromatic feedstock is converted into lighter products under a wide range of very high pressures (6895–13,790 kPa) and fairly high temperatures (672–1090 K) in the presence of hydrogen and special catalysts. When the feedstock has a high paraffinic content, the primary function of hydrogen is to prevent the formation of polycyclic aromatic compounds. Another important role of hydrogen in the hydro-cracking process is to reduce tar formation and prevent the build-up of coke on the catalyst.

Hydrogenation also serves to convert the sulphur and nitrogen compounds present in the feedstock to hydrogen sulphide and ammonia. Hydro-cracking produces relatively large amounts of *iso*-butane for alkylation feedstock. Hydro-cracking also performs isomerisation for pour-point control and smoke-point control, both of which are important in high-quality jet fuel.

### 2.3.1.4 Alkylation

Alkylation is a reaction in which an *iso*-paraffin is combined with an olefin to produce a branched-chain *iso*-paraffin with molecular weight equal to the sum of the reactants. A gaseous *iso*-paraffin is reacted with an olefin (propene or butene) in the presence of sulphuric acid or hydrofluoric acid as catalyst. The conversion of *iso*-butane and *n*-butene into *iso*-octane is shown in Equations 2.3 and 2.4.

$$C_4H_{10} \quad + \quad C_4H_8 \quad \rightarrow \quad C_8H_{18} \qquad (2.3)$$
$$\text{\textit{iso}-butane} \quad \text{\textit{n}-butene} \quad \text{\textit{iso}-octane}$$

$$
\begin{array}{ccc}
\quad\quad\text{CH}_3 & & \text{CH}_3 \quad \text{CH}_3 \\
\quad\quad| & & | \quad\quad | \\
\text{CH}_3\text{--CH} + \text{CH}_2\text{=CH--CH}_2\text{--CH}_3 & \rightarrow & \text{CH}_3\text{--C--CH}_2\text{--CH--CH}_3 \quad (2.4) \\
\quad\quad| & & | \\
\quad\quad\text{CH}_3 & & \text{CH}_3
\end{array}
$$

$$\text{\textit{iso}-butane} \qquad\qquad \text{olefin} \qquad\qquad \text{\textit{iso}-octane}$$

### 2.3.1.5 Polymerisation

Polymerisation is a method of converting the $C_3$ and $C_4$ olefins produced in the thermal and catalytic cracking processes into high-octane gasoline. Polymerisation in the petroleum industry is the process of converting light olefin gases, including ethylene, propylene and butylene, into hydrocarbons of higher molecular weight and higher octane number that can be used as gasoline blending stocks (Equation 2.5). Polymerisation combines two or more identical olefin molecules to form a single molecule with the same elements in the same proportions as the original molecules. Polymerisation may be accomplished thermally or in the presence of a catalyst at lower temperatures.

$$2(CH_3-CH_2-CH=CH_2) \rightarrow CH_3-\overset{\overset{\displaystyle CH_3}{|}}{CH}-CH_2-\overset{\overset{\displaystyle CH_3}{|}}{\underset{\underset{\displaystyle CH_3}{|}}{C}}=CH_2 \qquad (2.5)$$

butene                  *iso*-octene

### 2.3.1.6 Isomerisation

Isomerisation is the process employed for converting *n*-butane into *iso*-butane for alkylation. *n*-Pentane and *n*-hexane are converted into *iso*-paraffin for improving the octane number of the gasoline fuel (Equation 2.6).

$$CH_3-CH_2-CH_3-CH_3 \rightarrow \overset{\overset{\displaystyle CH_3}{|}}{\underset{\underset{\displaystyle CH_3}{|}}{CH}}-CH_3 \qquad (2.6)$$

*n*-butane                  *iso*-butane

### 2.3.1.7 Reforming

Reforming is a process of producing high-octane gasoline by heating, with or without a catalyst, for the naphtha fractions of the gasoline boiling range. Catalytic reforming is a chemical process used to convert petroleum refinery naphthas, typically having low octane ratings, into high-octane gasoline. Basically, the process rearranges or restructures the hydrocarbon molecules in the naphtha feedstocks as well as breaks some of the molecules into smaller molecules. The process is done by rearranging or reforming the molecules without disturbing their average molecular weight.

The main reactions are the formation of aromatics by the dehydrogenation of naphthenes and the dehydrocyclisation of paraffin. The process separates the hydrogen atoms from the hydrocarbon molecules and produces very significant amounts of by-product hydrogen gas for use in a number of other processes involved in a modern petroleum refinery.

## 2.4 Conventional Fuels for a Spark Ignition Engine

### 2.4.1 Motor Gasoline

Gasoline (gas), petroleum spirit (petrol) or petro-gasoline is a petroleum-derived liquid mixture consisting mostly of aliphatic hydrocarbons, enhanced with *iso*-octane or the aromatic hydrocarbons toluene and benzene to increase its octane rating. It is primarily used as a fuel in internal combustion engines. Gasoline is produced in oil refineries. It is separated from crude oil via distillation. Gasoline became the preferred automobile fuel because it releases a great deal of energy when burnt, as it is mixed readily with air in a carburettor.

### 2.4.1.1 Physico-Chemical Properties

#### 2.4.1.1.1 Gasoline Fuel Quality and Its Impact on Environment

Gasoline is obtained from the refinery and the properties that are to be considered are as follows.

#### 2.4.1.1.2 Octane Number

Octane number denotes the ability of the fuel to resist auto-ignition, which can cause engine knocks. The ignition quality of a gasoline is characterised by using its octane number. Octane number is determined in a cooperative fuel research (CFR) engine with *n*-heptane, which has an octane number of 0, and *iso*-octane, which has an octane number of 100 as per the ASTM method D2699 and D2700. The gasoline octane number is measured as the research octane number (RON) and motor octane number (MON). RON correlates best with low-speed, mild knocking conditions and MON correlates with high-temperature knocking conditions and a part throttle operation. RON values are higher than MON values, and the difference between these values is the sensitivity, which is less than 10. The average value of RON and MON is called the anti-knock index (AKI). The compound tetraethyl lead was used earlier as an additive to improve the octane number. However, owing to its toxic effect in terms of lead poisoning of the environment, it was subsequently banned.

Generally, vehicles are designed for a particular octane number. Gasoline with an octane number lower than required may result in knocking, which could lead to severe engine damage. Lower octane number can be handled by the engines equipped with knock sensors by retarding the spark timing. However, fuel consumption and maximum power may suffer, and at a very low octane number, knocks may still occur. Using a higher octane rating than that required will not improve the vehicle performance.

Octane number has no direct correlation to engine emissions. A higher octane number fuel permits the use of higher compression ratio engines. The higher thermal efficiency results in a better fuel economy. It may result in lower carbon dioxide emissions. Higher compression ratios may lead to higher cylinder gas temperatures, which may give rise to an increased level of nitrogen oxide emissions.

Lead alkyl additive was used worldwide to enhance the octane rating of gasoline. However, the health effect associated with the use of lead has led to the elimination of leaded gasoline in several countries. Vehicle emission control technologies like catalytic converters and oxygen sensors necessitate the use of unleaded petrol. The tolerance to lead contamination is very low as the efficiencies of the vehicle catalyst decrease since even a slight presence of lead can destroy a modern catalyst. Lead-free gasoline, therefore, is essential in the long term.

The octane number of gasoline could be enhanced up to about 95 by refinery chemical processes such as isomerisation, alkylation and catalytic reforming as shown in Figure 2.4. The research and motor octane numbers

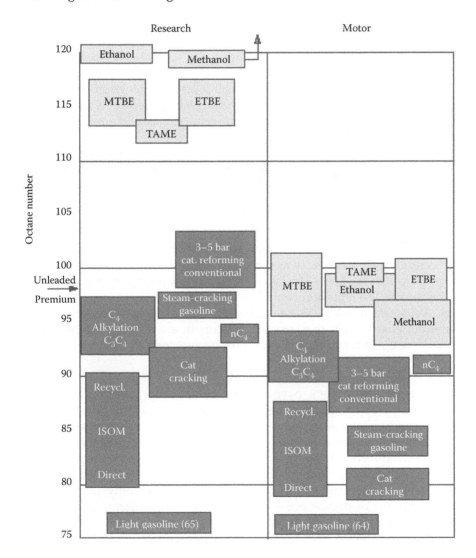

**FIGURE 2.4**

Octane number improvement with different refinery processes, additives and alternative fuels. (Adapted from D. Decroocq, Major Scientific and Technical Challenges about Development of New Processes in Refining and Petrochemistry Oil and Gas Science and Technology—Rev. IFP 52(5), September–October 1997.)

can be increased up to 115 and 100, respectively, by using additives such as methyl tetrabutyl ether (MTBE), tetra amine methyl ether and ethyl tetrabutyl ether. This is the major limitation of octane number increase for crude oil-based gasoline fuel. Alternative fuel such as methanol- and ethanol-blended gasoline is a viable and proven option to further increase the octane number to obtain the benefits of higher brake thermal efficiency from automobile vehicles.

### 2.4.1.1.3 Sulphur Content

Comprehensive testing done on the impact of sulphur on vehicular emissions indicates that emission reduction occurs with different technologies as sulphur is reduced from a higher to a lower level. Percentage emission reduction owing to lower sulphur in gasoline shall be high in case of vehicles complying with advanced vehicle emission technologies and low for vehicles with a lower level of technology.

Sulphur affects the efficiency of the catalyst used in after-treatment devices. It also adversely affects heated exhaust gas sensors. Reduced sulphur in fuel will provide a reduction of emissions from all catalyst-equipped vehicles. Hence, a low sulphur level in the fuel is an important parameter.

### 2.4.1.1.4 Ash-Forming Additives

To provide the precise closed-loop control of emission, sophisticated emission control equipments such as three-way catalysts and exhaust gas sensors are employed in modern vehicles. To maintain low emissions for the lifetime of the vehicle, the emission control systems must be kept in optimal condition. Ash-forming additives can adversely affect the operation of catalysts and other components (e.g., oxygen sensors) in a way that increases emissions. Hence, ash-forming additives must be avoided in gasoline.

### 2.4.1.1.5 Oxygenates

Owing to several reasons, oxygenated organic components such as ethers and ethanol may be added to petrol to increase the octane number. It may result in an increase in gasoline volume, induce a lean shift in engine stoichiometry and reduce carbon monoxide emissions. Carbon monoxide emissions are reduced during leaner operation, especially with carburettor vehicles without electronic feedback-controlled operation. In modern electronic feedback-controlled vehicles, the leaning effect only occurs during cold operations or during accelerations. Hence, the emission benefits from leaner operation are not realised.

Fuel leaning caused by oxygenates can affect drivability. In few cases, the over-leaning can also cause emissions to increase. As ethanol has a higher heat of vapourisation than ethers, some of the driveability degradation of petrol oxygenated with ethanol can be attributed to the additional heat needed to vapourise petrol.

### 2.4.1.1.6 *Olefin Content*

Olefins are unsaturated hydrocarbons and, in many cases, are also good octane components of petrol. Higher olefins in petrol may lead to deposit formation and increased emissions of reactive hydrocarbons and undesirable compounds. In two- and three-wheelers, they have been reported to lead to pre-ignition. Being thermally unstable, higher olefins may lead to gum formation and deposits in an engine intake system. Their evaporation into the atmosphere as chemically reactive species contributes to ozone formation.

### 2.4.1.1.7 *Aromatic Content*

Generally, aromatics are good octane components of petrol and higher-energy-density fuel molecules. A higher level of aromatics can enhance engine deposits and increase tailpipe emissions, including $CO_2$. Combustion chamber deposits have been linked to heavy aromatics and other high molecular weight compounds and these deposits increase tailpipe emissions, including HC and $NO_x$. Tailpipe $CO_2$ emission is a direct effect of petrol content. The linear relationship between $CO_2$ emissions and aromatic content was demonstrated by the EPEFE (European Program on Emissions, Fuels and Engine Technologies).

### 2.4.1.1.8 *Benzene Content*

Benzene is a natural constituent of crude oil. It is a product of catalytic reforming that produces high-octane petrol streams. It is known to be a human carcinogen. Evaporative and exhaust emissions of benzene from automobiles can be directly controlled by the benzene levels in petrol. An effective way to reduce human exposure to benzene is to control benzene in petrol.

### 2.4.1.1.9 *Volatility*

During the operation of spark ignition engines, both performance and emissions are affected by the volatility of gasoline. Hence, proper volatility of petrol has to be fixed. It is characterised by vapour pressure and distillation. The volatility needs of vehicles at different temperatures differ to a large extent. Hence, the vapour pressure should be controlled seasonally.

To reduce the possibility of hot fuel-handling problems, such as vapour lock or carbon canister overloading at high temperatures, the vapour pressure must be controlled precisely. At high temperatures, the control of vapour pressure is also important for the reduction of evaporative emissions. Higher vapour pressure at lower temperature is needed to allow good starting and warm-up performance.

Hot weather fuel-handling problems such as vapour lock, canister loading and higher emissions can be caused by excessively high petrol volatility.

Vapour lock occurs when too much vapour forms in the fuel system and fuel flow decreases to the engine. This can result in the loss of power, rough engine operation or engine stalls. A vapour/liquid ratio specification is necessary as the vapour pressure and distillation properties are not sufficient to ensure good vehicle performance.

### 2.4.1.1.10 Paraffins

The general formula of paraffin is $C_nH_{2n+2}$ ($n$ is a whole number, usually from 1 to 20). Straight- or branched-chain molecules can be gases or liquids at room temperature, depending upon the molecule. The examples are methane, ethane, propane, butane, *iso*-butene, pentane and hexane.

### 2.4.1.1.11 Aromatics

The general formula of aromatics is $C_nH_{2n-6}$. Ringed structures with one or more rings contain six carbon atoms, with alternating double and single bonds between the carbon atoms, typically liquids, for example, benzene and naphthalene.

### 2.4.1.1.12 Naphthenes or Cycloalkanes

The general formula is $C_nH_{2n}$ ($n$ is a whole number usually from 1 to 20). Ringed structures with one or more rings contain only single bonds between the carbon atoms, typically liquids at room temperature, for example, cyclohexane and methyl cyclopentane.

### 2.4.1.1.13 Alkenes

The general formula of alkenes is $C_nH_{2n}$ ($n$ is a whole number, usually from 1 to 20). Linear- or branched-chain molecules containing one carbon–carbon double bond can be liquid or gas, for example, ethylene, butane and *iso*-butene.

### 2.4.1.1.14 Alkynes

The general formula of alkynes is $C_nH_{2n-2}$ ($n$ is a whole number, usually from 1 to 20). Linear- or branched-chain molecules containing two carbon–carbon double bonds can be liquid or gas, for example, acetylene and butadienes.

A brief summary of the important characteristics of gasoline fuel is given in Table 2.4.

To minimise a significant change in the design of a spark ignition engine, it is desirable that the properties of any selected alternative fuel be preferably closer to that of the conventional gasoline fuel.

## 2.4.2 Liquefied Petroleum Fuels

LPG is a mixture of hydrocarbons, mainly propane ($C_3H_8$) and butane ($C_4H_{10}$). The composition of LPG variation depends on its end use and a country's geographical profile such as seasonal atmospheric temperature and the type

**TABLE 2.4**

Important Characteristics of Gasoline Fuels

| Fuel Characteristics | Comments |
| --- | --- |
| Octane number | a. Measure of anti-knocking quality |
| | b. High octane number implies short ignition delay |
| | c. *Iso*-paraffins and aromatics generally have higher octane numbers |
| | d. Influences both gaseous and particulate emissions |
| | e. Fuels with low auto-ignition temperatures are more likely to cause knocking |
| Distillation range | a. Affects fuel performance and safety |
| | b. Important to an engine's start and warm-up |
| | c. Presence of high-boiling components affects the degree of formation of solid combustion deposits |
| Specific gravity | Required for the conversion of measured volumes to volumes at a temperature of 15°C |
| Heat of combustion | a. A measure of energy available in a fuel |
| | b. A critical property of fuel intended for use in weight-limited vehicles |
| Flash point | a. Indicates the presence of highly volatile and flammable materials |
| | b. Measures the tendency of oil to form a flammable mixture with air |
| | c. Used to assess the overall flammability hazard of a material |
| Viscosity | a. Appropriate viscosity of fuel required for proper operation of an engine |
| | b. Important for flow of oil through pipelines, injector nozzles and orifices |
| | c. Effective atomisation of fuel in the cylinder requires a limited range of viscosity of the fuel to avoid excessive pumping pressures |
| Contamination (water/sediment) | a. Causes corrosion of equipment |
| | b. Causes problems in processing |
| | c. Required to accurately measure net volumes of actual fuel in sales, taxation, exchanges and custody transfer |
| Copper-strip corrosion | a. A measure to assess the relative degree of corrosivity |
| | b. Indicates the presence of sulphur compounds |
| Carbon residue | a. Correlates with the amount of carbonaceous deposits in the combustion chamber |
| | b. Greater carbon deposits expected for higher values of carbon residue |
| Ash | a. Results from oil, water-soluble metallic compounds or extraneous solids, such as dirt and rust |
| | b. Can be used to decide product's suitability for a given application |
| Sulphur | a. Controlled to minimise corrosion, wear and tear |
| | b. Causes environmental pollution from their combustion products |
| | c. Corrosive in nature and causes physical problems to engine parts |

of crude that is processed by refineries for LPG production. LPG is a by-product that is derived from natural gas extraction (55%) and petroleum refineries (45%). The hydrocarbon ($C_3$ and $C_4$) emanated during the atmospheric distillation refinery process is known as petroleum gas. Then the gas is liquefied by refrigerating it to below 400 K to condense gas into a liquid. The liquefaction process removes most of the water vapour, butane, propane and other trace gases that are usually included in ordinary natural gas. The typical composition of LPG is shown in Table 2.5 and the properties of propane, butane and LPG are given in Table 2.6.

Colder-climate countries use higher propane percentages due to easy vapourisation at low atmospheric temperature, whereas warmer countries use higher butane percentages. For example, different countries use various propane–butane percentages by volume such as Belgium (50%–50%), Denmark (50%–50%), France (35%–65%), Greece (20%–80%), Ireland (100%–0%), Italy (25%–75%), Netherlands (50%–50%), Spain (30%–70%), Sweden (95%–5%), United Kingdom (100%–0%), Germany (90%–10%) [4].

The LPG components are gases at room temperature but they will be in liquid state when they are compressed. LPG is stored in special tanks that keep it under pressure about 1379 kPa. An LPG-fuelled engine emits lesser carbon-based emissions (CO, HC and $CO_2$) as the carbon-to-hydrogen ratio of LPG (propane: $C_3H_8$; butane: $C_4H_{10}$) is lesser than that of gasoline ($C_8H_{18}$). Thermal efficiency of the engine increases due to the higher rate of mixture formation in view of its gaseous state.

The thermal output of a heat engine/combustion system depends on the calorific value. If inert gases such as $CO_2$, $N_2$, $H_2S$ and inorganic components are present, the calorific value of the fuel decreases and it affects the power output of an internal combustion engine.

**TABLE 2.5**

Composition of LPG (Meeting IS 4576 Specifications)

| S. No. | Component | Unit | Specification |
|--------|-----------|------|---------------|
| 1 | C2 (ethane) | Vol% | 0.2–0.1 |
| 2 | C3 (propane) | Vol% | 0.2–0.1 |
| 3 | C4 (butane) | Vol% | 37–47 |
| 4 | C5 (pentane) and heavier | Vol% | 1.5–2.5 (maximum) |
| 5 | Total olefins | Vol% | Nil |
| 6 | Corrosive compounds copper strip | | Not worse than 1 |
| 7 | $H_2S$ | ppm | Nil |
| 8 | Sulphur (volatile) | ppm | Nil |
| 9 | Total water content | | None |
| 10 | Calorific value (gross) | kcal/kg | 12,500 |
| 11 | Moisture | | Nil |

*Source:* Adapted from http://www.gailonline.com/gailnewsite/businesses/lpgproperties.html (GAIL).

**TABLE 2.6**

Properties of Propane, Butane and LPG

| S. No. | Particulars | Propane | Butane | LPG |
|---|---|---|---|---|
| 1 | Chemical formulae | $C_3H_8$ | $C_4H_{10}$ | 60% butane, 40% propane mix |
| 2 | Maximum vapour pressure saturated in kg/cm² at 65°C | 22.66 | 6.32 | 16.87 |
| 3 | Gross calorific value in kJ/kg | 49,813 | 49,359 | 49,526 |
| 4 | Specific gravity (liquid) at 15°C water = 1 | 0.504 | 0.582 | 0.543 |
| 5 | Specific gravity (vapour) at 15°C air = 1 | 1.50 | 2.01 | 1.75 |
| 6 | Ideal combustion ratio (air to gas) | 24 to 1 | 31 to 1 | 28 to 1 |
| 7 | Flammability limits (upper) | 9.60% | 8.60% | 9.1% |
| 8 | Flammability limits (lower) | 2.15% | 1.55% | 1.90% |
| 9 | Ignition temperature (°C) | 493–504 | 482–537 | 488–502 |
| 10 | Volume of gas produced per unit volume of liquid | 274 | 233 | 250 |
| 11 | Volume of air required to burn unit volume of gas | 23 | 30 | 26 |
| 12 | Volume of oxygen required to burn unit volume of gas | 4.8 | 6.25 | 5.5 |
| 13 | Maximum flame temperature (°C) | 1979.44 | 1990 | 1985 |
| 14 | Volatility: evaporation temperature in °C for 95 (°C) by volume at 760 mm Hg pressure maximum | –38 | 2 | 2 |
| 15 | Boiling points (°C) | –45 | –2 | –22 |
| 16 | Percent gas in air for maximum flame temperature | 4.4 | 3.5 | 3.9 |
| 17 | Limits of flammability (lower) (% gas in gas/air mixture) (upper) | 2.0; 11.0 | 1.9; 8.5 | 1.95; 9.75 |

*Source:* Adapted from http://www.gasindia.in/technical-specification.html (Indian Oil).

The inert gas alters the specific gravity of the mixer gas. The Wobbe index or Wobbe number is a function of the calorific value and the specific gravity of gas as shown in Equation 2.7. The Wobbe index is a measure of the amount of heat released by a gas burner with a constant orifice, which is equal to the gross calorific value of the gas at standard temperature and pressure divided by the square root of the specific gravity of the gas as shown in Equation 2.7.

Wobbe index or Wobbe number = CV/square root of specific gravity  (2.7)

### 2.4.2.1 Availability of LPG

Gas Authority of India Limited (GAIL) produces LPG at its seven fractionating units. The details of its locations and their production capacities are given in Table 2.7. The yearwise production is shown in Table 2.8.

**TABLE 2.7**

Production of Petroleum Products in India

| S. No. | Location | LPG Production (MT/annum) |
|--------|----------|--------------------------|
| 1 | Vijaipur (2 nos.), Madhya Pradesh | 406,000 |
| 2 | Auraiya Pata, Uttar Pradesh | 258,250 |
| 3 | Gandhar and Vaghodia, Gujarat | 207,000 and 73,000 |
| 4 | Usar, Maharashtra | 139,500 |
| 5 | Lakwa, Assam | 85,000 |

*Source:* Adapted from Basic statistics on petroleum and natural gas, Ministry of petroleum and natural gas, Government of India, http://petroleum.nic.in/petstat.pdf, 2011.

### 2.4.2.2 Cost of LPG in India

The international prices of crude oil and major petroleum products are given in Table 2.9 and the growth rate of the LPG market in India is shown in Table 2.10. The retail selling prices of petroleum products in the metropolitan cities of India are shown in Table 2.11.

### 2.4.2.3 Production Process of LPG

Because LPG has natural origins, it is not 'made' of other raw materials; instead, it is 'found' in petroleum chemical mixtures deep within the earth. These petroleum mixtures are literally rock oil, combinations of various hydrocarbon-rich fluids that accumulate in subterranean reservoirs made of porous layers of sandstone and carbonate rock. Petroleum is derived from various living organisms buried with sediments of early geological eras. The organisms were trapped between rock layers without oxygen and could not break down, or oxidise, completely. Instead, over tens of millions of years, the residual organic material was converted to propane-rich petroleum via two primary processes, diagenesis and catagenesis. Diagenesis occurs below 50°C when the organic 'soup' undergoes microbial action (and some chemical reactions), which results in dehydration, condensation, cyclisation and polymerisation. Catagenesis, on the other hand, occurs under high temperatures of 50–200°C and causes the organic materials to react via thermocatalytic cracking, decarboxylation and hydrogen disproportionation. These complex reactions form petroleum in the sedimentary rocks. A schematic diagram of the production process of LPG is shown in Figure 2.5.

LPG manufacture involves the separation and collection of gases from its petroleum sources. LPGs are isolated from petrochemical mixtures in one of two ways—by separation from the natural gas phase of petroleum and by refinement of crude oil.

1. Both processes begin when underground oilfields are tapped by drilling oil wells. The gas/oil hydrocarbon mixture is piped out of the well and into a gas trap, which separates the stream into

**TABLE 2.8**

Yearwise Production of LPG in India

| Products | 2005–2006 | 2006–2007 | 2007–2008 | 2008–2009 | 2009–2010 | 2010–2011 ('000 tonne) |
|---|---|---|---|---|---|---|
| *a. From Crude Oil* | | | | | | |
| **1. Light distillates of which** | **32,427** | **38,104** | **40,111** | **40,222** | **51,197** | **55,197** |
| LPG | 5525 | 6315 | 6732 | 6996 | 8091 | 7538 |
| Mogas | 10,502 | 12,539 | 14,167 | 16,020 | 22,537 | 26,135 |
| Naphtha | 14,509 | 16,660 | 16,440 | 14,826 | 17,105 | 17,531 |
| Others LD | 1891 | 2590 | 2772 | 2380 | 3464 | 3994 |
| **2. Middle distillates of which** | **64,432** | **71,225** | **76,649** | **80,309** | **93,790** | **99,776** |
| Kerosene | 9078 | 8491 | 7794 | 8223 | 8545 | 7702 |
| ATF/RTF/jet A-1 | 6196 | 7805 | 9107 | 8071 | 9296 | 9570 |
| HSD | 47,572 | 53,465 | 58,361 | 62,889 | 73,281 | 78,053 |
| LDO | 923 | 803 | 671 | 606 | 472 | 578 |
| Others MD | 663 | 661 | 716 | 520 | 2196 | 3874 |
| **3. Heavy end of which** | **22,891** | **25,931** | **28,170** | **29,985** | **34,782** | **35,391** |
| Furnace oil | 10,320 | 12,325 | 12,638 | 14,749 | 15,828 | 18,659 |
| LSH/HHS/RFO | 3985 | 3372 | 3166 | 2935 | 2518 | 1860 |
| Lube oils | 677 | 825 | 881 | 874 | 950 | 737 |
| Bitumen | 3576 | 3891 | 4507 | 4713 | 4889 | 4478 |
| Petroleum coke | 3182 | 3779 | 4129 | 4241 | 3709 | 2632 |
| Paraffin wax | 63 | 63 | 64 | 69 | 64 | 179 |
| Others waxes | 3 | 5 | 7 | 5 | 3 | 6 |
| **Total (1 + 2 + 3)** | **119,750** | **135,260** | **144,930** | **150,516** | **179,769** | **190,364** |
| *b. From Natural Gas* | | | | | | |
| LPG | 2185 | 2093 | 2060 | 2162 | 2243 | 2168 |

*Source:* Adapted from Basic statistics on petroleum and natural gas, Ministry of petroleum and natural gas, Government of India. http://petroleum.nic.in/petstat.pdf, 2011.

*Notes:* Includes RIL SEZ production in 2009–2010 and 2010–2011.

*Light distillate:* Includes propylene, C-3, propane, hexane, special boiling point spirit, benzene, toluene, petroleum hydrocarbon solvent, natural heptanes, methyl tertiary butyl ether, poly *iso*-butene, poly butadiene feedstock and methyl ethyl ketone feedstock.

*Middle distillate:* Includes mineral turpentine oil, JP-5, linear alkyl benzene feedstock, aromex, jute batching oil, solvent 1425, low sulphur, heavy fuel HSD, desulphurisation hydro-cracker bottom and special kerosene.

*Heavy ends:* Includes carbon black feedstock, sulphur, solar oil, light aluminium rolling oil and extracts.

[a] Provisional. The "'000 tonne" means the values in table are multiplied by 1000.

**TABLE 2.9**

International Prices of Crude Oil and Major Petroleum Products

| Period | Crude Oil (Indian Basket) | Petrol | Diesel | Kerosene | LPG (US$/Barrel & LPG US$/MT) |
|---|---|---|---|---|---|
| 2002–2003 | 26.65 | 30.04 | 28.86 | 29.24 | 279.67 |
| 2003–2004 | 27.97 | 35.01 | 30.39 | 31.11 | 277.02 |
| 2004–2005 | 39.21 | 48.97 | 46.91 | 49.51 | 368.57 |
| 2005–2006 | 55.72 | 64.51 | 64.7 | 69.43 | 481.04 |
| 2006–2007 | 62.46 | 72.62 | 74.12 | 77.03 | 499.67 |
| 2007–2008 | 79.25 | 90.76 | 92.91 | 94.33 | 683.49 |
| 2008–2009 | 83.57 | 89.42 | 101.75 | 104.37 | 688.00 |
| 2009–2010 | 69.76 | 76.23 | 74.67 | 75.35 | 582.69 |
| 2010–2011 | 85.09 | 92.43 | 95.66 | 96.79 | 746.20 |

*Source:* International Trade Department, IOCL; Adapted from Basic statistics on petroleum and natural gas, Ministry of petroleum and natural gas, Government of India. http://petroleum.nic.in/petstat.pdf, 2011.

*Note:* The composition of the Indian basket of crude represents the average of Oman and Dubai for sour grades and brent (dated) for sweet grade in the ratio of 64:37 for 2009–2010; 62:38 for 2008–2009; 61:39 for 2007–2008; 60:40 for 2006–2007; 58:42 for 2005–2006 and 57:43 for the prior periods.

Price of kerosene and diesel (0.5% sulphur) is for the Arab market. LPG price is Saudi Aramco CP based on 60:40 butane:propane ratio. Price of petrol is 92 RON unleaded for the Singapore market.

crude oil and 'wet' gas, which contains natural gasoline, LPGs and natural gas. Crude oil is heavier and sinks to the bottom of the trap; it is then pumped into an oil storage tank for later refinement. (Although propane is most easily isolated from the 'wet gas' mixture, it can be produced from crude oil. Crude oil undergoes a variety of complex chemical processes, including catalytic cracking, crude distillation and others. While the amount of propane produced by refinery processing is small compared to the amount separated from natural gas, it is still important because propane produced in this manner is commonly used as a fuel for refineries or to make LPG or ethylene.)

2. The 'wet' gas comes off the top of the trap and is piped to a gasoline absorption plant, where it is cooled and pumped through an absorption oil to remove the natural gasoline and LPGs. The remaining dry gas, about 90% methane, comes off the top of the trap and is piped to towns and cities for distribution by gas utility companies.

3. The absorbing oil, saturated with hydrocarbons, is piped to a still where the hydrocarbons are boiled off. This petroleum mixture is known as 'wild gasoline'. The clean absorbing oil is then returned to the absorber, where it repeats the process.

**TABLE 2.10**

Growth in LPG Marketing in India

| Item | Unit | 2005–2006 | 2006–2007 | 2007–2008 | 2008–2009 | 2009–2010 | 2010–2011* (US$/Barrel &LPG US$/MT) |
|---|---|---|---|---|---|---|---|
| Consumption (PSU/PVT) | TMT | 10,456 | 10,849 | 12,010 | 12,191 | 13,135 | 14,328 |
| LPG customers (PSUs)@ | Lac | 891 | 949 | 1018 | 1068 | 1163 | 1269 |
| LPG distributors (PSUs)@ | No. | 9270 | 9363 | 9365 | 9366 | 9686 | 10,541 |
| LPG markets (PSUs)@ | No. | 4288 | 4359 | 4393 | 4420 | 4599 | 4866 |
| Enrolment | Lac | 46.6 | 53.8 | 64.9 | 53.2 | 86.2 | 104.2 |
| Indigenous production (PSU/Pvt) of which | TMT | 7717 | 8454 | 8868 | 9170 | 10,338 | 9685 |
| a. Refineries | TMT | 5532 | 6359 | 6743 | 7008 | 8087 | 7518 |
| b. Fractionators | TMT | 2185 | 2095 | 2125 | 2162 | 2249 | 2167 |
| Imports (PSU/Pvt) | TMT | 2883 | 2278 | 2833 | 2360 | 2718 | 4338 |
| Bottling capacity@ | TMTPA | 8122 | 8448 | 8697 | 8967 | 11,569 | 11,889 |

*Source:* PPAC; Adapted from Basic statistics on petroleum and natural gas, Ministry of petroleum and natural gas, Government of India. IOCL, http://petroleum.nic.in/petstat.pdf, 2011.

*Note:* *provisional; @year-end position.

4. The 'wild gasoline' is pumped to stabiliser towers, where the natural liquid gasoline is removed from the bottom and a mixture of LPGs is drawn off the top.

5. This mixture of LP gases, which is about 10% of the total gas mixture, can be used as a mixture or further separated into its three parts—butane, *iso*-butane and propane (about 5% of the total gas mixture).

## 2.4.3 Compressed Natural Gas

CNG is a gaseous fuel consisting of a mixture of hydrocarbons, mainly methane. Natural gas is compressed to a pressure of 200–250 bars, to form CNG. Colourless, odourless, non-carcinogenic and non-toxic CNG has a limited flammability range and is lighter than air. Commonly referred to as the green fuel because of its lead-free characteristic, it reduces harmful emissions and is non-corrosive, thereby enhancing the lifespan of spark plugs. Another practical advantage observed in countries where CNG is already in vogue is the extension of the life of lubricating oils as the fuel does not contaminate and dilute the crank-case oil.

**TABLE 2.11**

Retail Selling Prices of Selected Petroleum Products in Metropolitan Cities in India

| | | | | | LPG (Rs per 14.2 kg Cylinder) | (ATF) | |
|---|---|---|---|---|---|---|---|
| | | | | | | Domestic Airlines* | International Airlines |
| | MS | HSDO | LDO$ | FO$ | | | |
| Cities/Town | (Rs/L) | | (Rs/kL) | | | (Rs/kL) | |
| *As on 1.10.2007* | | | | | | | |
| Mumbai | 48.41 | 34.96 | 30692.14 | 22003.71 | 298.00 | 29891.56 | 37390.00 |
| Kolkata | 46.90 | 32.88 | 32722.14 | 21128.32 | 300.50 | 33366.70 | 29100.59 |
| Delhi | 42.85 | 30.25 | 30899.64 | N.A. | 294.75 | 30124.20 | 27622.93 |
| Chennai | 47.48 | 33.31 | 34588.27 | 22637.16 | 288.10 | 30606.35 | 27038.37 |
| *As on 1.4.2008* | | | | | | Rs/kL | $/kL |
| Mumbai | 50.51 | 36.08 | 35079.15 | 30148.86 | 298.00 | 44153.26 | 1016.67 |
| Kolkata | 48.95 | 33.92 | 38462.88 | 30148.86 | 298.00 | 44153.26 | 1016.67 |
| Delhi | 45.52 | 31.76 | 35555.03 | N.A. | 294.75 | 44424.41 | 1023.40 |
| Chennai | 49.61 | 34.40 | 39454.13 | 28410.49 | 288.10 | 44822.97 | 1006.81 |
| *As on 1.4.2009* | | | | | | | |
| Mumbai | 44.55 | 34.45 | 34821.70 | 20404.58 | 312.05 | 30784.81 | 452.53 |
| Kolkata | 44.05 | 32.92 | 36765.92 | 19927.38 | 327.25 | 37744.02 | 499.19 |
| Delhi | 40.62 | 30.86 | 34413.45 | N.A. | 279.70 | 29925.97 | 458.44 |
| Chennai | 44.24 | 32.82 | 38166.91 | 19557.12 | 314.55 | 33306.77 | 453.22 |
| *As on 1.4.2010* | | | | | | | |
| Mumbai | 52.20 | 39.88 | 46586.94 | 32555.98 | 313.45 | 42159.48 | 689.78 |
| Kolkata | 51.67 | 37.99 | 50145.03 | 31155.92 | 328.70 | 49294.36 | 733.61 |
| Delhi | 47.93 | 38.10 | 46195.19 | 27920.00 | 310.35 | 40841.40 | 694.62 |
| Chennai | 52.13 | 38.05 | 51239.39 | 30766.62 | 315.95 | 45151.97 | 692.63 |
| *As on 1.4.2011* | | | | | | | |
| Mumbai | 63.08 | 42.06 | 61840.56 | 43214.20 | 348.45 | 47920.21 | 1011.81 |
| Kolkata | 62.51 | 40.06 | 66118.90 | 41002.11 | 365.10 | 53721.59 | 1056.37 |
| Delhi | 58.37 | 37.75 | 65614.63 | N.A. | 345.35 | 49297.77 | 1020.66 |
| Chennai | 61.93 | 40.16 | 68187.86 | 40583.86 | 352.35 | 49320.21 | 1016.40 |

*Source:* Adapted from Basic statistics on petroleum and natural gas, Ministry of petroleum and natural gas, Government of India. http://petroleum.nic.in/petstat.pdf, 2011; HPCL's website.

*Note:* $: No real sale; N.A.: not available; *posted airfield price, exclusive of sales and other levies. Retail selling prices of ATF for international airlines based on $/kL from 1.4.2008.
1 US$ = INR 54.6 as on 20 September 2012.

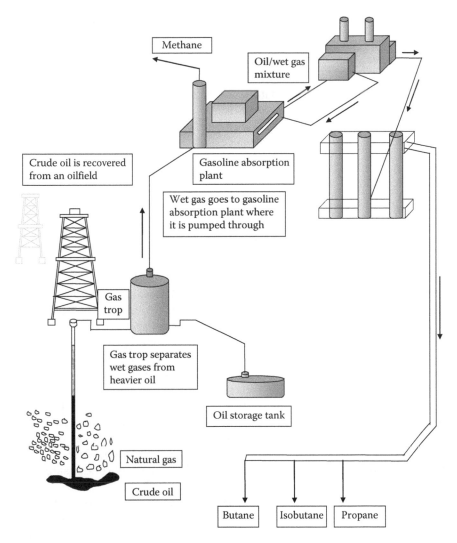

**FIGURE 2.5**
Schematic diagram of the production process of LPG.

### 2.4.3.1 Resources

Natural gas is defined as the gas obtained from a natural underground reservoir. It generally contains a large quantity of methane along with heavier hydrocarbons such as ethane, propane, *iso*-butene and normal butane. It contains a considerable amount of non-hydrocarbons such as nitrogen, hydrogen sulphide and carbon dioxide. Some traces of helium, carbonyl sulphide and various mercaptans are also present in natural gas.

**TABLE 2.12**

Compositions of Natural Gas

| S. No. | CNG Composition | % by Vol. |
|--------|-----------------|-----------|
| 1 | Methane | 91.9 |
| 2 | Ethane | 3.7 |
| 3 | Carbon dioxide | 2 |
| 4 | Propane | 1.2 |
| 5 | *iso*-Butane | 0.4 |
| 6 | *iso*-Pentane | 0.2 |
| 7 | *n*-Pentane | 0.2 |
| 8 | Nitrogen | 0.2 |
| 9 | *n*-Butane | 0.1 |

*Source:* Adapted from P. Goel and Sidhartha, *International Journal of Atmospheric Environment*, 37(38), 5423–5431, 2003.

The resources of CNG are oilfields and natural gas fields. Natural gas is a mixture of hydrocarbons—mainly methane ($CH_4$)—and is produced either from gas wells or in conjunction with crude oil production. Natural gas is consumed in the residential, commercial, industrial and utility markets. The interest in natural gas as an alternative fuel stems mainly from its clean burning qualities, its domestic resource base and its commercial availability to end users. Owing to the gaseous nature of this fuel, it must be stored on board a vehicle either in a compressed gaseous state (CNG) or in a liquefied state (LNG).

Gas streams produced from reservoirs contain natural gas, liquids and other materials. Processing is required to separate the gas from petroleum liquids and to remove the contaminants. In addition, natural gas (methane) can also come from landfill gas and water/sewage treatment. First, the gas is separated from free liquids such as crude oil, hydrocarbon condensate, water and entrained solids.

The separated gas is further processed to meet specified requirements. For example, natural gas for transmission companies must generally meet certain pipeline quality specifications with respect to the water content, hydrocarbon dew point, heating value and hydrogen sulphide content. A dehydration plant controls the water content; a gas processing plant removes certain hydrocarbon components to hydrocarbon dew point specifications; and a gas sweetening plant removes hydrogen sulphide and other sulphur compounds (when present). A typical gas composition of CNG is shown in Table 2.12 and the classification of natural gas based on composition is shown in Table 2.13.

### 2.4.3.2 Availability of Natural Gas: World View

Natural gas can be hard to find since it can be trapped in porous rocks deep underground. However, various methods have been developed to find out natural gas deposits. The methods employed are as follows:

**TABLE 2.13**

Classification of Natural Gas Composition

| Classification Based on Composition | Components |
|---|---|
| Lean gas | Methane |
| Wet gas | Considerable amounts of the higher-molecular-weight hydrocarbons |
| Sour gas | Hydrogen sulphide |
| Sweet gas | Little, if any, hydrogen sulphide |
| Residue gas | Natural gas from which the higher-molecular-weight hydrocarbons have been extracted |
| Casing head gas | Derived from petroleum but is separated at the separation facility at the wellhead |

1. Looking at surface rocks to find clues about underground formations
2. Setting off small explosions or dropping heavy weights on the surface and recording the sound waves as they bounce back from the rock layers underground
3. Measuring the gravitational pull of rock masses deep within the earth

Scientists are also researching new ways to obtain natural (methane) gas from biomass as a fuel source derived from plant and animal wastes. Methane gas is produced naturally whenever organic matter decays. Coal beds and landfills are other sources of natural gas. The world status of reserves of natural gas is shown in Table 2.14. The oil and gas reserves in India are shown in Table 2.15.

### 2.4.3.3 Cost Analysis of CNG (United States)

The annual average lower 48 wellhead and Henry Hub spot market prices for natural gas (1990–2035) are shown in Figure 2.6 and the Europe price chart for natural gas is depicted in Figure 2.7.

### 2.4.3.4 Natural Gas Production Process

Natural gas from high-pressure wells is usually passed through field separators at the well to remove hydrocarbon condensate and water [9]. Natural gas, butane and propane are usually present in the gas, and gas processing plants are required for the recovery of these liquefiable constituents as shown in Figure 2.8.

Natural gas is considered 'sour' if hydrogen sulphide ($H_2S$) is present in amounts greater than 5.7 mg per normal cubic metre ($mg/Nm^3$). The $H_2S$ must be removed (called 'sweetening' the gas) before the gas can be utilised. If $H_2S$ is present, the gas is usually sweetened by absorption of the $H_2S$ in an amine solution. Other methods, such as carbonate processes, solid bed

**TABLE 2.14**

World Status of Sources of Natural Gas

| | At End 1990 (Trillion Cubic Metres) | At End 2000 (Trillion Cubic Metres) | At End 2009 (Trillion Cubic Metres) | At End 2010 (Trillion Cubic Metres) |
|---|---|---|---|---|
| United States | 4.8 | 5.0 | 7.7 | 7.7 |
| Canada | 2.7 | 1.7 | 1.7 | 1.7 |
| Mexico | 2.0 | 0.8 | 0.5 | 0.5 |
| Total North America | 9.5 | 7.5 | 9.9 | 9.9 |
| Argentina | 0.7 | 0.8 | 0.4 | 0.3 |
| Bolivia | 0.1 | 0.7 | 0.7 | 0.3 |
| Brazil | 0.1 | 0.2 | 0.4 | 0.4 |
| Colombia | 0.1 | 0.1 | 0.1 | 0.1 |
| Peru | 0.3 | 0.2 | 0.4 | 0.4 |
| Trinidad and Tobago | 0.3 | 0.6 | 0.4 | 0.4 |
| Venezuela | 3.4 | 4.2 | 5.1 | 5.5 |
| Other South and Central America | 0.2 | 0.1 | 0.1 | 0.1 |
| Total South and Central America | 5.2 | 6.9 | 7.5 | 7.4 |
| Azerbaijan | n/a | 1.2 | 1.3 | 1.3 |
| Denmark | 0.1 | 0.1 | 0.1 | 0.1 |
| Germany | 0.2 | 0.2 | 0.1 | 0.1 |
| Italy | 0.3 | 0.2 | 0.1 | 0.1 |
| Kazakhstan | n/a | 1.8 | 1.9 | 1.8 |
| Netherlands | 1.8 | 1.5 | 1.2 | 1.2 |
| Norway | 1.7 | 1.3 | 2.0 | 2.0 |
| Poland | 0.2 | 0.1 | 0.1 | 0.1 |
| Romania | 0.1 | 0.3 | 0.6 | 0.6 |
| Russian Federation | n/a | 42.3 | 44.4 | 44.8 |
| Turkmenistan | n/a | 2.6 | 8.0 | 8.0 |
| Ukraine | n/a | 1.0 | 1.0 | 0.9 |
| United Kingdom | 0.5 | 1.2 | 0.3 | 0.3 |
| Uzbekistan | n/a | 1.7 | 1.6 | 1.6 |
| Other Europe and Eurasia | 49.7 | 0.5 | 0.4 | 0.3 |
| Total Europe and Eurasia | 54.5 | 55.9 | 63.0 | 63.1 |
| Bahrain | 0.2 | 0.1 | 0.2 | 0.2 |
| Iran | 17.0 | 26.0 | 29.6 | 29.6 |

**TABLE 2.14 (continued)**

World Status of Sources of Natural Gas

| | At End 1990 (Trillion Cubic Metres) | At End 2000 (Trillion Cubic Metres) | At End 2009 (Trillion Cubic Metres) | At End 2010 (Trillion Cubic Metres) |
|---|---|---|---|---|
| Iraq | 3.1 | 3.1 | 3.2 | 3.2 |
| Kuwait | 1.5 | 1.6 | 1.8 | 1.8 |
| Oman | 0.3 | 0.9 | 0.7 | 0.7 |
| Qatar | 4.6 | 14.4 | 25.3 | 25.3 |
| Saudi Arabia | 5.2 | 6.3 | 7.9 | 8.0 |
| Syria | 0.2 | 0.2 | 0.3 | 0.3 |
| United Arab Emirates | 5.6 | 6.0 | 6.1 | 6.0 |
| Yemen | 0.2 | 0.5 | 0.5 | 0.5 |
| Other Middle East | † | 0.1 | 0.1 | 0.2 |
| Total Middle East | 38.0 | 59.1 | 75.7 | 75.8 |
| Algeria | 3.3 | 4.5 | 4.5 | 4.5 |
| Egypt | 0.4 | 1.4 | 2.2 | 2.2 |
| Libya | 1.2 | 1.3 | 1.5 | 1.5 |
| Nigeria | 2.8 | 4.1 | 5.3 | 5.3 |
| Other Africa | 0.8 | 1.1 | 1.2 | 1.2 |
| Total Africa | 8.6 | 12.5 | 14.7 | 14.7 |
| Australia | 0.9 | 2.2 | 2.9 | 2.9 |
| Bangladesh | 0.7 | 0.3 | 0.4 | 0.4 |
| Brunei | 0.3 | 0.4 | 0.3 | 0.3 |
| China | 1.0 | 1.4 | 2.8 | 2.8 |
| India | 0.7 | 0.8 | 1.1 | 1.5 |
| Indonesia | 2.9 | 2.7 | 3.0 | 3.1 |
| Malaysia | 1.6 | 2.3 | 2.4 | 2.4 |
| Myanmar | 0.3 | 0.3 | 0.3 | 0.3 |
| Pakistan | 0.6 | 0.7 | 0.8 | 0.8 |
| Papua New Guinea | 0.2 | 0.4 | 0.4 | 0.4 |
| Thailand | 0.2 | 0.4 | 0.3 | 0.3 |
| Vietnam | † | 0.2 | 0.7 | 0.6 |
| Other Asia Pacific | 0.3 | 0.3 | 0.4 | 0.4 |
| Total Asia Pacific | 9.9 | 12.3 | 15.8 | 16.2 |
| Total world | 125.7 | 154.3 | 186.6 | 187.1 |

*Note:* n/a: not available; †: less than 0.05.

**TABLE 2.15**

Oil and Gas Reserves in India

| Area | 2005 | 2006 | 2007 | 2008 | 2009 | 2010 | 2011 |
|---|---|---|---|---|---|---|---|
| 1 | 2 | 3 | 4 | 5 | 6 | 7 | 8 |
| *Crude Oil (Million Metric Tonnes)* | | | | | | | |
| Onshore | 376 | 387 | 357 | 404 | 406 | 403 | 403 |
| Offshore | 410 | 369 | 368 | 367 | 368 | 371 | 354 |
| Total | 786 | 756 | 725 | 771 | 774 | 774 | 757 |
| *Natural Gas (Billion Cubic Metres)* | | | | | | | |
| Onshore | 340 | 330 | 270 | 304 | 328 | 334 | 394 |
| Offshore | 761 | 745 | 785 | 786 | 788 | 815 | 847 |
| Total | 1101 | 1075 | 1055 | 1090 | 1116 | 1149 | 1241 |

*Source:* ONGC, OIL and DGH; Adapted from Basic statistics on petroleum and natural gas, Ministry of petroleum and natural gas, Government of India. http://petroleum.nic.in/petstat.pdf, 2011.

*Note:* The oil and natural gas reserves (proved and indicated) data relate to 1 April of each year.

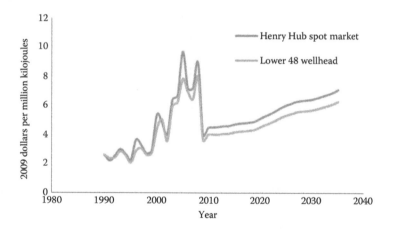

**FIGURE 2.6**
Annual average lower 48 wellhead and Henry Hub spot market prices for natural gas, 1990–2035 (2009 dollars per million kilojoules). (Adapted from http://www.mongabay.com/images/commodities/charts/chart-ngeu.html.)

absorbents and physical absorption, are employed in the other sweetening plants.

Many chemical processes are available for sweetening natural gas [10,11]. At present, the amine process (also known as the Girdler process) is the most widely used method for $H_2S$ removal. The process is summarised in reaction 2.8

$$2\ RNH_2 + H_2S \rightarrow (RNH_3)_2S \tag{2.8}$$

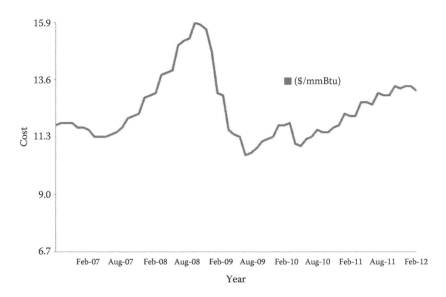

**FIGURE 2.7**
Natural gas, Europe price chart (one thousand US dollars per thousand British thermal units). (Adapted from http://www.mongabay.com/images/commodities/charts/chart-ngeu.html.)

where
R = mono-, di- or tri-ethanol
N = nitrogen
H = hydrogen
S = sulphur

The recovered hydrogen sulphide gas stream may be (1) vented, (2) flared in waste gas flares or modern smokeless flares, (3) incinerated or (4) utilised for the production of elemental sulphur or sulphuric acid. If the recovered $H_2S$ gas stream is not to be utilised as a feedstock for commercial applications, the gas is usually passed to a tail gas incinerator in which the $H_2S$ is oxidised to $SO_2$ and is then passed to the atmosphere out a stack.

### 2.4.3.5 Natural Gas Purification

Natural gas produced at the well contains contaminants and natural gas liquids that have to be removed before sending to the consumers. These contaminants can cause an operation problem, pipe rupture or pipe deterioration. The natural gas purification process is shown in Figure 2.9.

### 2.4.3.6 Advantages of Natural Gas as Transportation Fuel

- The octane ratings are higher than that of gasoline with resultant anti-knock protection.

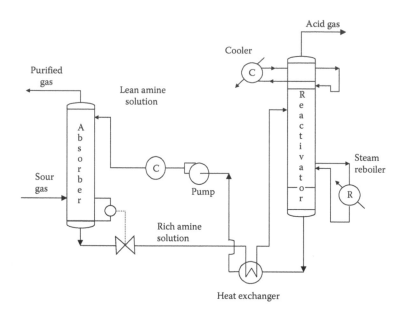

**FIGURE 2.8**
Flow diagram of the amine process for gas sweetening. (Adapted from R. R. Maddox, *Gas and Liquid Sweetening*, 2nd Ed. Campbell Petroleum Series, Norman, OK, 1974; R. E. Kirk and D. F. Othmer (eds.), *Encyclopedia of Chemical Technology*. Vol. 7, Interscience Encyclopedia, Inc., New York, 1951.)

- There is no tetraethyl lead (anti-knock agent) to foul spark plugs.
- It mixes thoroughly with air, burning more completely, quietly and smoothly than liquid fuel–air mixtures.
- It is generally uniform in composition and quality with consistent specific gravity and heating value, in contrast to the possible deterioration of stored liquid fuel or adulteration of the same.
- It requires only a simple carburetion system.
- The gas does not collect in the combustion chamber as an unburnt liquid to find its way past the piston rings to dilute the lubricating oil.
- Less sludge accumulates in the lubricating oil because of the higher combustion temperature of natural gas.
- Natural gas burns completely, leaving little or no fuel soot.
- Gas engines require only a simple fuel filtering system.

### 2.4.3.7 Comparison between CNG and LPG

Various parameters such as safety, burning quality and properties of CNG and LPG are compared in Table 2.16. The fuel qualities of natural gas for engine application are discussed below.

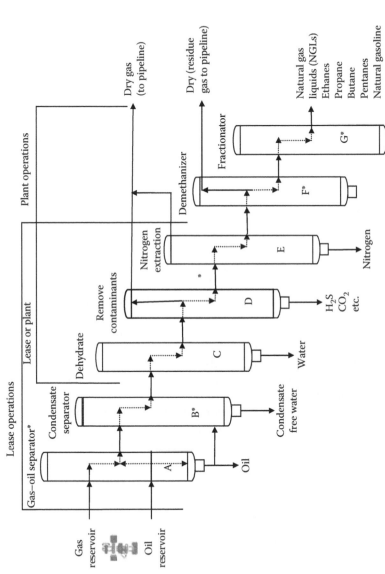

**FIGURE 2.9**
Natural gas purification process. (Adapted from Annual Energy Outlook 2011, U.S. Energy Information Administration, Report Number: DOE/EIA-0383, http://205.254 135.7/forecasts/archive/aeo11/forecasts/archive/aeo11/source_natural_gas.cfm, 2011.)

**TABLE 2.16**

Comparison between CNG and LPG

| S. No. | Features | CNG | LPG |
|---|---|---|---|
| 1 | Safety | Very safe as it is lighter than air; when leaked, it will dissipate quickly | Less safe, as it is heavier than air; when leaked, it will pool on the ground |
| 2 | Readiness for use | Ready to use as it is in gas form | It is in a liquid form; when used, it needs to be converted to gas |
| 3 | Burning quality | Complete burning | Complete burning |
| 4 | Properties | Sulphurless, colourless and odourless gas with sootless burning | Colourless and odourless but odour is normally added for safety reasons |
| 5 | Expense | Storage tank is not needed | Storage tank and advance ordering are needed |

### 2.4.3.8 Methane Number

The main parameter for rating the knock resistance of gaseous fuels is the methane number (MN), which represents the knock resistance of gaseous fuel for the spark ignition engine. The MN must always be at least equal to the methane number requirement (MNR) of the gas engine for ensuring safe engine operation. The MN required by the engine is affected by the design and operating parameters, with the adjustment of the MN requirement being achieved by the changing engine operation. Measures such as optimisation of spark ignition timing could affect the MNR.

Different scales have been used to rate the knock resistance of CNG, including the MON and the MN. The differences in these ratings are the reference fuel blends used for comparing to natural gas.

MN uses a reference fuel blend of methane (methane number: 100) and hydrogen (methane number: 0). The correlations between the motor octane and methane numbers are shown in Equations 2.9 and 2.10 [12].

$$MON = -406.14 + 508.04 * (H/C) - 173.55 * (H/C)^2 + 20.17 * (H/C)^3 \quad (2.9)$$

$$MN = 1.624 * MON - 119.1 \quad (2.10)$$

If a gas mixture has an MN of 60, its knock resistance is equivalent to that of a gas mixture of 60.

### 2.4.4 Other Hydrocarbons

Heavier hydrocarbons could form liquid phases in the network, causing blockage problems. The control of the hydrocarbon dew point is the preferred method for controlling liquid formation.

## 2.4.5 Inert Gas

Typical inert gases are carbon dioxide, nitrogen, helium and argon. The inert gas reduces the calorific value of natural gas as they are non-combustible. However, they are normally present in relatively small amounts. If it is present in higher amounts (e.g., of biogas: 40% $CO_2$), it reduces the combustion efficiency of the engine. Both carbon dioxide and nitrogen can be used to lower the calorific value of a gas for industrial burner application if it has a high Wobbe index of mixture. As far as the internal combustion engine application is concerned, a higher calorific value is always advantageous as it relates directly to the power output of the engine.

## 2.4.6 Contaminants

The contaminant, such as mercury, if present in very low concentrations, affects downstream operations such as corrosion of the pipeline network and restricts the gas flow.

At very high temperature conditions, metal impurities, such as mercury, may form amalgams with the engine components, causing embrittlement, cracking and premature failure.

## 2.4.7 Water

Hydrates are ice-like solids containing hydrocarbons that can form if the temperature of the gas decreases. The operating parameters such as temperature, pressure and composition of natural gas in the transmission pipeline need to be controlled to prevent the formation of water droplets and hydrates. The presence of water in natural gas can cause corrosion of the pipeline. Hydrate formation can block valves and the pipeline flow.

## 2.4.8 Oxygen

Oxygen in gas promotes pipeline corrosion in the presence of water and sulphur. In underground storage sites, oxygen promotes bacterial activity, which produces hydrogen sulphide. So the oxygen percentage should be minimal.

## 2.4.9 Hydrogen

Hydrogen in a gas mixture leads to corrosion, which cracks the steel pipeline.

## 2.4.10 Hydrogen Sulphide

As hydrogen sulphide is toxic, it needs to be removed on health and safety grounds. It reacts with iron oxide, resulting in the corrosion of ferrous metals. It also reacts with copper piping used in domestic systems to form copper sulphide flakes, which form a black dust, resulting in the blockage of filters.

### 2.4.11 Sulphur

Organic sulphur compounds are mercaptans and sulphides that may be present naturally in the gas. Sulphur needs to be limited due to its highly unpleasant odour.

---

## 2.5 Conventional Fuels for Compression Ignition Engines/ Vehicles

### 2.5.1 Diesel

Diesel is used as a primary fuel in a CI engine. The CI engine gives a higher fuel economy than the spark ignition engine as it operates at a higher compression ratio (16:1 to 22:1). To improve the engine performance and emission reduction, the fuel quality of diesel needs to be upgraded. Parameters that affect the fuel quality of diesel are given below.

#### 2.5.1.1 Cetane Number

Cetane number is a measure of the compression ignition quality of diesel fuel. It is measured using the CFR engine. It influences cold start-ability, exhaust emissions and combustion noise. This fuel property is most influential for cold start and white smoke emissions. A high cetane number implies less knocking tendency in a CI engine. A typical cetane number of diesel fuel lies in the range of 45–51.

#### 2.5.1.2 Density

The density of diesel should be in the optimum range. If it is higher than the optimum value, it affects the atomisation of fuel particles, resulting in poorer combustion. The core of the injector is the main cause for soot emission. Higher fuel density increases the Sauter mean diameter (SMD) and penetration and decreases the spray cone angle and air entrainment of spray, resulting in a poorer combustion.

#### 2.5.1.3 Viscosity

It affects the pumping work, leakage in the pump and the spray characteristics. If the viscosity is higher than the optimum value, it increases the input power requirement of the injection system. It reduces the spray cone angle, which affects the mixture formation process. If it is lesser than the specified value, fuel leakage would occur in an injection pump. Hence, it should be in the optimal range.

### 2.5.1.4 Sulphur Content

The presence of sulphur in diesel fuel contributes to particulate matter emissions. It can lead to corrosion and wear of the engine system. Sulphur in fuel reduces the catalytic converter efficiency and also induces the sulphur poisoning problem.

### 2.5.1.5 Distillation Characteristics

Distillation characteristics relate to fuel evaporation during engine operation. The distillation properties will vary depending upon the climate (atmospheric temperature) of a country. Diesel fuel has different hydrocarbon blends, which have different boiling points. T10 refers to the temperature at which 10% of the fuel evaporates that is important for the starting of the system. If it is higher than the specified value, the engine will experience a cold start-ability problem. T50 refers to the cruising speed of the vehicle, whereas T90 indicates the deposit of the fuel on the combustion chamber of the system. The fuel should have correct distillation properties. To minimise the design modifications of an engine, the properties of a selected alternative fuel should be comparable with the properties given in Table 2.17.

**TABLE 2.17**

Typical Properties of Conventional Fuels

| S. No. | Properties | Diesel | Gasoline | Compressed Natural Gas | Liquefied Natural Gas |
|---|---|---|---|---|---|
| 1 | Molecular formula | $C_{10}$–$C_{18}$ | $C_4$–$C_{12}$ | $CH_4$ | $C_3H_8$ and $C_4H_{10}$ |
| 2 | Research octane number | — | 91 | 110 | 95 |
| 3 | Cetane number | 51 | — | — | — |
| 4 | Density 15°C, kg/m³ | 820–845 | 720–775 | — | — |
| 5 | Sulphur, total, maximum (mg/kg) | 50 | 50 | 7.5 mg/m³ | — |
| 6 | Calorific value (MJ/kg) | 44.5 | 44.5 | 48 | 42.5 |
| 7 | Reid vapour pressure (RVP) kPa | — | 60 | — | — |
| 8 | Olefin content, maximum (% volume) | — | 21 | — | — |
| 9 | Aromatic content, maximum (% volume) | | 35 | — | — |
| 10 | Polyaromatic hydrocarbon (PAH) | 11 | — | — | — |
| 11 | Mercaptans (grains) | | | <200 | |
| 12 | $CO_2$ (mol%) | — | — | 3 | — |
| 13 | Oxygen (mol%) | — | — | <3% | — |
| 14 | Nitrogen (mol%) | — | — | <3% | — |

*Source:* Adapted from *Guidebook to Gas Interchangeability and Gas Quality*, BP, IGU. http://www.igu.org/igu-publications/Gas%20Interchangeability%202011%20v6%20HighRes.pdf; www.siamindia.com.

## 2.6 Conclusion

Diesel or gasoline derived from crude oil has high sulphur, aromatics and olefins content, which are not desirable in view of the stringent fuel quality and emission norms. As the fossil fuel resource gets depleted at a faster rate, alternative fuel studies have to be explored. The technology that is adapted for producing alternative fuels is discussed in the next chapter.

**PROBLEMS**

**Example 2.1**

Determine the stoichiometric air–fuel ratio by volume and by mass, wet and dry products of combustion, and mass of dry products per unit mass of fuel $CH_4$ (assume the system is with complete combustion).

**Solution to Example 2.1**

$$CH_4 + m_s(O_2 + 3.76N_2) \rightarrow n_1CO_2 + n_2H_2O + n_3N_2$$

| | |
|---|---|
| Carbon balance: | $n_1 = 1$ |
| Hydrogen balance: | $n_2 = 2$ |
| Oxygen balance: | $m_s = n_1 + 0.5n_2 = 2$ |
| Nitrogen balance: | $n_3 = 3.76\ ms = 7.52$ |
| Volumetric (A/F)s | $= 4.76\ ms = 9.52$ |
| Gravimetric (A/F)s | $= (137.9 \times 2)/(12 + 4) = 17.24$ approximately |

| Product | | Mol | Vol.% Wet | Vol.% Dry |
|---------|------|------|-----------|-----------|
| $CO_2$ | | 1 | 9.51 | 11.74 |
| $N_2$ | | 7.52 | 71.48 | 88.26 |
| | Σdry | 8.52 | | 100.00 |
| $H_2O$ | | 2 | 19.01 | |
| | Σwet | 10.52 | 100.00 | |

Mass of dry products/unit mass fuel $= (A/F)_s + 1 - \dfrac{18\ n_2}{RMM_{fuel}\ (mol\ fuel)}$

$$= 17.24 + 1 - \frac{18 \times 2}{16 \times 1}$$

$$= 15.99\ kg\ air/kg\ fuel$$

**Example 2.2**

Determine the approximate air–fuel ratio by mass and the wet and dry products of combustion, for lean and rich mixtures of ethanol ($C_2H_5OH$) in air.

Calculate the air–fuel ratio for (a) 10% excess air assuming no CO in products and (b) 10% deficient air assuming no $O_2$ in the products.

**Solution to Example 2.2**

In this problem, $x = 2$, $y = 6$ and $z = 1$.

$$m_s = (x + y/4 - 0.5z) = 2 + 1.5 - 0.5 = 3.0$$

RMM of fuel $= 24 + 6 + 16 = 46$

Gravimetric $(A/F)_s = (137.9 \times 3)/46 = 8.99$ approximately

1. With 10% excess air; $m = 1.1\ m_s$, so that since $\varphi = m_s/m = 1/1.1 = 0.91$, so $m = 3.3$

   Gravimetric $(A/F) = 1.1(A/F)_s = 1.1 \times 8.99 = 9.89$ approximately

   The general combustion equation on a molar basis is

   $$C_2H_5OH + 3.3(O_2 + 3.76N_2) \rightarrow n_1CO_2 + n_2H_2O + n_5O_2 + n_6N_2$$

   | Carbon balance: | $n_1 = 2$ |
   |---|---|
   | Hydrogen balance: | $n_2 = 3$ |
   | Oxygen balance: | $n_1 + 0.5\ n_2 + n_5 = 3.3 + 0.5$ |
   | | $n_3 = 0.3$ |
   | Nitrogen balance: | $n_4 = 3.76 \times 3.3 = 12.41$ |

   | Product | | Mol | Vol.% Wet | Vol.% Dry |
   |---|---|---|---|---|
   | $CO_2$ | | 2 | 11.29 | 13.60 |
   | $O_2$ | | 0.3 | 1.69 | 2.04 |
   | $N_2$ | | 12.41 | 70.07 | 84.36 |
   | | $\Sigma dry$ | 14.71 | | 100.00 |
   | $H_2O$ | | 3 | 16.95 | |
   | | $\Sigma wet$ | 17.71 | 100.00 | |

2. With 10% deficient air; $m = 0.9\ m_s$ so that since $\varphi = m_s/m = 1/0.9 = 1.11$, so $m = 2.7$

   Gravimetric $(A/F) = 0.9\ (A/F)_s = 0.9 \times 8.99 = 8.09$ approximately.

   The general combustion equation on a molar basis is

   $$C_2H_5OH + 2.7(O_2 + 3.76N_2) \rightarrow n_1CO_2 + n_2H_2O + n_3CO + n_4N_2$$

   | Carbon balance: | $n_1 + n_3 = 2$ |
   |---|---|
   | Hydrogen balance: | $n_2 = 3$ |
   | Oxygen balance: | $n_1 + 0.5\ n_2 + 0.5\ n_3 = 2.7 + 0.5$ |
   | | $n_3 = 0.6$ |
   | | $n_1 = 1.4$ |
   | Nitrogen balance: | $n_4 = 3.76 \times 2.7 = 10.15$ |

| Product | | Mol | Vol.% Wet | Vol.% Dry |
|---------|---------|-------|-----------|-----------|
| $CO_2$ | | 1.4 | 9.24 | 11.52 |
| CO | | 0.6 | 3.96 | 4.94 |
| $N_2$ | | 10.15 | 67.00 | 83.54 |
| | Σdry | 12.15 | | 100.00 |
| $H_2O$ | | 3 | 19.80 | |
| | Σwet | 15.15 | 100.00 | |

## Example 2.3

Determine the adiabatic flame temperature of the hydrocarbon fuel $C_{10}H_{22}$ for $\Phi = 1$.

## Solution to Example 2.3

$$C_{10}H_{22} + 15.5(O_2 + 3.76N_2) \rightarrow 10CO_2 + 11H_2O + 58.28N_2$$

$$\text{We know that } H_{reactant} = H_{product}$$

$H_{reactant}$

$$= [h_f^\circ\, C_{10}H_{22}] + 15.5[h_f^\circ\, O_2 + C_{pO_2}\, (T_2 - T_{ref})] + 58.28[h_f^\circ\, N_2 + C_{pN_2}(T_2 - T_{ref})]$$

$$= [-1.755644 * 10^3 * 142] + 15.5[0.972(726.879 - 298)] * 32$$

$$+ 58.28[1.056(726.879 - 298)] * 28$$

$$= 696.4 * 10^3\, kJ$$

$H_{product}$

$$= 10[h_f^\circ\, CO_2 + C_{pCO_2}(T_{ad} - T_{ref})] + 11[h_f^\circ\, H_2O + C_{pH_2O}\, (T_{ad} - T_{ref})]$$

$$+ 58.28[h_f^\circ N_2 + C_{pN_2}\, (T_{ad} - T_{ref})]$$

$$= 10 * 44[-8946.8 + 1.348(T_{ad} - 298)] + 11 * 18[-13430.8$$

$$+ 2.711(T_{ad} - 298)] + 58.28 * 28[0 + 1.263(T_{ad} - 298)]$$

$$= -7.546E6 + 3190.9119\, T_{ad}$$

$H_{reactant} = H_{product}$

$696.4 * 103 = -7.546E6 + 3190.9119\, T_{ad}$

$T_{ad} = 2583.086\ K$

**Example 2.4**

A petrol engine has the following ultimate analysis: carbon, C, 83.7 and hydrogen, $H_2$, 16.3. If the dry products analysis by volume is $CO_2$, 11.8%, $O_2$, 3.7% and nitrogen, 84.5%. Determine the air–fuel ratio used by the (1) ratio method, (2) carbon balance method and (3) oxygen–hydrogen balance method.

**Solution to Example 2.4**

1. Ratio method. The general combustion equation on a molar basis is

$$C_xH_y + m(O_2 + 3.76N_2) \rightarrow n_1CO_2 + n_2H_2O + n_3O_2 + n_4N_2$$

Product analysis gives $\dfrac{n_3}{n_1} = \dfrac{3.7}{11.8} = 0.314$; thus, $n_3 = 0.314\, n_1$

and $\dfrac{n_4}{n_1} = \dfrac{84.5}{11.8} = 7.16$; thus, $n_4 = 7.16\, n_1$

Carbon balance: $\qquad\qquad x = n_1$

Hydrogen balance: $\qquad\quad y = 2\, n_2$

Oxygen balance: $\qquad\quad m = n_1 + 0.5\, n_2 + n_5$

$\qquad\qquad\qquad\qquad\qquad = 1.314\, n_1 + 0.5 n_2$

Nitrogen balance: $\qquad\quad m = \dfrac{n_6}{m} = \dfrac{7.16 n_1}{3.76} = 1.9\, n_1$

Thus, $1.9\, n_1 = 1.314\, n_1 + 0.5\, n_2$

So $n_2 = 1.172\, n_1$ and hence $y = 2.344\, n_1$

$$\text{Gravimetric air–fuel ratio} = \frac{137.9}{12x + y}$$

$$= \frac{137.9 \times 1.9\, n_1}{12\, n_1 + 2.344\, n_1}$$

$$= 18.27$$

2. Carbon balance method

   Volume fractions are the same as molar fractions. 1 mol of dry products contains 0.845 mol of nitrogen. Therefore, 0.845 mol of nitrogen is brought in with $0.845 \times 28$ $(100/76.8) = 30.81$ kg air.

   1 mol of dry products contains 0.118 mol of carbon dioxide, which in turn contains $0.118 \times 12$ kg of carbon. This amount of carbon comes from the combustion of carbon in the fuel.

   Since 0.837 kg carbon is contained in (or associated with) 1 kg of fuel, $0.118 \times 12$ kg carbon is associated with $0.118 \times 12/0.837 = 1.692$ kg fuel.

   Thus, 1 mol of dry products is produced when 1.692 kg fuel burns with 30.81 kg of air, so

$$\text{Gravimetric air fuel ratio} = \frac{30.81}{1.692} = 18.21$$

The carbon balance method using the carbon balance between reactants and products is most satisfactory when carbon is predominant in the fuel. The result may be inaccurate, however, if the exhaust is smoky, indicating unburnt carbon that is not included in the analysis. An oxygen–hydrogen balance can be made as a check, although, since the relevant quantities are relatively small, the result is more open to error from inaccuracies of measurement.

3. Oxygen–hydrogen balance method
One mol of dry products contains oxygen from carbon dioxide and free oxygen, which amounts to

$$32(0.118 + 0.037) = 4.96 \text{ kg oxygen}$$

But the amount of oxygen brought in with air for combustion (accompanying the nitrogen) is

$$0.845 \times 28 \ (23.2/76.8) = 7.147 \text{ kg oxygen}$$

Now, during combustion, only 4.96 kg of oxygen has been consumed to form carbon dioxide and generate free oxygen. The rest has been used up to form water vapour, which is given by

$$7.147 – 4.96 = 2.187 \text{ kg oxygen}$$

From the equation for the oxidation of hydrogen

$$H_2 + 1/2O_2 \leftrightarrow H_2O$$

We know that for every unit mass of oxygen required, the mass of hydrogen required is 1/8. Hence, the mass of hydrogen involved is 2.187/8 kg.

Since 0.163 kg hydrogen is contained in 1 kg fuel, 2.187/8 kg hydrogen is contained in $2.187/(8 \times 0.163) = 1.677$ kg fuel.

Thus, 1 mol of dry products is produced when 1.677 kg fuel burns with 30.81 kg of air so

$$\text{Gravimetric air–fuel ratio} = \frac{30.81}{1.677} = 18.37$$

Results are quite similar for all the three methods.

**Example 2.5**

*Iso*-octane at 2500 K burns with air at an equivalence ratio of 1.2. Find the products of combustion.

$$C_8H_{18} + m(O_2 + 3.76N_2) \rightarrow n_1CO_2 + n_2H_2O + n_3CO + n_4H_2 + n_5N_2$$

**Solution to Example 2.5**

Here, $m_s = 12.5$ and since $\varphi = m_s/m$, $m = m_s/\phi = \dfrac{12.5}{1.2} = 10.42$

| | | |
|---|---|---|
| Carbon balance: | $n_1 + n_3 = 8$ | (1) |
| Hydrogen balance: | $n_2 + n_4 = 9$ | (2) |
| Oxygen balance: | $n_1 + 0.5\, n_2 + 0.5\, n_3 = m = 10.42$ | (3) |
| Nitrogen balance: | $n_5 = 3.76\, m = 3.76 \times 10.42 = 39.18$ | (4) |

At 2500 K

$$K_{WG} = \frac{n_2 n_3}{n_1 n_4} = 6.0814 \tag{5}$$

| | | |
|---|---|---|
| From (1), we have | $n_3 = 8 - n_1$ | (6) |
| Substituting for $n_3$ in (3), we get | $n_1 + 0.5n_2 + 0.5(8 - n_1) = 10.42$ | (7) |
| Rearranging, we get | $0.5\, n_1 + 0.5\, n_2 = 6.42$ | (8) |
| which can be written as | $n_2 = 12.84 - n_1$ | (8a) |
| From (5), we have | $n_2 n_3 = 6.0814\, n_1 n_4$ | (9) |
| And from (2), we have | $n_4 = 9 - n_2$ | (10) |
| Substituting for $n_3$ and $n_4$ in (9), we get | $n_2(8 - n_1) = 6.0814\, n_1(9 - n_2)$ | (11) |
| Thus, | $8\, n_2 - n_1 n_2 = 54.7326\, n_1 - 6.0814\, n_1 n_2$ | (12) |

We can substitute for $n_2$ in (12) from (8a) to get

$$8(12.84 - n_1) - n_1(12.84 - n_1) = 54.7326\, n_1 - 6.0814\, n_1(12.84 - n_1) \tag{13}$$

This can be written as

$$102.72 - 8\, n_1 - 12.84\, n_1 + n_{12} = 54.7326\, n_1 - 78.085\, n_1 + 6.0814\, n_{12}$$

and when this is rearranged after simplification, we get

$$-5.0814 n_{12} + 2.5124 n_1 + 102.72 = 0$$

This is a quadratic equation with two roots, which are

$$n_1 = \frac{2.5124 \pm \sqrt{2.5124^2 + 4 \times 5.0814 \times 102.72}}{-2 \times 5.0814}$$

The two roots for $n_1$ are $-4.75$ and $+4.25$. We take the positive root so $n_1 = 4.25$, $n_3 = 3.75$, $n_2 = 8.59$, $n_4 = 0.41$. We can obtain the mole fractions also.

**UNSOLVED PROBLEMS**

1. Find a relationship between the partial pressures of water vapour, hydrogen and oxygen for an equilibrium mixture at 400 K and 1 atmosphere absolute.

2. Find the equilibrium concentration in moles of a water vapour, oxygen and hydrogen mixture at 3500 K and 1 atm. Initially, 1 mol $H_2$, 3 mol of $O_2$ and 6 mol of $H_2O$ were present. The temperature of the reactants is not specified.

3. The *number of moles* of products of combustion of a hydrocarbon fuel and air when at 1 atm pressure and a certain temperature are as follows: $CO_2$ 9.01, $H_2O$ 5.72, CO 2.99, $H_2$ 0.28, $O_2$ 1.635 and $N_2$ 56.4. Determine the following:

   a. The chemical formula for the fuel

   b. The stoichiometric air–fuel ratio by mass for the fuel

   c. The equivalence ratio that gives the above moles

   d. The equilibrium constants based on pressures in atmospheres for the reactions

   $$H_2 + \tfrac{1}{2}O_2 \Leftrightarrow H_2O$$

   $$CO + \tfrac{1}{2}O_2 \Leftrightarrow CO_2$$

   e. What would be the number of moles of the species listed above if the mixture of products is at a pressure of 10atm without change in temperature?

4. Calculate the gross calorific value and Wobbe index of a fuel gas having the following data:

| Gas | Composition (% Vol.) | CV Gross (kcal/ Nm³) (Average) |
|---|---|---|
| Methane ($CH_4$) | 87.0 | 9500 |
| Ethane ($C_2H_6$) | 9.0 | 16,644 |
| Propane ($C_3H_6$) | 2.6 | 23,688 |
| Butane ($C_4H_{10}$) | 1.4 | 30,714 |

5. Determine the exhaust gas analysis and air–fuel ratio by weight when a medium-viscosity fuel oil with 84.9% carbon, 11.4% hydrogen, 3.2% sulphur, 0.4% oxygen and 0.1% ash content is burnt with 20% excess air.

6. A furnace burns producer gas with 10% excess air at a rate of 7200 Nm³/h and discharge flue gas at 760 mm and 400°C. Calculate the flue gas composition, air requirement and volume of flue gas per hour.

7. The raw biogas obtained from the biogas plant was purified by water scrubbing technology. The compositions of raw and purified biogas obtained are tabulated below.

| Type of Gas Composition | Raw Biogas (% Vol.) | Purified Biogas (% Vol.) | Density (kg/m³) | Calorific Value (kJ/kg) |
|---|---|---|---|---|
| (Methane) $CH_4$ | 60% | 95% | 0.668 | 39,820 |
| (Carbon dioxide) $CO_2$ | 38% | 3% | 1.842 | — |
| (Nitrogen) $N_2$ | 1.6% | 1.6% | 1.165 | — |
| (Hydrogen) $H_2$ | 0.4% | 0.4% | 0.0899 | 120,210 |

Calculate the density and calorific value of pure and raw biogas. Find the total volume of the pure gas obtained.

## References

1. D. Decroocq, Major scientific and technical challenges about development of new processes in refining and petrochemistry oil and gas science and technology—Rev. IFP 52(5), September–October 1997.
2. http://www.gailonline.com/gailnewsite/businesses/lpgproperties.html (GAIL).
3. http://www.gasindia.in/technical-specification.html (Indian Oil).
4. M. Masi, Experimental analysis on a spark ignition petrol engine fuelled with LPG (liquefied petroleum gas). *Energy*, 41(1), 1–9, 2011.
5. Basic statistics on petroleum and natural gas, Ministry of petroleum and natural gas, Government of India. http://petroleum.nic.in/petstat.pdf, 2011.
6. P. Goel and Sidhartha, Present scenario of air quality in Delhi—A case study of CNG implementation. *International Journal of Atmospheric Environment*, 37(38), 5423–5431, 2003.
7. Annual Energy Outlook 2011, U.S. Energy Information Administration, Report Number: DOE/EIA-0383, http://205.254.135.7/forecasts/archive/aeo11/source_natural_gas.cfm, 2011.
8. http://www.mongabay.com/images/commodities/charts/chart-ngeu.html.
9. D. K. Katz et al., *Handbook of Natural Gas Engineering*, McGraw-Hill Book Company, New York, 1959.
10. R. R. Maddox, *Gas and Liquid Sweetening*, 2nd Ed. Campbell Petroleum Series, Norman, OK, 1974.
11. R. E. Kirk and D. F. Othmer (eds.), *Encyclopedia of Chemical Technology*. Vol. 7, Interscience Encyclopedia, Inc., New York, 1951.
12. *Guidebook to Gas Interchangeability and Gas Quality*, BP, IGU. http://www.igu.org/igu-publications/Gas%20Interchangeability%202011%20v6%20HighRes.pdf
13. www.siamindia.com.

# 3

# *Alternative Fuels for Land Transportation*

## 3.1 Introduction

Several alternative fuels are being considered to replace conventional gasoline and diesel in the transportation sector. They consist of liquid and gaseous fuels. Among the liquid fuels, biofuels are receiving greater attention due to their renewable nature.

## 3.2 Alternative Fuels for Spark Ignition Engines

### 3.2.1 Fossil-Based Fuels

#### 3.2.1.1 Gas to Liquid Fuel

*Resources*: Natural gas (the final liquid fuel being gas to liquid (GTL)) and coal (coal to liquid (CTL)).

*Introduction*: As the percentage of raw material cost is about 50–75%, the petroleum fuel cost increases steeply with crude oil price increase, which results in instability in economic development. If cheaper raw materials such as agro-waste, refinery effluents, petroleum coke, coal, naphtha, residual oil and natural gas are used, the end product price will be lower, resulting in energy sustainability.

*Production technology*: Methanol fuel can be produced using the GTL process, which comprises two chemical processes: gasification and synthesis. The process layout is shown in Figure 3.1. First, the cheaper raw material is converted into carbon monoxide (CO) and hydrogen, known as synthesis gas through the gasification process. The syngas is converted into valuable end products such as Fisher–Tropsch synthesis (FTS) diesel fuel, dimethyl ether (DME), methanol and various kinds of valuable liquid fuels using the synthesis process.

The fundamental reactions of synthesis gas chemistry are methanol synthesis, FTS, oxo synthesis and methane synthesis. For example, FTS produces

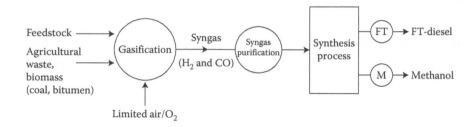

**FIGURE 3.1**
Process layout of methanol production through GTL route.

hydrocarbons of different lengths from a gas mixture of $H_2$ and CO (syngas) from biomass gasification, called bio-syngas, whereas oxo synthesis (hydroformylation) is an important industrial process for the production of aldehydes from alkenes.

Methanol is used as a fuel in a spark ignition engine. A detailed discussion is given in Section 3.2.3.1.

### 3.2.2 Hydrogen from Fossil Fuel

*Resources*: Coal and natural gas

*Hydrogen from fossil fuels*: Hydrogen can be produced from most of the fossil fuels. Since carbon dioxide is produced as a by-product, the $CO_2$ could be captured to ensure a sustainable (zero-emission) process.

*Production from natural gas*: Hydrogen can currently be produced from natural gas by means of three different chemical processes:

- Steam reforming method (SRM)
- Partial oxidation (POX)
- Auto-thermal reforming (ATR)

*Steam reforming method (SRM)*: Steam reforming involves the endothermic conversion of methane and water vapour into hydrogen and carbon monoxide as shown in Equation 3.1. The process typically occurs at temperatures of 700–850°C and pressures of 3–25 bar. The product gas contains approximately 12% CO, which can be further converted to $CO_2$ and $H_2$ through the water gas shift reaction as shown in Equation 3.2

$$CH_4 + H_2O + heat = CO + 3H_2 \qquad (3.1)$$

$$CO + H_2O = CO_2 + H_2 + heat \qquad (3.2)$$

*Partial oxidation*: Partial oxidation of natural gas is a process whereby hydrogen is produced through the partial combustion of methane with oxygen gas to yield carbon monoxide and hydrogen (Equation 3.3). In this process, heat is produced in an exothermic reaction, and hence a more compact design is possible as there is no need for any external heating of the reactor. The CO produced is further converted to $H_2$ as described in Equation 3.4.

$$CH_4 + 1/2O_2 = CO + 2H_2 + \text{heat} \qquad (3.3)$$

$$CO + H_2O = CO + H_2 + \text{heat} \qquad (3.4)$$

*Auto-thermal reforming*: ATR is a combination of both steam reforming (1) and partial oxidation (3). The total reaction is exothermic. The outlet temperature from the reactor is in the range of 950–1100°C, and the gas pressure can be as high as 100 bar. Again, the CO produced is converted to $H_2$ through the water gas shift reaction (Equation 3.2). The detailed discussion on the production of hydrogen from renewable fuel resources is given later.

### 3.2.3 Biofuels: Alcohol Fuels

#### 3.2.3.1 Methanol

Methanol or methyl alcohol ($CH_3OH$) is a colourless liquid with a boiling point of 65°C. Methanol will mix with a wide variety of organic liquids as well as with water, and accordingly it is often used as a solvent for domestic and industrial applications. It is most familiar in the home as one of the constituents of methylated spirits. Methanol is the raw material for many chemicals, formaldehyde, dimethyl terephthalate, methylamines and methyl halides, methyl methacrylate, acetic acid and gasoline.

Methanol is produced from the raw materials of coal, natural gas and biomass through the SRM, steam gasification method, photosynthesis process and carnol process as shown in Equations 3.5 through 3.8.

SRM:

$$CH_4 + 0.67CO_2 = 0.67CH_3OH + 0.67H_2O + C \qquad (3.5)$$

SRM with $CO_2$ addition:

$$CH_4 + 0.67H_2O + 0.33CO_2 = 1.33CH_3OH \qquad (3.6)$$

Biomass steam gasification process:

$$CH_{1.4}O_{0.7} + 0.3H_2O = 0.5CH_3OH + 0.5CO_2 \qquad (3.7)$$

Photosynthesis process:

$$CO_2 + 0.7H_2O = CH_{1.4}O_{0.7} + O_2 \qquad (3.8)$$

As the $CO_2$ mitigation gets high momentum due to global warming and climate change, methanol production using natural gas along with $CO_2$ emission gets more attention. $CO_2$ emission is extracted from the stack gases of a coal-fired power plant using mono-ethanolamine (MEA) solvent. Methanol production with natural gas and carbon dioxide is shown in Equation 3.9.

$$CH_4 + 0.67CO_2 = 0.67CH_3OH + 0.68H_2O + C \qquad (3.9)$$

The present study had indicated that methanol can be produced from $CO_2$ and $H_2$ using the synthesis process as shown in Equations 3.10 and 3.12. If hydrogen is produced from renewable energy sources such as solar, wind and biomass, the yield from the reaction is bio-methanol.

$$CO_2 + 3H_2 = CH_3OH + H_2O \qquad (3.10)$$

$$CO_2 + H_2 = CO + H_2O \qquad (3.11)$$

$$CO + 2H_2 = CH_3OH \qquad (3.12)$$

A methanol production process is shown in Figure 3.1. The catalyst used for the methanol synthesis process was $Cu/ZnO/ZrO_2/Al_2O_3/SiO_2$. The reactor temperature and pressure were about 525 K and 50 bar, respectively. The methanol yield was about 99.95% with traces of methyl format (460 ppm) and higher alcohols $(C_2–C_4)$ (70 ppm) [1].

Methanol was first produced as a by-product in the manufacture of charcoal through the destructive distillation of wood, with yields of 12–24 L per ton of wood. Most of the methanol today is produced from natural gas. In principle, many carbon-containing materials may be substituted for natural gas as the starting material. These include (in addition to wood) coal, lignite and even municipal wastes. Each of these raw materials, however, must first be converted to syngas; for this step, each alternative feedstock requires process modifications that increase capital investment costs over those required for natural gas.

The descriptions of the processes for converting natural gas and wood to methanol follow.

*Methanol production from wood*: The production of methanol through the conversion of wood to syngas is being examined in several countries. In terms of converting carbon to methanol, wood is inherently less efficient than natural gas. The initial gasification step in producing syngas from wood yields a mixture of CO and $H_2$ deficient in hydrogen; to bring the ratio of $H_2$ to CO to 2:1, a part of the CO is reacted with steam to yield additional hydrogen:

$$CO + H_2O \rightarrow CO_2 + H_2 \qquad (3.13)$$

This carbon dioxide is then removed from the process stream and discarded. Approximately 50% of the carbon in the wood entering the process is non-productively released to the atmosphere. This also means that 50% of all the forestry and transport activities to provide the wood to the plant are wasted. About 2.25 kg of natural gas are required to produce 4 L of methanol as opposed to about 9 kg of dry wood.

Figure 3.2 shows the sequence of operation suggested by Hokanson and Rowell [2] for the production of methanol from wood.

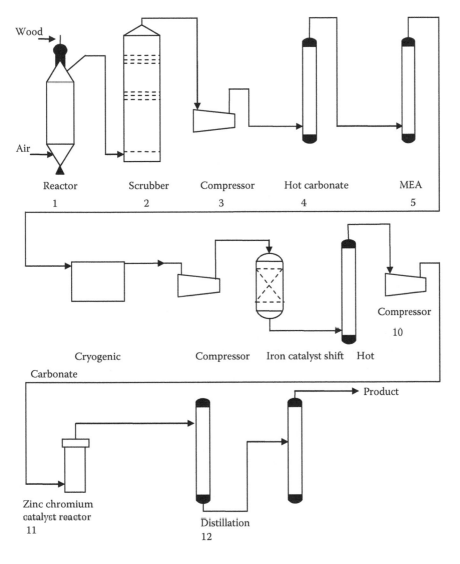

**FIGURE 3.2**
Block diagram of methanol synthesis from wood waste (Adapted from R. Feachem et al., 1983. *Sanitation and Disease: Health Aspects of Excreta and Waste Water Management.* J. Wiley and Sons, Chichester, UK.)

In the first step, wood is charged at the top of the reactor and ash discharged from the bottom. Air and steam are charged near the base of the reactor.

The sequence of reactions occurring in wood gasification is as follows:

Drying (100–200°C)

Moist wood and heat → Dry wood and water vapour

Pyrolysis (200–500°C)

Dry wood and heat → Char + CO + $CO_2$ + $H_2$ + $CH_4$ + tars and pyroligenous acids

Gasification (500°C+)

Char + $O_2$ + $H_2O$ → CO + $H_2$ + $CO_2$

The raw gas typically contains hydrogen (18%), carbon monoxide (22.8%), carbon dioxide (9.2%), methane (2.5%), other hydrocarbons (0.9%), oxygen (0.5%) and nitrogen (45.8%).

*Gas purification and shift conversion*: The raw gas is then purified to remove all but hydrogen and carbon monoxide. This mixture undergoes a reaction to convert part of the CO to $H_2$ so that the final mixture contains a 2:1 ratio of $H_2$ to CO. In this conversion, additional $CO_2$ is formed and must be removed before methanol synthesis. In detail, the raw gas from the reactor passes through a scrubber (2) cooling the gas to about 32°C and removing the tars and acid.

The gas is then compressed to about 7 bar (3) and treated in two stages to remove the carbon dioxide. In the first stage (4), a hot potassium carbonate solution reduces the $CO_2$ content to about 300 ppm. In the second stage (5), MEA is used to reduce the $CO_2$ content to about 50 ppm.

The gas is then passed through a cryogenic system (6), which removes the residual $CO_2$ and water vapour, plus methane and other hydrocarbons and finally nitrogen.

The purified gas is a mixture of hydrogen (approximately 44%) and carbon monoxide (approximately 56%). It requires further processing to provide the 2:1 ratio of $H_2$ to CO needed to produce methanol.

Following cryogenic purification, the gas is compressed (7) to 28 bar for shift conversion. Part of the CO reacts with water vapour in the presence of an iron catalyst to form additional hydrogen (8) so that the exit gas contains a 2:1 ratio of $H_2$ to CO.

Since the shift reaction (Equation 3.3) also produces $CO_2$, it is necessary to rescrub the gas with a potassium carbonate absorption system (9).

### 3.2.3.1.1 Mono-ethanolamine

Process steps: (1) partial oxidation of wood waste, (2) clean and cool crude gas, (3) compress to 7 bar, (4) remove residual carbon dioxide formed in shift, (6) remove nitrogen and hydrocarbons, (7) compress to 28 bar psig, (8) shift gas to two parts hydrogen and one part carbon monoxide, (9) remove carbon dioxide formed in shift, (10) compress to 140 bar, (11) convert hydrogen and carbon monoxide to methanol and (12) refine crude methanol into a specification-grade product.

*Methanol synthesis*: Figure 3.2 shows the block diagram of methanol synthesis from wood waste. The syngas is then compressed (10) to 140–280 bar and passed into the methanol synthesis reactor (11). In the reactor, approximately 95% of the gas is converted to methanol over a zinc–chromium catalyst. The unreacted gases are separated and recycled and the methanol purified by distillation (12).

*Integrated gasification combined cycle (IGCC)*: IGCC [3] facilities such as the cool water plant have proven to be a clean, efficient and economic means of generating electric power from coal. The IGCC facility has environmental advantages over conventional pulverised coal or fluidised bed combustion, especially with the more stringent air pollution controls now being contemplated because of acid rain concerns. Although IGCC facilities would be roughly equivalent in capital investment to conventional coal-fired steam plants, the resulting cost of power would be lower due to higher efficiency. Flexibility in the IGCC facility, however, may be somewhat more expensive. One efficient way to provide flexibility is to convert some of the energy from the gasifier into a storable liquid such as methanol. The once-through methanol (OTM) concept is being developed with this application in mind.

Figure 3.3 shows a general diagram of an IGCC/OTM plant. The IGCC plant is composed of a gasifier and its waste heat recovery (WHR) unit, an acid gas removal (AGR) unit and a combustion turbine with a WHR system. The IGCC could be modified by adding an OTM process after the AGR system to prevent

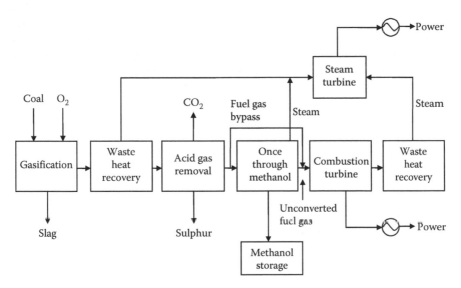

**FIGURE 3.3**
A schematic diagram of an IGCC. (Adapted from Minowa et al. *Fuel*, 74(12) 1995, 1735–1738.)

the methanol catalyst from being poisoned by sulphur compounds. The OTM unit would be composed of a guard bed system, a methanol synthesis and recovery section and a methanol storage area, and could include a peaking combustion turbine. In the methanol reactor, 2 mol of $H_2$ and 1 mol of CO react over the copper-based catalyst to form 1 mol of methanol ($CH_3OH$), which is condensed as a liquid. Up to 25% of the energy in the fuel gas can be converted to methanol. The unconverted fuel gas (depleted synthesis gas) can be saturated with water vapour and then burned in the combustion turbine of the conventional combined cycle plant.

Conventional methanol produces one mole methanol by its conversion to the reactants (two mole hydrogen and one mole carbon monoxide). It also recycles the unconverted synthesis gas at a high ratio to feed gas to maximise methanol production, since only partial methanol conversion occurs during each methanol reactor pass. Thus, recycling the unconverted synthesis gas allows almost complete conversion of the $H_2$ and CO to methanol in a conventional plant.

### 3.2.3.2 Ethanol

The resources required for ethanol ($C_2H_5OH$) production are sugarcane, molasses, barley, seed corn, wood biomass, beverage waste, waste sugars/ starches and wheat starch.

Ethanol is an alcohol-based alternative fuel produced by fermenting and distilling starch crops that have been converted into simple sugars. The feedstock for this fuel includes corn, barley and wheat. Ethanol can also be produced from "cellulosic biomass" such as trees and grasses and is called bio-ethanol. It is commonly used to increase the octane number, substitute gasoline and improve the emission quality of gasoline blended with ethanol to form an E10 blend (10% ethanol and 90% gasoline), but it can be used in higher concentrations such as E85 or E95. Ethanol is an octane booster for gasoline fuel and it enhances the thermal efficiency of a spark ignition engine by increasing the engine's compression ratio. Apart from this, ethanol has several other advantages. It is a sustainable and renewable fuel resulting in less dependence on oil imports. If it is produced from sugarcane molasses, it is a sustainable fuel since as the sugar production increases, the ethanol production also goes up. It reduces air pollution and trade deficit. It has enormous potential for rural economic development and creation of hundreds of jobs (direct and indirect). The world ethanol production is shown in Table 3.1. It is clearly seen from the table that ethanol production had increased by about 75,568 million litres in 2009 as compared to the previous year of 2008.

*Production technology:* Biological feedstocks that contain appreciable amounts of sugar or materials that can be converted into sugar such as starch or cellulose are examples of feedstocks that contain sugar. Corn contains starch that can be relatively easily converted into sugar. A significant percentage of trees and grasses are made up of cellulose, which can also be

**TABLE 3.1**

World Ethanol Production by Country (Million Litres)

| Country | 2007 | 2008 | 2009 |
|---|---|---|---|
| United States | 24,600 | 34,065 | 40,120 |
| Brazil | 19,000 | 25,500 | 24,900 |
| Europe | 2160 | 2780 | 3940 |
| China | 1840 | 1900 | 2050 |
| Canada | 800 | 900 | 1100 |
| Thailand | 300 | 340 | 1650 |
| Colombia | 285 | 302 | 315 |
| India | 200 | 250 | 350 |
| Australia | 100 | 100 | 216 |
| Other | 310 | 490 | 934 |
| World | 49,595 | 66,627 | 75,575 |

*Source:* Adapted from F. O. Licht, cited in Renewable Fuels Association, Ethanol Industry Outlook 2008, 2009, and 2010, p. 16, 29, and 22. Available at www.ethanolrfa.org.

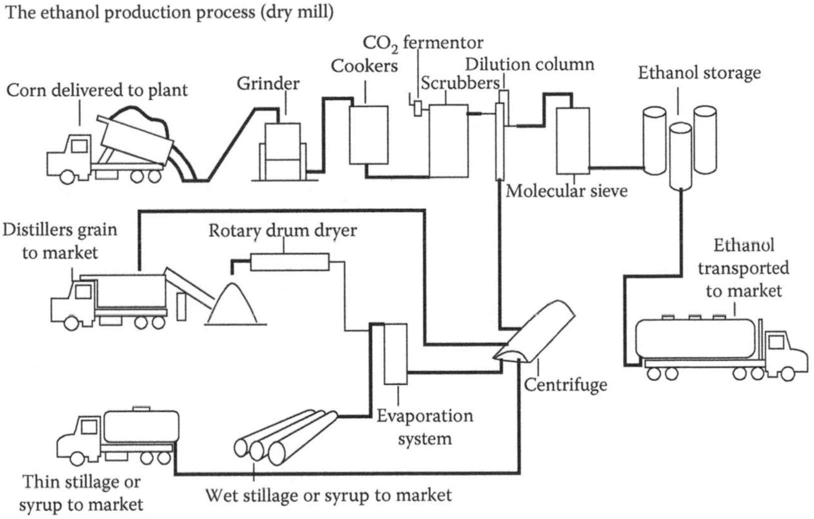

**FIGURE 3.4**

Production process for ethanol. (Adapted from F. O. Licht, cited in Renewable Fuels Association, Ethanol Industry Outlook 2008, 2009, and 2010, p. 16, 29, and 22. Available at www.ethanolrfa.org.)

converted to sugar. Cellulose has three main uses (industrial, beverage and fuel) and the production process varies slightly for each of them. The ethanol production process from corn feedstock is shown in Figure 3.4. The production process consists of seven different stages. They are milling, liquefaction, saccharification, fermentation, distillation, dehydration and denaturing.

*Milling process*: Corn (or barley or wheat) will pass through the hammer mills, which grind it into a fine powder called meal.

*Liquefaction*: The meal is then mixed with water and alpha-amylase and will pass through cookers where the starch is liquefied. Heat will be applied at this stage to enable liquefaction in the cookers with a high-temperature stage (120–150°C), and a low-temperature holding period (95°C) will be used. These high temperatures reduce bacteria levels in the mash.

*Saccharification*: Mash from the cookers will then be cooled and the secondary enzyme (gluco-amylase) will be added to convert the liquefied starch to fermentable sugars (dextrose), a process called saccharification.

*Fermentation*: Yeast will then be added to the mash to ferment the sugars to ethanol and carbon dioxide. Using a continuous process, the fermenting mash will be allowed to flow or cascade through several fermentation units until the mash is fully fermented and then leaves the final tank. In one batch fermentation process, the mash stays in one fermentation unit for about 48 h before the distillation process is started.

*Distillation*: The fermented mash, now called beer, will contain about 10% alcohol, as well as the non-fermentable solids from the corn and yeast cells. The mash will then be pumped to the continuous-flow, multi-column distillation system where the alcohol will be removed from the solids and the water. Alcohol will leave the top of the final column at about 96% strength and the residue mash, called stillage, will be transferred from the base of the column to the co-product processing area.

*Dehydration*: The alcohol from the top of the column will then pass through a dehydration system where the remaining water will be removed. Most ethanol plants use a molecular sieve to capture the last bit of water in the ethanol. The alcohol product at this stage is called anhydrous (pure without water) ethanol.

*Denaturing*: Ethanol that is to be used as a fuel is then denatured with a small amount (2–5%) of some product like gasoline to make it unfit for human consumption.

*Cellulosic ethanol production*: Cellulosic ethanol has not yet been produced commercially. However, several commercial cellulosic ethanol production plants are under construction, and intensive research and development is rapidly advancing the state of cellulosic ethanol technology, including biochemical and thermo-chemical conversion processes. A schematic diagram of a biochemical cellulosic ethanol production process is shown in Figure 3.4.

*Biochemical conversion*: Because cellulosic feedstock is more difficult to break down into fermentable sugars than starch- and sugar-based feedstock, the cellulosic biochemical conversion process requires additional steps (see diagram below). Two key steps are biomass pre-treatment and cellulose hydrolysis. During pre-treatment, the hemicellulose part of the biomass is broken down into simple sugars and removed for fermentation. During cellulose hydrolysis, the cellulose part of the biomass is broken down into simple sugar glucose as shown in Figure 3.5.

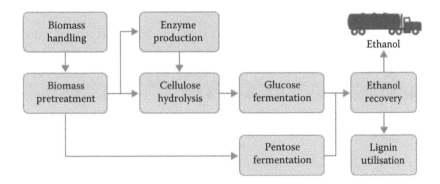

**FIGURE 3.5**
Schematic of a biochemical cellulosic ethanol production process. (Adapted from DOE Biomass Program, www1.eere.energy.gov.)

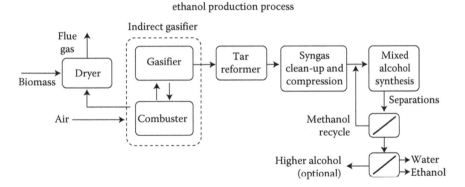

**FIGURE 3.6**
Cellulosic ethanol production. (http://www.afdc.energy.gov/afdc/ethanol/production_cellulosic.html.)

*Thermo-chemical conversion*: Ethanol can also be produced using thermo-chemical processes. In this approach, heat and chemicals are used to break biomass into syngas (a mixture of carbon monoxide and hydrogen) and reassemble it into products such as ethanol (Figure 3.6).

### 3.2.3.3 Propanol

Propanol ($C_3H_7OH$) has the potential to be used as a fuel and its characteristics are similar to methanol and ethanol. Propanol has a higher energy density than methanol and ethanol but lower than butanol, and a flashpoint that is lower than butanol but higher than ethanol and methanol.

The two forms of propanol are 1-propanol and 2-propanol. 2-Propanol is produced from the hydration of propene, which is extracted during oil refining.

The production of 1-propanol is a more complicated process as two steps are required—catalytic hydroformylation of ethylene to produce propanol and then catalytic hydrogenation of the propanol. 1-Propanol is primarily used as a solvent in the pharmaceutical, paint and cosmetic industries, whereas 2-propanol has received attention for use in direct 2-propanol fuel cells. There is little information on the utilisation of propanol as a fuel in a spark ignition engine.

### 3.2.3.4 Butanol

Butanol ($C_4H_9OH$), also commonly known as butyl alcohol, *n*-butanol or methylolpropane, is a linear four-carbon aliphatic alcohol (primary alcohol) having the molecular formula $C_4H_9OH$ (MW 74.12 g/mol). Butanol is a colourless, flammable, slightly hydrophobic liquid with a distinct banana-like aroma and strong alcoholic odour. In direct contact, it may irritate the eyes and skin. Its vapour has an irritant effect on the mucous membranes and a narcotic effect when inhaled in high concentrations.

One of the major pre-eminent roles of biobutanol (bio-based butanol) is its application in the next generation of motor fuels. While ethanol has received most of the attention as a fuel additive for many reasons, butanol could be a better direct option due to its own intrinsic physical and chemical properties and energy content as compared to ethanol. This means butanol consumption is close to that of pure gasoline, whereas ethanol–gasoline blends are consumed much faster to obtain the same power input. Additionally, butanol can be mixed with common gasoline at any percentage ratio in a similar way as with existing gasoline–ethanol blends (e.g., 23% in Brazil and 10% in the United States and some parts of Europe). Also, butanol usage does not require any modification in car engines or substitutions, producing similar mileage performance to gasoline.

Besides the expected role as an engine biofuel, butanol is actually an important bulk chemical with a broad range of industrial uses. Almost half of the worldwide production is used in the form of butyl acrylate and methacrylate esters used in the production latex surface coatings, enamels, nitrocellulose lacquers, adhesives/scalants, elastomers, textiles, super absorbents, flocculants, fibres and plastics. Other important butanol-derived compounds are butyl glycol ether, butyl acetate and plasticisers. Compounds of minor applicability are butyl amines and amino resins. Butanol and derived compounds are excellent diluents in paint thinners, hydraulic and brake fluid formulations. It is also used as a solvent in the perfume industry and for the manufacture of antibiotics, vitamins and hormones. Other applications include the manufacture of safety glass, detergents, flotation aids (e.g., butyl xanthate), de-icing fluids, cosmetics (eye make-up, nail care products, shaving and personal hygiene products). It is also commonly used as an extracting agent and in food and flavour industries).

*Butanol production process flow diagram*: Figure 3.7 shows a schematic of the batch fermentation process. Molasses, containing 55 wt% fermentable sugars and 30 wt% non-fermentable solids, is diluted to 60 g/L sugar and

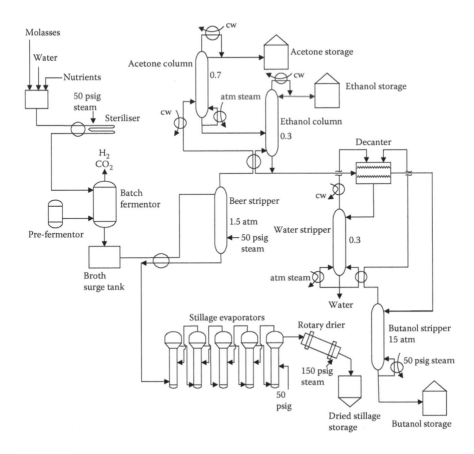

**FIGURE 3.7**
Process flow diagram of batch fermentation for production of butanol. (Adapted from Y. D. Park, J. J. Park and J. H. Lim, 1979. Research Reports of the Office of Rural Development, Suweon, Korea.)

mixed with nutrients in the feed mix tank. Butanol inhibition prevents the use of higher sugar concentrations in the fermenter. The diluted feed is continuously sterilised by direct steam injection and fed into batch fermenters. Fermenters are inoculated with actively growing cells of a strain of *Clostridium acetobutylicum* produced in smaller seed fermenters. After 30 h of fermentation, the broth, containing 13.7 g/L butanol, 5.4 g/L acetone, 1.5 g/L ethanol, 0.2 g/L butyric acid, 0.3 g/L acetic acid and 3.0 g/L cells, is discharged into the broth surge tank. The batch fermenters are operated on a staggered schedule so that downstream processing is continuous.

Butanol, acetone and ethanol are stripped from the broth with 35 bar steam in the beer stripper after being heated to 100°C by heat exchange with the stripper bottom product. The stripped broth, containing acetic and butyric acids, cells, proteins and non-fermentable molasses solids, is evaporated to

50 wt% solids in the multiple-effect stillage evaporators and then dried to 85 wt% solids in a rotary dryer to give a dried stillage product that can be used as an animal feed supplement.

The overhead vapour from the beer stripper, containing approximately 70 wt% water and 30 wt% acetone, butanol and ethanol, is separated in a series of four distillation columns. 99.5 wt% acetone is taken overhead from the first column. This column is operated at 0.7 atm so that low-pressure steam from the last effect of the stillage evaporators can be used in the reboiler. The bottom product from the acetone column is fed to the ethanol column, which operates at 0.3 atm pressure.

Vacuum operation reduces the reflux needed to produce the 95 wt% ethanol overhead product and allows the total reboiler duty to be met by condensing the overhead vapours from the beer stripper in the ethanol column reboiler. The bottom product from the ethanol still and the overhead streams from the water and butanol strippers are fed to a decanter where an aqueous-rich phase is allowed to separate from a butanol-rich phase. The water-rich phase, containing approximately 9.5 wt% butanol, is refluxed to the water stripper, which produces water containing less than 0.01 wt% butanol. The operating pressures of the stripping columns are set such that about half of the heat duty in the water stripper reboiler is met by condensing the overhead vapours from the butanol stripper.

Butanol can also be produced from the same agricultural feedstock as ethanol (e.g., sugarcane, corn, wheat, sorghum and, in future, dedicated energy grasses). The production process is similar to ethanol offering the possibility to retrofit the existing ethanol capacity to butanol production.

The existing infrastructure can be utilised for butanol production. It is a much superior fuel to ethanol in terms of toxicity, energy density and storage. Butanol can be blended with gasoline at higher concentrations than ethanol for use in standard vehicle engines. Currently, European regulations allow butanol to be blended up to 15% by volume and U.S. regulations allow 16% by volume. Its energy content is closer to that of gasoline and is higher than that of ethanol.

It can be easily blended to conventional gasoline, and therefore it can use the industry's existing distribution infrastructure without requiring modifications in the blending facilities, storage tanks or retail station pumps. It can be more easily transported through pipelines, unlike existing biofuels. When butanol is added to gasoline, it does not increase the vapour pressure, which results in the decrease of emission of volatile organic compounds (VOCs).

*Raw material required for producing 1 L of alcohol based on EU report*: The renewable ethanol produced in Europe is almost 100% produced from European-grown raw material. The most commonly used feedstock is wheat, although other cereals can be used such as barley, rye, maize and triticale. Close to two-thirds of all raw material used are cereal-based with the rest of the feedstock being mainly derived from sugar beet [8].

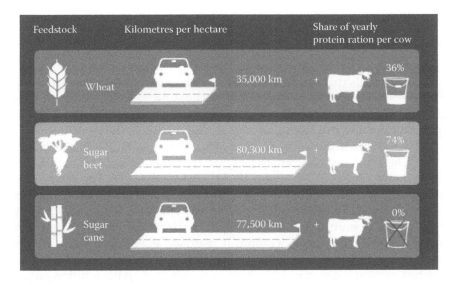

**FIGURE 3.8**
Bio-ethanol output per hectare. (Adapted from Microalgae for biodiesel production and other application: A review material, *A Renewable and Sustainable Energy Reviews*, 14(1); January 2010, 217–232.)

On an average, 2.7 kg of grain produces 1 L of ethanol and 1 kg of protein-rich animal feed [8]. The same applies to sugar beet. When 7.9 kg of sugar beet is used to produce 1 L of ethanol, 600 g of a co-product called vinasse is produced at the same time. Vinasse can be used as a rich fertiliser, animal feed or as a source of biogas production. Additionally, 600 g of carbohydrate-rich dried beet pulp are finally left over from the process and used as an animal-feed concentrate. Bio-ethanol output per hectare is shown in Figure 3.8.

### 3.2.3.5 Dimethyl Carbonate

Dimethyl carbonate ($C_3H_6O_3$) is an oxygenated fuel that is produced from methanol as shown in Equation 3.14. It is not getting importance as an auto fuel due to the constraint in the need of a specific raw material of methanol. It is a good strategy if methanol is directly used as a fuel in a spark ignition engine. Otherwise, it is converted into dimethyl carbonate and then can be used as a fuel in the engine resulting in lower energy content than methanol, which is due to a loss in conversion efficiency. So there is no report on the use of dimethyl carbonate as a fuel for a spark ignition engine due to its poor energy sustainability as a fuel. The production of DMC is shown in Equation 3.14.

$$2CH_3OH + CO + 1/2O_2 \longrightarrow (CH_3O)_2C=O + H_2O \tag{3.14}$$

| Methanol | Carbon monoxide | Oxygen | | Dimethyl carbonate | Water |

### 3.2.4 Biofuels: Gaseous Fuels

#### 3.2.4.1 Producer Gas

Producer gas is normally generated by using waste wood and waste biomass. Other fuels that can be obtained from biomass gasification are shown in Figure 3.9. The producer gas primarily contains methane and inert gases ($CO_2$ and $N_2$). It is produced using the gasification process under a limited air environment with application of heat. The carbon in the feedstock is converted to CO due to the lesser amount of air. However, $CO_2$ will also be generated. The producer gas contains a large fraction of inert gas, such as $CO_2$ and $N_2$, in the range of 12–15% and 48–50%, respectively [9]. The octane number of producer gas is higher than that of gasoline and CNG due to the presence of inert gas. However, it has a lower calorific value of about 4.7 MJ/m$^3$ as compared to natural gas (30 MJ/m$^3$) and affects the power output of an engine, resulting in power deration. If methane is enriched from the producer gas by removing the inert gas, including inorganic compounds, the enriched methane gas could be used as a fuel in a spark ignition engine. However, the economics of the enrichment of the gas may be an impediment to implement in a transport fleet. Figure 3.9 gives an overview about the possible applications of the producer gas from a steam-blown gasifier.

A schematic diagram of the two-stage gasifier is shown in Figure 3.9 and the operational diagram is given in Figure 3.10. The plant had two reactors in a series that were made of STS-316 and heated indirectly using electrical furnaces. The bottom reactor had a height of 390 mm and an inner diameter

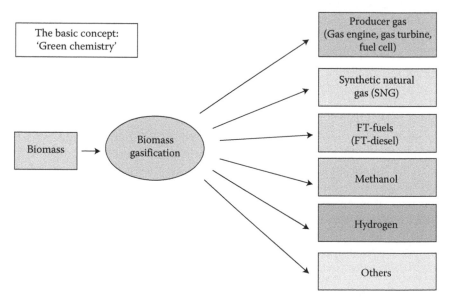

**FIGURE 3.9**
Schematic diagram of the two-stage gasifier.

of 110 mm. The bottom reactor had four thermocouples and was filled with silica sand. The upper reactor was 340 mm in height and its inner diameter was 160 mm. The two reactors were separated using a tuyere distributor having 29 bubble caps. The activated carbon was applied in the upper reactor to reduce the tar yield. The reaction temperatures of the two reactors were controlled separately. At first, the reaction temperature was set at a value below the aimed temperature.

During experiments, it rose to some extent and stayed at this level. The reaction temperatures of the two reactors were defined as the average values taken from the thermocouples. The woody waste was placed in the sand in the bottom reactor through two screw feeders. As a fluidising medium, air was heated at 750°C using a pre-heater to reduce heat loss in the bottom reactor. The producer gas yielded by the two-stage gasifier was immediately passed through a two-stage cyclone and a hot filter to collect its particles (Figure 3.10). The cyclone and the hot filter were designed to capture particles bigger than 100 and 10 µm, respectively. After the particle separation, the condensate liquid was captured in a series of stainless and glass condensers that were cooled up to 10°C, using water as the cooling medium. Finally, to capture the aerosols and untreated particles in the producer gas,

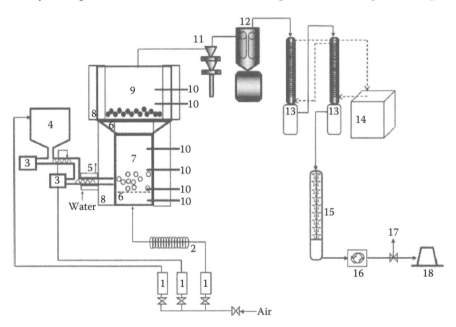

**FIGURE 3.10**
Operational diagram of two stage gasification plant. (1) Flow meter, (2) pre-heater, (3) screw feeders, (4) silo, (5) water jacket, (6) tuyere cap distributor, (7) fluidised bed, (8) electric furnace, (9) activated carbon zone, (10) thermocouples, (11) two-stage cyclone, (12) hot filter, (13) condensers, (14) chillers, (15) electrostatic precipitator, (16) gas meter, (17) gas sampler, (18) flare stack. (Adapted from Ben-Zion Ginzburg, 1993, *Renewable Energy*, 3(2–3), 249–252.)

an electrostatic precipitator was used in the experiments. The remaining producer gas was either burnt in a flare stack or sampled using Teflon gas bags at 3 min intervals, to analyse their composition.

### 3.2.4.2 Biogas

The resources that are normally used are manure (animal, human sewage), straw, grass, maize, seed cake, sugar beet, barley, hemp, rice straw and agro-waste.

Biogas is produced through a biological process. Extracellular enzyme degrades complex carbohydrates, proteins and lipids into their constituent units during the hydrolysis process. It is a fermentation process where hydrolysis products are converted to acetic acid, hydrogen and carbon dioxide. The methanogenic bacteria mediate these reactions in the digester, thus producing suitable conditions for methane gas production. The anaerobic bacteria control methane production from acidogenesis products. Anaerobic digesters are typically designed to operate in the mesophilic (20–40°C) or thermophilic (above 40°C) temperature zones. Sludge produced from the anaerobic digestion of liquid biomass is often used as a fertiliser [11].

Biogas typically contains 50–70% methane and 30–50% carbon dioxide, depending on the feedstock [12]. As the energy content of the biogas is about 25 MJ/kg as compared to natural gas (48 MJ/kg), the power out drops significantly. Methane enrichment of biogas can be obtained by using a water scrubbing system for the removal of $CO_2$ content [12]. Methane content is enriched from 65% by volume with biogas to 95% with the water scrubbing system. The property of enriched biogas is comparable to base CNG fuel and can be used as a fuel in a spark ignition engine. However, the economics of the enrichment cost would play a vital role in the implementation in vehicle fleets.

Anaerobic treatment is the use of biological processes, in the absence of oxygen, for the breakdown of organic matter and the stabilisation of these materials, by conversion to methane and carbon dioxide gases and a nearly stable residue. As early as the eighteenth century, the anaerobic process of decomposing organic matter was known, and in the middle of the nineteenth century, it became clear that anaerobic bacteria are involved in the decomposition process. But it has only been a century since anaerobic digestion was reported to be a useful method for the treatment of sewage and offensive material. Since that time, the applications of anaerobic digestion have grown steadily in both its microbiological and chemical aspects.

The environmental aspect and the need for renewable energy are receiving interest and considerable financial support in both the developed and developing countries. The research and application work in these directions have expanded, and many systems using anaerobic digestion have been constructed in many countries. Anaerobic digestion provides some exciting possibilities and solutions to global concerns such as alternative energy production; handling human, animal, municipal and industrial wastes safely; controlling environmental pollution and expanding food supplies.

Most technical data available on biogas plants relate primarily to two digester designs, the floating cover and fixed-dome models. Promising new techniques such as bag, dry fermentation, plug flow, filter and anaerobic baffled reactors should be explored to establish a firmer technical base on which to make decisions regarding the viability of biogas technology. Along with this increase in interest, several newer processes have been developed that offer promise for more economical treatment and for stabilising other than sewage materials—agricultural and industrial wastes, solid, organic municipal residues and so on—and generating not only an alternative energy source but also materials that are useful as fodder substitutes and substrates for the mushroom and greenhouse industries, in addition to their traditional use as organic fertilisers. Other benefits of anaerobic digestion include reduction of odours, reduction or elimination of pathogenic bacteria (depending upon the temperature of the treatment) and the use of the environmentally acceptable slurry.

The technology of anaerobic digestion has not yet realised its full potential for energy production. In most industrialised countries, biogas programmes (except for sewage treatments) are often hindered by operational difficulties, high costs of plants and as yet low energy prices. In most developing countries, the expansion of biogas programmes has been hindered because of the need for better economic initiatives, organised supervision and initial financial help, while in other developing countries, slow development has been observed, along with a lack of urgency, because of readily available and inexpensive non-commercial fuels, such as firewood.

Biogas technology is also potentially useful in the recycling of nutrients back to the soil. Burning non-commercial fuel sources, such as dung and agricultural residues, in countries where they are used as a fuel rather than as a fertiliser, leads to a severe ecological imbalance, since the nutrients, nitrogen, phosphorus, potassium and micro-nutrients, are essentially lost from the ecosystem. Biogas production from organic materials not only produces energy but also preserves the nutrients, which can, in some cases, be recycled back to the land in the form of slurry. The organic digested material also acts as a soil conditioner by contributing humus. Manure could be used as raw material for fermentation process for producing valuable products instead of less valuable utilization such as fertilization to soil. Chinese workers report that digested biomass increases agricultural productivity by as much as 30% over farmyard manure, on an equivalent basis [13]. This is due in part to the biochemical processes occurring during digestion, which cause the nitrogen in the digested slurry to be more accessible for plant utilisation, and to the fact that less nitrogen is lost during digestion than in storage or composting. The stability of the digested slurry and its low BOD (biological oxygen demand) and COD (chemical oxygen demand) are also of great importance. This aspect of biogas technology may in fact be more important than the gas produced [14,15].

In the area of public health and pollution control, biogas technology can solve another major problem: that of the disposal of sanitation wastes.

Digestion of these wastes can reduce the parasitic and pathogenic bacterial counts by over 90% [13,16], breaking the vicious circle of reinfection via drinking water, which in many rural areas is untreated. Industrial waste treatment, using anaerobic digestion, is also possible.

To draw conclusions about the feasibility of the anaerobic digestion process, it can be examined in one of two ways: a strictly financial approach, involving the analysis of monetary benefits such as the sale or reuse of products (methane, carbon dioxide and slurry with all its applications) and the costs of constructing and maintaining facilities; or as a social assessment of input and output, including intangibles such as improvements in public health, reduced deforestation and reduced reliance on imported fossil fuels, in a social cost–benefit analysis.

There is an increased recognition, in both developing and industrialised countries, of the need for technical and economical efficiency in the allocation and exploitation of resources. Systems for the recovery and utilisation of household and community wastes are gaining a more prominent place in the world community. During the past few years, anaerobic fermentation has developed from a comparatively simple technique of biomass conversion, with the main purpose of energy production, into a multifunctional system:

1. Treatment of organic wastes and wastewaters in a broad range of organic loads and substrate concentrations
2. Energy production and utilisation
3. Improvement of sanitation; reduction of odours
4. Production of high-quality fertiliser

Research and development (R&D) has shifted from basic studies on the anaerobic fermentation of quasi-homogeneous substrates, with contents of organic solids in the range of about 5–10%, to the digestion of more complex materials that need modified digester designs. The main fields of R&D activities are

1. Fermentation at high organic loadings
2. High-rate digestion of diluted wastewaters of agro-industries, including substrate separation during fermentation; immobilisation of the microorganisms
3. Fermentation and reuse of specific materials in integrative farming systems
4. Biogas purification
5. Simple but effective digested design/construction of standardised fermenters
6. Domestic wastewater treatment

*Anaerobic processes, plant types and control*: Batch and dry fermentation: This is the simplest of all the processes. The operation involves merely charging an airtight reactor with the substrate, a seed inoculum and, in some cases, a chemical (regularly a base) to maintain almost neutral pH. The reactor is then sealed, and fermentation is allowed to proceed for 30–180 days, depending on ambient temperature. During this period, the daily gas production builds up to a maximum and then declines. This fermentation can be conducted at a "normal" solids content (6–10%) or at high concentrations (>20%), which is then known as "dry" fermentation. Its main components are shown in Figure 3.11.

*Fixed dome (Chinese)*: A fixed-dome biogas digester (see Figure 3.12) was built in Jiangsu, China as early as 1936, and since then considerable research

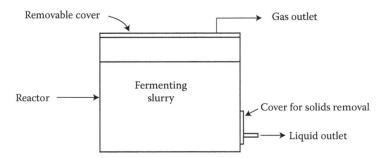

**FIGURE 3.11**
Batch digester.

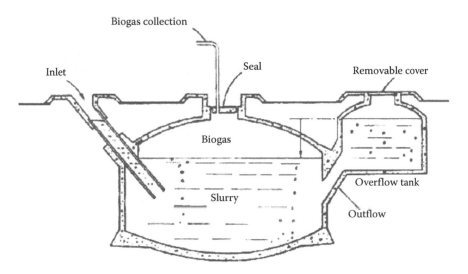

**FIGURE 3.12**
Fixed-dome (Chinese) digester.

has been carried out in China on various digester models. The water pressure digester was developed in the 1950s. In one variation, the displaced effluent flows on to the roof of the reactor, and thus it enables the roof to withstand the gas pressure. In terms of absolute numbers, the fixed dome is by far the most common digester type in developing countries. This reactor consists of a gas-tight chamber constructed of bricks, stones or poured concrete. Both the top and bottom of the reactor are hemispherical, and are joined together by straight sides.

The digester is fed semi-continuously (i.e., once a day); the inlet pipe is straight and ends at the mid-level in the digester. There is a manhole plug at the top of the digester to facilitate the entrance for cleaning, and the gas outlet pipe exits from the manhole cover. The gas produced during digestion is stored under the dome and displaces some of the digester contents into the effluent chamber, leading to gas pressures in the dome of between 1 and 1.5 m of water. This creates quite high structural forces and is the reason why the reactor has a hemispherical top and bottom.

The design of fixed-dome digesters has been developed in China based on (1) four important horizontal lines; (2) gas pressure; (3) average rate of gas production; (4) gas storage; (5) digester size and (6) geometric forms, loads and forces. The inside water level at ambient pressure is 95% of the total volume. The gas pressure in fixed-dome digesters is equal to, or below, 120 cm of water. The ratios of key dimensions are kept constant, for example, diameter to height ratio of the cylinder is 2:1. The HRT, for both cow and pig manure, is 35–40 days at total solids concentrations of 5–8% and 4–7%, respectively. Gas production varies between 0.15 and 0.6 $m^3$ per day, depending on ambient temperature. The state of development of fixed-dome digesters is quite advanced, and much is known about the material, methods of construction, cost and suitable digester feedstock and gas production rates.

*Floating dome (Indian or KVIC design)*: In India, the history of biogas technology has developed since 1937. In 1950, Patel designed a plant with a floating gas holder, which caused renewed interest in biogas in India. The Khadi and Village Industries Commission (KVIC) of Mumbai began using the Patel model biogas plant in a planned programme in 1962, and since then it has made a number of improvements in the design. The floating dome digester (see Figure 3.13) is disseminated by KVIC and workshops recognised by KVIC. Those most commonly constructed are of 6 and 8 $m^3$ gas production capacities. The digester is designed for 30, 40 and 55 days' retention time: the lowest time applies to the hot southern states and the highest to the cooler northern states. Construction costs vary according to ambient temperatures, for which partial compensation is allowed for by subsidies. The main material fed is cattle manure. At the community plant level, nightsoil is digested in a mixture with cattle dung, and at the large farm level, other types are being introduced to digest materials such as water hyacinth. The drum was

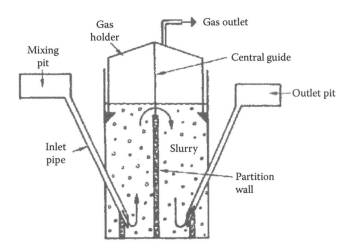

**FIGURE 3.13**
Floating dome (Indian) digester.

originally made of mild steel, until fibreglass reinforced plastic (FRP) was introduced successfully to overcome the problem of corrosion. Nearly all new digesters are equipped with FRP gas holders. The cost of a mild steel gas holder is approximately 40–50% of the total cost of the plant. FRP gas holders are 5–10% more expensive than the steel drum.

A typical KVIC design is shown schematically in Figure 3.13. The reactor wall and bottom are usually constructed of brick, although reinforced concrete is sometimes used. The gas produced in the digester is trapped under a floating cover that rises and falls on a central guide. The volume of the gas cover is approximately 50% of the total daily gas production. The pressure of the gas available depends on the weight of the gas holder per unit area, and usually varies between 4 and 8 cm of water pressure.

The reactor is fed semi-continuously through an inlet pipe, and displaces an equal amount of slurry through an outlet pipe. When the reactor has a high height:diameter ratio, a central baffle is included to prevent short-circuiting. Most of the KVIC-type digesters are operated at ambient temperatures, so that retention times depend on local variations. Typical retention times are 30–40 days in warm climates, such as southern India, where ambient temperatures vary from 20–40°C, 40–50 days in moderate climates, such as the central and plain areas of India, where the minimum temperatures go down to 5°C, and 50–80 days in cold climates, such as the hilly areas of northern India, where the minimum temperatures go below 0°C.

The typical feedstock is cattle dung, although substrates such as agricultural residues, nightsoil and aquatic plants have been used. Cattle manure, generally about 20% solids, is diluted to 10% total solids before feeding, by adding an equal quantity of water. The daily average gas yield varies

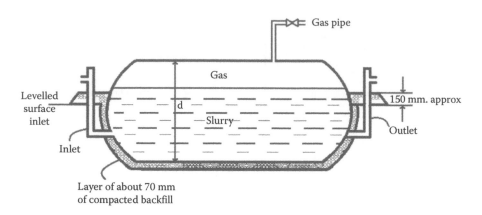

**FIGURE 3.14**
Bag-red mud (Taiwan, China) digester.

from 0.20 to 0.60 volume of gas per volume of digester ratio in cold to warm climates.

*Bag design (Taiwan, China):* The bag digester is essentially a long cylinder (length:diameter 3:14) made of PVC, a neoprene-coated nylon fabric, or "red mud plastic" (RMP), a proprietary PVC, to which wastes from aluminium production are reported to be added. Integral with the bag are the feed and outlet pipes and a gas pipe (Figure 3.14). The feed pipe is arranged so that a maximum water pressure of approximately 40 cm is maintained in the bag. The digester acts essentially as a plug flow (unmixed) reactor, although it can be stored in a separate gas bag [17].

The basic design originated in Taiwan, China, in the 1960s [18] due to problems experienced with brick and metal digesters. The original material used, a neoprene-coated nylon, was expensive and did not weather well. In 1974, a new membrane, RMP, was produced from the residue from aluminium refineries. Owing to its availability, PVC is also starting to be used extensively, especially in central America. The membrane digester is extremely light (e.g., a 50 m³ digester weighs 270 kg), and can be installed easily by excavating a shallow trench, slightly deeper than the radius of the digester.

*Plug flow design:* The plug flow reactor, while similar to the bag reactor, is constructed of different materials and classified separately. A typical plug flow reactor consists of a trench lined with either concrete or an impermeable membrane (Figure 3.15). To ensure true plug flow conditions, the length has to be considerably greater than the width and depth.

The reactor is covered with either a flexible cover gas holder, anchored to the ground, or with a concrete or galvanised iron top. In the latter type, a gas storage vessel is required. The inlet and outlet to the reactor are at opposite ends, and feeding is carried out semi-continuously, with the feed displacing an equal amount of effluent at the other end.

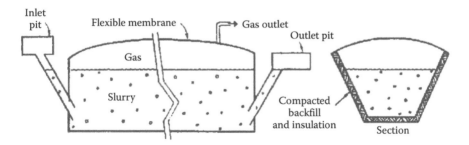

**FIGURE 3.15**
Plug flow digester.

### 3.2.4.3 Bio-Syngas

Resources that are normally used include biomass and agro-waste. Synthesis gas is a mixture of carbon monoxide and hydrogen. It is produced by using the gasification process under a limited air environment with the application of heat. The feedstock is cheaper materials such as biomass, agro-waste and grass. The syngas consists of about 40% combustible gases such as carbon monoxide (CO), hydrogen ($H_2$), methane ($CH_4$) and the remainder of primary non-combustible gases such as nitrogen ($N_2$) and carbon dioxide ($CO_2$) [19]. The typical composition of synthesis gas is shown in Table 3.2. It has a lower calorific value of 4.13 MJ/kg as compared to natural gas (48 MJ/kg) and this is a problem similar to that of fuels such as biogas and producer gas. So the fuel quality of raw syngas needs to be upgraded for better performance of a spark ignition engine. It is worth mentioning that syngas is a feedstock for the F–T or methanol synthesis process for producing valuable end products such as methanol, F–T diesel and DME. Thus, syngas as a fuel may not be a good option to utilise in an internal combustion engine.

*Production of syngas*: The coal and methane co-conversion experiments were carried out on a 45 mm (I.D.) auto-thermal fluidised bed reactor. The height of the reactor is about 1500 mm. The system includes coal feeding, gas distributing, reactor, de-dusting, flow measurement and controlling (see the flowchart in Figure 3.16).

**TABLE 3.2**

Composition of a Typical Synthesis Gas

| Gas | Composition (%) |
| --- | --- |
| Hydrogen | 13.7 |
| Carbon monoxide | 22.3 |
| Methane | 1.9 |
| Carbon dioxide | 16.8 |
| Nitrogen | 45.3 |

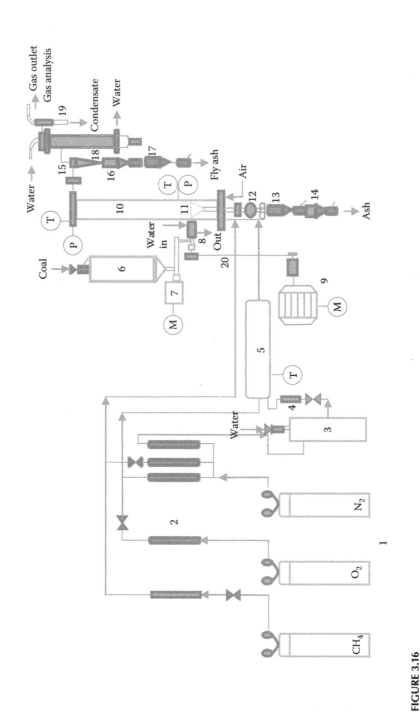

**FIGURE 3.16**

Flow sheet of the bench-scale fluidised bed coal and methane co-conversion system. (1) Gas cylinder, (2) gas rotameters, (3) distilled water tank, (4) water rotameter, (5) steam evaporator and mixer, (6) coal hopper, (7) measuring screw feeder, (8) pushing feeder, (9) motor, (10) gasifier, (11) gas distributor, (12) ash discharger, (13) 1st ash hopper, (14) 2nd ash hopper, (15) cyclone, (16) 1st fly ash hopper, (17) 2nd fly ash hopper, (18) gas cleaner/cooler, (19) liquid and gas separator, (20) driving band. (Adapted from T. M. Mata et al., 2010, *Renewable and Sustainable Energy Reviews,* 14(1), 217–232.)

- Coal preparation

  Raw coal is crushed to a size of 0~1 mm diameter with a jaw crusher and then put into the coal hopper and dried.

- Feeding system

  The dried coal is continuously fed by an adjustable screw feeder, and then it is pushed into the reactor by another screw feeder.

- Reaction gas supply unit

  The reaction gases (oxygen, steam and methane) are fed through the gas distributor and ash discharging pipe.

- Co-conversion reactor

  The coal and methane co-conversion reactor is a fluidised bed. Dry coal reacts with methane, oxygen and steam in the reactor, producing $CO$, $H_2$, $N_2$ $CH_4$, $CO_2$ and $H_2S$. In the reaction region, a part of the oxygen and steam enters the reactor through the inverse-cone-shaped gas distributor to keep the particle fluidising, and the remaining part and methane enters through the ash discharging pipe into the reactor. A certain amount of $N_2$ enters the reactor also through the ash discharging pipe to adjust the temperature and to prevent ash sintering. The ash is removed through the ash discharging pipe to the ash hopper system and taken out of the reactor. The produced gas exits from the top of the reactor and then enters into the cyclone.

- De-dusting, gas cooling/cleaning system

  High-temperature gas from the top of the reactor enters into the cyclone which separates the solid fine particles from the gas. Then it passes through the water scrubbing system. The remaining fly ash in the gas is further captured, and in the mean time the gas is cooled. The cleaned gas is taken out of the system after measuring and analysing.

- Operating and controlling system

  The co-conversion system is equipped with the necessary measuring instruments (temperature, pressure, flow rate, etc.). The reaction temperature is controlled manually by adjusting the reacting agent flow rate or coal feed rate.

### 3.2.4.4 Hydrogen

The resources needed for producing hydrogen comprise natural gas, coal, renewable resources such as biomass, water, sunlight, wind, wave or hydropower. A variety of process technologies can be employed, including chemical, biological, electrolytic, photolytic and thermo-chemical. Production cost influences the choice of the various options for hydrogen production. An overview of the various feedstocks and process technologies is presented in Figure 3.17.

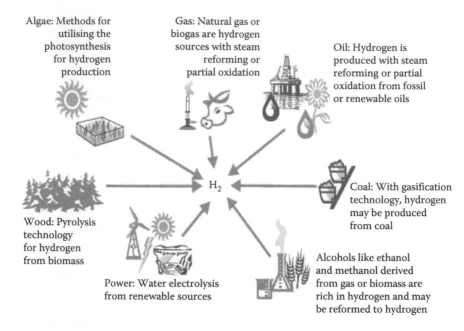

**FIGURE 3.17**
Some feedstock and process alternatives. (www.doe.com.)

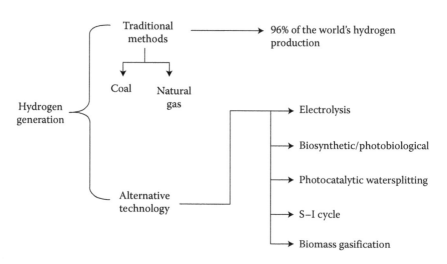

**FIGURE 3.18**
Hydrogen production methods.

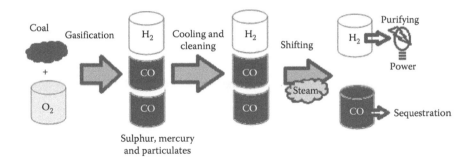

**FIGURE 3.19**
Coal gasification process.

*Hydrogen generation*: The various methods used for producing hydrogen are shown in Figure 3.18.

*Production of hydrogen from coal*: Hydrogen can be produced from coal through gasification processes. Gasification is a process in which coal or biomass is converted into gaseous components by applying heat under pressure and in the presence of steam (Figure 3.19). A subsequent series of chemical reactions produces a synthesis gas, which is reacted with steam to produce more hydrogen than that can be separated and purified.

Underground coal gasification is the conversion of the coal itself to a usable syngas consisting of hydrogen, carbon monoxide and methane. The conversion is achieved by introducing oxygen and steam into the coal seam, and igniting the coal. It can be used at the surface for heating, power generation, hydrogen production or the manufacture of key liquid fuels such as diesel fuel or methanol. It consists of mixing coal with oxygen, air or steam at very high temperatures without letting combustion occur (partial combustion).

A chemical reaction process is given in Equation 3.15, in which carbon is converted to carbon monoxide and hydrogen.

$$C(s) + H_2O + heat = CO + H_2 \qquad (3.15)$$

Since this reaction is endothermic, additional heat is required. The CO is further converted to $CO_2$ and $H_2$ through the water gas shift reaction described in Equation 3.2.

*Hydrogen from splitting of water*: Hydrogen can be produced from the splitting of water through water electrolysis, photo-electrolysis, photo-biological production and high-temperature water decomposition.

*Water electrolysis*: Water electrolysis is the process whereby water is split into hydrogen and oxygen through the application of electrical energy, as in Equation 3.16. The total energy that is needed for water electrolysis increases slightly with temperature, while the required electrical energy decreases. Future potential costs for electrolytic hydrogen are presented in Figure 3.20, where the possibilities to considerably reduce the production cost are evident.

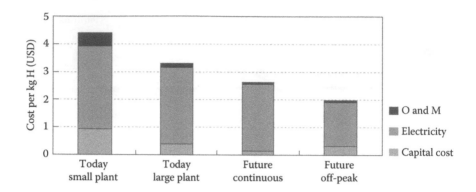

**FIGURE 3.20**
Future potential costs of electrolytic hydrogen. (Adapted from US DoE; www.doe.com.)

$$H_2O + \text{electricity} = H_2 + 1/2O_2 \qquad (3.16)$$

*Photo-electrolysis (photolysis)*: Photovoltaic (PV) systems coupled to electrolysers are commercially available. The systems offer some flexibility, as the output can be electricity from photovoltaic cells or hydrogen from the electrolyser. Direct photo-electrolysis represents an advanced alternative to a PV-electrolysis system by combining both processes in a single apparatus. This principle is illustrated in Figure 3.21. Photo-electrolysis of water is the process whereby light is used to split water directly into hydrogen and oxygen. Such systems offer great potential for the cost reduction of electrolytic hydrogen, as compared with conventional two-step technologies.

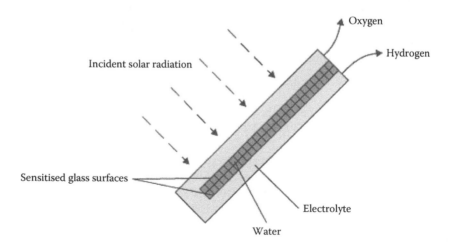

**FIGURE 3.21**
Principle of a photo-electrolytic cell. (Adapted from Hydrogen Solar Production Company, www.hydrogensolar.com.)

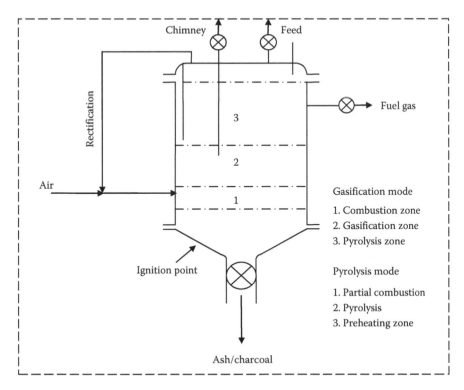

**FIGURE 3.22**
Gasification process.

*Hydrogen production from biomass*: Biomass has the potential to become an important $H_2$ source via different processes as shown in Figures 3.22 and 3.23.

### 3.2.4.4.1 By Thermo-Chemical Process

- Gasification
- High-pressure conversion in the presence of water
- Pyrolysis

*Photo-biological production (biophotolysis)*: The photo-biological production of hydrogen is based on two steps: photosynthesis (Equation 3.17) and hydrogen production catalysed by hydrogenases (Equation 3.18), for example, green algae and cyanobacteria. Long-term basic and applied research is needed in this area, but if successful, a long-term solution for renewable hydrogen production will result. Figure 3.24 shows hydrogen production from a thermo-biological process. It is of vital importance to understand the natural processes and the genetic regulations of $H_2$ production. Metabolic and genetic engineering may be used to demonstrate the process in larger

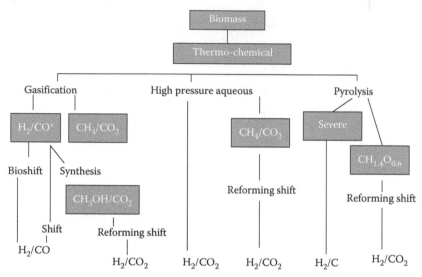

*In boxes-storable intermediates.

**FIGURE 3.23**
Hydrogen production from thermo-chemical processes.

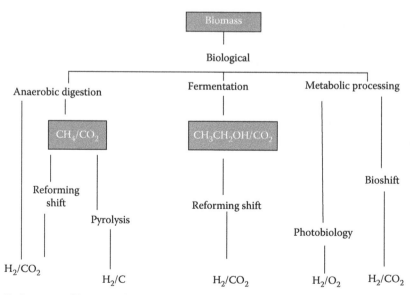

*In boxes-storable intermediates.

**FIGURE 3.24**
Hydrogen production from thermo-biological processes.

bioreactors. Another option is to reproduce the two steps using artificial photosynthesis.

Photosynthesis:

$$2H_2O = 4H^+ + 4e^- + O_2 \tag{3.17}$$

Hydrogen production:

$$4H^+ + 4e^- = 2H_2 \tag{3.18}$$

The principal biological processes are

- Anaerobic digestion
- Fermentation

Anaerobic digestion directly yields an $H_2$ and $CO_2$ mixture or via the reforming shift of storable intermediate products, which give rise to $CH_4$ and $CO_2$. The pyrolysis of $CH_4$ yields $H_2$ and carbon. Fermentation yields ethanol and $CO_2$ and via the reforming shift, $H_2$ and $O_2$ are produced. Hydrogen can be produced using different feed stocks and processes as given below:

- Hydrogen produced by biological processes has the advantages of:
  - Environment-friendly process
  - Sustainable production processes
- Biohydrogen is produced from variety of processes as given below:
  - Thermophilic fermentation
  - Photo-fermentation
    - Photo-fermentation requires energy input
    - Dark fermentation does not require energy input
  - Gas cleaning
  - Bio-photolysis requires energy input
- Photosynthetically using algae
  - Biological hydrogen production is done in a bioreactor based on the production of hydrogen by algae. Algae produce hydrogen under certain conditions
- Fermentatively
  - Produced by many bacteria, particularly Clostridia
  - Through dark fermentation either by
    - A mixed culture of hydrogen producing sludge or
    - A pure culture of anaerobic bacteria such as *Clostridium butyricum*

- Bio-photolysis of water by microalgae and cyanobacteria
- Cyanobacteria, formerly known as blue-green algae, are capable of oxygenic photosynthesis according to the following reaction:

$$CO_2 + H_2O \rightarrow 6[CH_2O] + O_2$$

Photosynthesis consists of two processes: light energy conversion to biochemical energy by a photochemical reaction, and $CO_2$ reduction to organic compounds such as sugar phosphates, through the use of this biochemical energy by Calvin cycle enzymes. Hydrogenase and nitrogenase enzymes are both capable of hydrogen production.

### 3.2.4.4.2 Dark Fermentation

- Carried out under anaerobic conditions.
- Carbohydrates (mainly glucose) are the preferred carbon source.
- Mainly gives rise to acetic and butyric acids along with hydrogen gas.
- Heterotrophic bacteria that grow on organic substrates by oxidation produce the metabolic energy. This leads to the generation of electrons, which need to be neutralised.
- In anaerobic conditions, components like protons that are reduced to molecular hydrogen act as electron acceptors.

$$2e^- + 2H^+ \rightarrow H_2$$

In essence, dark fermentation is a concerted action of a well-studied metabolic pathway and the hydrogenase enzymes.

### 3.2.4.4.3 Plasma Reforming

- Thermal plasma technology can be used in the production of hydrogen and hydrogen-rich gases from a variety of fuels.
- Hydrogen-rich gas (50–75% $H_2$, with 25–50% CO for steam reforming) can be efficiently produced in plasma reformers with a variety of hydrocarbon fuels (gasoline, diesel, oil, biomass, natural gas, jet fuel, etc.) with conversion efficiencies to hydrogen-rich gas close to 100%.
- Plasma conditions (high temperatures and a high degree of ionisation) can be used to accelerate thermodynamically favourable chemical reactions without a catalyst or to provide the energy required for endothermic reforming processes.

## 3.3 Alternative Fuels for a Compression Ignition Engine

### 3.3.1 Biofuels

#### 3.3.1.1 Introduction

Biofuels are liquid or gas fuels that are not derived from petroleum-based fossils fuels or that contain a proportion of non-fossil fuels. Biofuels can be broadly classified as first-generation, second-generation and third-generation biofuels.

*First-generation biofuels*: First-generation biofuels are the more common fuels that are produced from food crops and animal fats. Some examples include biodiesel, vegetable oil and biogas.

*Second-generation biofuels*: Second-generation biofuels are made from waste biomass, making them a more sustainable solution as compared to their first-generation counterparts. They include various alcohols (such as ethanol) and diesel derived from wood.

*Third-generation biofuels*: Third-generation biofuels are generally made from algae that are formed on a massive scale. By way of photosynthesis and the breaking down of carbon dioxide, the carbohydrates extracted from these microorganisms are used to make various fuels.

#### 3.3.1.2 Biodiesel

Biodiesel refers to a vegetable oil- or animal fat-based diesel fuel consisting of long-chain alkyl (methyl, propyl or ethyl) esters. Biodiesel is typically made by chemically reacting lipids (e.g., vegetable oil, animal fat [with alcohol-producing fatty acid esters]). Biodiesel is meant to be used in standard diesel engines and is thus distinct from the vegetable and waste oils used to fuel *converted* diesel engines. Biodiesel can be used alone, or it can be blended with petrodiesel. Biodiesel can also be used as a low-carbon alternative to heating oil.

Biodiesel has been an area of considerable interest in terms of environment-friendliness, biodegradability, by-products and properties close to diesel, which are briefly given below:

1. *Environment friendliness*: Biodiesel has very less environmental impact as compared to the present diesel analogues. It is renewable and non-toxic; therefore, its production is rapidly increasing, especially in Europe, United States and Asia. It contributes no net carbon dioxide or sulphur to the atmosphere and emits less gaseous pollutants than normal diesel. Non-renewable fuels emit pollutants in the form of oxides of nitrogen, oxides of sulphur, carbon dioxide, carbon monoxide, lead, hydrocarbons and so on during their processing and use. It is a carbon-neutral fuel because all the carbon dioxide

($CO_2$) released during consumption had been sequestered from the atmosphere for the growth of vegetable oil crops.

2. *Biodegradability*: Biodiesel is highly degradable. It degrades about four times faster than petrodiesel. Its oxygen content improves the biodegradation process, leading to a decreased level of quick biodegradation. 90–98% of biodiesel is mineralised in 21–28 days under aerobic as well as anaerobic conditions. Biodiesel increases the biodegradability of crude oil by means of co-metabolism. The time taken to reach 50% biodegradation reduced from 28 to 22 days in 5% biodiesel mixture and from 28 to 16 days in case of 20% biodiesel mixture at room temperature. The risks of handling, transporting and storing biodiesel are much lower than those associated with petrodiesel. Biodiesel is safe to handle and transport because it is as biodegradable as sugar and has a high flashpoint compared to petroleum diesel fuel [13].

3. *Products from by-products of biodiesel*: Glycerol can be chemically converted to glycol, propionic acid, acrylic acid, propanol, acrolein, propanediol and so on. Steam gasification of glycerol leads to the formation of hydrogen and syngas. It can also be biologically converted to citric acid, sophorolipids, 1,3-propanediol and so on. Glycerol is also used as a feed for certain animals.

4. *Properties close to petroleum diesel*: Biodiesel can be used directly in diesel engines without any requirements as other alternative fuels require specific engines. Its combustion properties are very close to petroleum diesel. At present, it can be used as additive petroleum diesel, to improve the low lubricity of pure ultra-low sulphur petrodiesel fuel. It can be blended at any ratio with petroleum diesel.

### 3.3.1.2.1 Fatty Acid

A fatty acid is a carboxylic acid with a long untraced chain. Carboxylic acids are organic acids characterised by the presence of a carboxyl group, which has the formula –COOH.

The simplest series of carboxylic acids is the alkanoic acids

R–COOH

where R is a hydrogen or an alkyl group.

The alkyl group is a group of carbon and hydrogen atoms derived from an alkane molecule by removing one hydrogen atom such as methyl and ethyl. Saturated and unsaturated components of a feedstock are shown in Table 3.3.

*Production process of biodiesel*: Transesterification is the process (Figure 3.25) of exchanging the alkoxy group of an ester compound with another alcohol (Equation 3.19 and Figure 3.8). Alkoxy group are generally alkyl (C and H) group linked with oxygen R–O. These reactions are often catalysed by the addition of an acid or a base. Acids can catalyse the reaction by donating a proton to the carbonyl group, thus making it more reactive. Bases can catalyse

**TABLE 3.3**
Saturated and Unsaturated Components of Feedstock

| Carbon Number | Saturated Fatty Acid | | | | | | | Monounsaturated Acid | | | | | |
|---|---|---|---|---|---|---|---|---|---|---|---|---|---|
| | 8 | 10 | 12 | 14 | 16 | 18 | >18 | <16 | 16 | 18 | >18 | Di | Tri |
| Beef tallow | — | — | 0.2 | 2–3 | 21–26 | 21–26 | 0.4–1 | 0.5 | 2–3 | 39–42 | 0.3 | 2 | — |
| Butter | 1–2 | 2–3 | 1–4 | 8–13 | 25–30 | 8–13 | 0.4–2 | 1–2 | 2–5 | 22–29 | 0.2–1.5 | 3 | — |
| Coconut | 5–9 | 4–10 | 44–51 | 13–18 | 7–10 | 1–4 | — | — | — | 5–8 | — | 1–3 | — |
| Cod liver | — | — | — | 2–6 | 7–14 | 0–1 | — | 0–2 | 10–20 | 25–31 | 35–52 | — | — |
| Corn | — | — | — | 0–2 | 8–10 | 1–4 | — | — | 1–2 | 30–50 | 0–2 | 34–56 | — |
| Cottonseed | — | — | — | 0–3 | 17–23 | 1–3 | — | — | — | 23–41 | 2–3 | 34–55 | — |
| Lard | — | — | — | 1 | 25–30 | 12–16 | — | 0.2 | 2–5 | 41–51 | 2–3 | 4–22 | — |
| Linseed | — | — | — | 0.2 | 5–9 | 0–1 | — | — | — | 9–29 | — | 8–29 | 45–67 |
| Palm | — | — | — | 1–6 | 32–47 | 1–6 | — | — | — | 40–52 | — | 2–11 | — |
| Palm kernel | 2–4 | 3–7 | 45–52 | 14–19 | 6–9 | 1–3 | 1–2 | — | 0–1 | 10–18 | — | 1–2 | — |
| Peanut | — | — | — | 0.5 | 6–11 | 3–6 | 5–10 | — | 1–2 | 39–66 | — | 17–38 | — |
| Rapeseed | — | — | — | — | 2–5 | 1–2 | 0.9 | — | 0.2 | 10–15 | 50–60 | 10–20 | 5–10 |
| Safflower | — | — | — | — | 5.2 | 2.2 | — | — | — | 76.3 | — | 16.2 | — |
| Soya bean | — | — | — | 0.3 | 7–11 | 3–6 | 5–10 | — | 0–1 | 22–34 | — | 50–60 | 2–10 |
| Sunflower | — | — | — | — | 6.0 | 4.2 | 1.4 | — | — | 18.7 | — | 69.3 | 0.3 |
| Tung | — | — | — | — | — | — | — | — | — | 4–13 | — | 8–15 | Bulk |

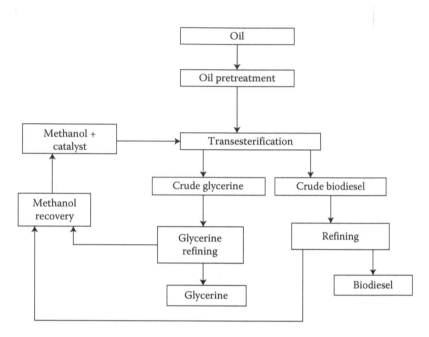

**FIGURE 3.25**
Transesterification process.

the reaction by removing a proton from the alcohol, thus making it more reactive. The cetane number of pure fatty acid ester is shown in Table 3.4.

Vegetable oil + methanol/ethanol → esters + glycerine (triglyceride)

$$
\begin{array}{ll}
CH_2 - OCOR & CH_2OH \\
| & | \\
CH - OCOR + 3CH_3OH \rightarrow 3RCOOCH_3 \; CHOH \\
| & | \\
CH_2 - OCOR & CH_2OH
\end{array}
\tag{3.19}
$$

The crude straight vegetable oil (SVO) is pre-heated in the temperature range of 90–100°C for removal of water content. The acid value of the oil was found to be about 34% by the titrimetric method. In the first stage, the high-acid-value crude oil was treated with an acid catalyst ($H_2SO_4$ concentration of 0.5% w/w of oil) and alcohol in the temperature range of 55–60°C to avoid soap formation during the transesterification process. After 3 h of reaction in the reactor, the mixture was allowed to settle for 2 h. Two layers, such as the alcohol fraction (top layer) and transesterified oil (bottom layer), were formed due to their phase difference. Then transesterified oil was separated for further processing with the base catalyst. The reaction was repeated till the acid value was brought down to less than 1%.

**TABLE 3.4**
Cetane Number of Pure Fatty Acid Esters

| Fuel | Cetane Number |
| --- | --- |
| Caprylic acid methyl ester | 18.0 |
| Caprylic acid methyl ester | 33.6 |
| Capric acid methyl ester | 47.9 |
| Capric acid methyl ester | 47.2 |
| Capric acid ethyl ester | 51.2 |
| Lauric acid methyl ester | 60.8 |
| Lauric acid methyl ester | 61.4 |
| Myristic acid methyl ester | 73.5 |
| Myristic acid methyl ester | 66.2 |
| Myristic acid ethyl ester | 66.9 |
| Palmitic acid methyl ester | 74.3 |
| Palmitic acid methyl ester | 74.5 |
| Stearic acid methyl ester | 75.6 |
| Stearic acid methyl ester | 86.9 |
| Stearic acid ethyl ester | 76.8 |
| Oleic acid methyl ester | 55.0 |
| Oleic acid ethyl ester | 53.9 |
| Linoleic acid methyl ester | 42.2 |
| Linoleic acid ethyl ester | 37.1 |
| Linolenic acid methyl ester | 22.7 |
| Linolenic acid ethyl ester | 26.7 |

In the second stage, the transesterification process was done with a base catalyst (KOH). The methanol and KOH were added into the acid-treated transesterified oil, which was treated in the first stage. The reaction was carried out with methanol/oil at a molar ratio of 6:1 with a base catalyst (KOH concentration of 1% w/w of oil) at 65°C for 3 h. The processed oil was kept in a conical flask for the separation of the monoester (biodiesel) and glycerol. In the third stage, the biodiesel was washed for a number of times using hot distilled water until the complete removal of unreacted triglyceride, alcohol and salt. Finally, the oil was heated at a temperature range of 70–95°C to remove water moisture.

Today, making biodiesel is not just about making a renewable fuel. For biodiesel producers, this is the challenge of producing high-quality biodiesel with consistent characteristics, regardless of feedstock at times, when feedstock prices increase with food and energy prices. 'In the current environment, only the least-cost producers are able to produce fuel at positive margin'.

Basically, making biodiesel from oil, methanol (or ethanol) and catalyst is a simple chemical process. The problem lies in the chemical reaction kinetics. The conventional transesterification of the triglycerides to fatty acid methyl ester (FAME) and glycerine is slow and not complete. During the conversion process, not all fatty acid chains are turned into alkyl esters (biodiesel). This reduces the biodiesel quality and yield significantly.

*Commercial biodiesel production technologies*: Current technologies for producing biodiesel can be classified into three categories:

1. Base-catalysed transesterification with refined oils
2. Base-catalysed transesterification with low fatty acid greases and fats
3. Acid esterification followed by transesterification of lower or high free fatty acid fats and oils

Other processes under development include biocatalysed transesterification, pyrolysis of vegetable oil/seeds and transesterification with supercritical methanol.

The goal of all technologies is to produce fuel-grade esters meeting standard specifications (e.g., ASTM/European/BIS).

The key quality control issues, including the complete (or nearly complete) removal of alcohol, catalyst, water, soaps, glycerine and unreacted or partially reacted triglycerides and free fatty acids (FFA), are major concerns for its use as an auto fuel.

Non-removal of these contaminants causes the biodiesel to fail in one or more fuel standards. There are numerous variations in the basic technology. Variations of basic technology in the production of biodiesel are listed below:

- Different catalysts, for example, NaOH, KOH, MeONa, non-alkaline catalysts, acids, metal complexes and biocatalysts, can be used.
- Anhydrous ethanol, isopropanol or butanol can be substituted for methanol.
- Alcohols other than methanol may require additional process steps and quality control.
- Basic transesterification is carried out at an atmospheric pressure and temperature around 60–70°C.
- Some technologies use higher temperatures and elevated pressure, typically in a supercritical range of methanol.
- For high FFA feedstock, acid-catalysed esterification followed by base-catalysed transesterification is used or FFA can be removed first and the purified oil then transesterified.

The production processes for biodiesel are well known. The three basic routes to biodiesel production from oils and fats are listed below:

- Base-catalysed transesterification of the oil
- Direct acid-catalysed transesterification of the oil
- Conversion of the oil to its fatty acids and then to biodiesel

Most of the biodiesel produced today is from the base-catalysed reaction for several reasons:

- It requires lower temperature and pressure
- It yields higher conversion (98%) with minimal side reactions and reaction time
- It is a direct conversion to biodiesel with no intermediate compounds
- No exotic materials of construction are needed

The chemical reaction for base-catalysed biodiesel production is depicted below. 45 kg of fat or oil (such as soya bean oil) are reacted with 4.5 kg of a short-chain alcohol in the presence of a catalyst to produce 4.5 kg of glycerine and 45 kg of biodiesel. The short-chain alcohol, signified by ROH (usually methanol, but sometimes ethanol), is charged in excess to assist in quick conversion. The catalyst is usually sodium or potassium hydroxide that has already been mixed with the methanol. R', R" and R"' indicate the fatty acid chains associated with the oil or fat, which are largely palmitic, stearic, oleic and linoleic acids for naturally occurring oils and fats.

*Stability of biodiesel*: Biodiesel is with ester functionality, which on exposure to air gets hydrolysed to alcohol and acid. The presence of this alcohol reduces the flashpoint and the acid increases the total acid number. This makes biodiesel unstable on storage. The presence of a large percentage of saturated fatty acids in the biomass makes the biodiesel more stable than the presence of unsaturated fatty acids. However, a large proportion of saturates lowers the cold-temperature properties of biodiesel such as cloud and pour points. Therefore, the limiting factor of biodiesel is its inverse relationship between the level of saturation and cold-temperature properties [13].

*Oxidative degradation of biodiesel*: As shown in Figure 3.26, fatty acid methyl esters form a radical next to the double bond during the oxidation process. This radical binds with the oxygen in the air, which is a bi-radical, to form a peroxide radical. A new radical is created from the fatty acid methyl ester by this peroxide radical, which binds with the oxygen in the air. This augments the auto-oxidation cycle at an exponentially rapid rate, whereby 100 new radicals are created quickly from one single radical, resulting in the formation of a series of by-products. The fuel thus gets deteriorated as there is a formation of sediment and gum [14]. Peroxide formation through this route leads to oligomerisation even at an ambient temperature. The greater the level of unsaturation in a fatty oil or ester, the more susceptible it will be to oxidation. Once the LOOHs have formed, they decompose and inter-react to form numerous secondary oxidation products, including higher-molecular-weight oligomers often called polymers. The primary oxidation products are hydroperoxides and conjugated dienes, and procedures to measure both are established for fatty oils and esters. The secondary oxidation products have been measured by many procedures, depending on the type of the compound of interest. The peroxidation process is inhibited by tocopherols, mannitol and

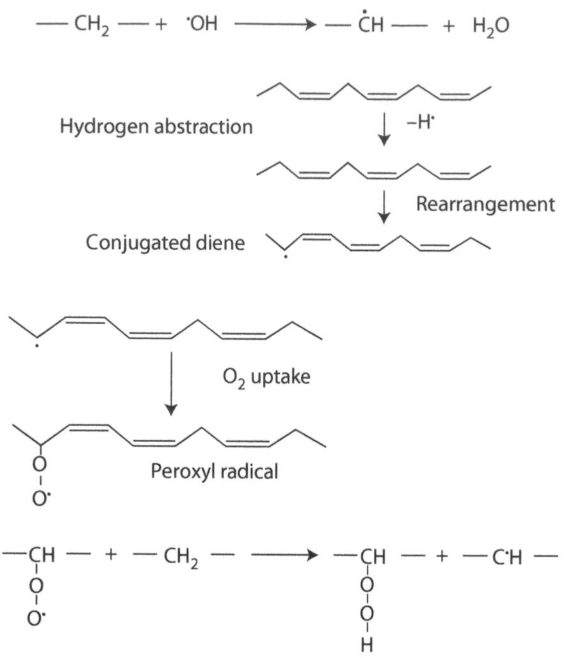

$$— CH_2 — + \ ^\cdot OH \longrightarrow — \overset{\cdot}{C}H — + \ H_2O$$

Hydrogen abstraction                    $\downarrow$ –H$^\cdot$

$\downarrow$ Rearrangement

Conjugated diene

$\downarrow$ O$_2$ uptake

Peroxyl radical

$$—\overset{|}{\underset{\overset{|}{O} \\ \overset{|}{O^\cdot}}{C}H— \ + \ — CH_2 — \longrightarrow —\overset{|}{\underset{\overset{|}{O} \\ \overset{|}{O} \\ \overset{|}{H}}{C}H— \ + \ —C^\cdot H —}$$

**FIGURE 3.26**
Mechanism of peroxy radical formation on methylene group.

formate [15]. Antioxidants have either been used or proposed for use to control fatty oil oxidation. Crude fatty oils contain the naturally occurring phenolic antioxidants—tocopherols. As the deterioration of biodiesel is attributed to the formation of peroxide in the initial step, the remedy suggested is to prevent peroxide formation during the stages of biodiesel manufacture, and throughout its distribution chain, synthetic antioxidants such as phenolic types or aminic types have to be added to make it stable and hence acceptable in the market.

*Thermal degradation of biodiesel*: Viscosity is one of the most important fuel properties. The effects of viscosity can be seen in the quality of atomisation and combustion as well as engine wears. The quality of fuel atomisation is significantly affected by viscosity [16]. The viscosity of biodiesel increases with the increase in the degree of thermal degradation due to the trans-isomer formation on double bonds. The decomposition of biodiesel and its corresponding fatty acids linearly increases from 293 to 625 K. The densities of biodiesel fuels decreased linearly with temperature from 293 to 575 K [17]. The combustion heat of biodiesel decreases with an increase in the degree of thermal degradation [18]. The thermal polymerisation of fatty oils and esters does not become important until temperatures of 525 K are reached. Thermal polymerisation occurs by the Diels–Alder reaction, and two fatty acid chains are linked by a cyclohexene ring. Thermal polymerisation of biodiesel causes an increase in its viscosity [15].

*Microalgae—Biofuels:* Third-generation biofuels derived from microalgae are considered to be a viable alternative energy resource that is devoid of the major drawbacks associated with first- and second-generation biofuels. Microalgae are able to produce 15–300 times more oil for biodiesel production than traditional crops on an area basis. Furthermore, compared with conventional crop plants that are usually harvested once or twice a year, microalgae have a very short harvesting cycle (≈1–10 days, depending on the process), allowing multiple or continuous harvests with significantly increased yields.

Fatty acids constitute 40% of the overall mass of algae. It can be grown throughout the year. Algae can grow in practically every place where there is enough sunshine. They are the principal producers of oxygen on the earth. They require less water than terrestrial crops. Some algae can grow in saline water and some even in wastewater on degraded land. They are considered the fastest-growing photosynthesising organisms, which can complete their life cycle within a few days. The oil yield per acre from microalgae can theoretically be 100 times higher than that of terrestrial oil crops [19]. Approximately 46 tons of oil/hectare/year can be produced from diatom algae. Some algae produce up to 50% oil by weight. They require very less area for their growth and they are mostly marine. Miao and Wu [20,22] reported a heterotrophic growth of *Chlorella* prototothecoides capable of yielding as high as a 55% lipid content and converting the lipid to biodiesel [23]. The lipid contents in different species of microalgae are given in Table 3.5. Microalgae have been suggested to be very good candidates for fuel production because of their advantages of higher photosynthetic efficiency, higher biomass production and faster growth compared to other energy crops [3,10,24]. No carbon source is required for the growth of algae and any carbon dioxide released on combustion will have been previously fixed so that the energy supply will be carbon dioxide neutral. Some microalgae, such as the *Chlorella* species, are unicellular with a cell size in the range of 3–10 µm, which is ideal for combustion in a diesel engine. Algae are considered as second-generation feedstock for biodiesel. Algae have an advanced photosynthetic capacity similar to higher-order plants, which have two photosystems and the ability to split water. According to Chisti [19], 1 kg of dry algae biomass utilises about 1.83 kg of $CO_2$; thus, the microalgae biomass production can help in bio-fixation of waste $CO_2$ with respect to air quality maintenance and improvement.

Combined with their ability to grow under harsher conditions and their reduced needs for nutrients, they can be grown in areas unsuitable for agricultural purposes independently of the seasonal weather changes, thus not competing for arable land use, and they can use wastewaters as the culture medium, not requiring the use of freshwater. Moreover, they use less water than terrestrial plants [25]. They also use less water for their growth than other traditional oil crops. They can double their biomass within a day. Thus, it was feasible to use algae oil production to completely replace fossil diesel. Microalgae are estimated to produce biomass at a rate 50 times greater than the fastest-growing terrestrial plant like switchgrass [25].

**TABLE 3.5**

Lipid Content and Productivities of Different Microalgal Species

| Marine and Freshwater Microalgae Species | Lipid Content (% Dry Weight Biomass) | Lipid Productivity (mg/L/day) | Volumetric Productivity of Biomass (g/L/day) | Areal Productivity of Biomass (g/m²/day) |
|---|---|---|---|---|
| *Ankistrodesmus* sp. | 24.0–31.0 | — | — | 11.5–17.4 |
| *Botryococcus braunii* | 25.0–75.0 | — | 0.02 | 3.0 |
| *Chaetoceros calcitrans* | 14.6–16.4/39.8 | 17.6 | 0.04 | — |
| *Chlorella emersonii* | 25.0–63.0 | 10.3–50.0 | 0.036–0.041 | 0.91–0.97 |
| *Chlorella protothecoides* | 14.6–57.8 | 1214 | 2.0–7.70 | — |
| *Chlorella sorokiniana* | 19.0–22.0 | 44.7 | 0.23–1.47 | — |
| *Chorella vulgaris* | 5.0–58.0 | 11.2–40.0 | 0.02–0.2 | 0.57–0.95 |
| *Chlorella* sp. | 10.0–48.0 | 42.1 | 0.02–2.5 | 1.61–16.47/25 |
| *Chlorella pyrenoidosa* | 2.0 | — | 2.90–3.64 | 72.5/130 |
| *Chlorella* | 18.0–57.0 | 18.7 | — | 3.5–13.9 |
| *Chlorococcum* sp. | 19.3 | 53.7 | 0.28 | — |
| *Crypthecodinium cohnii* | 20.0–51.1 | — | 10 | — |
| *Dunadiella salina* | 6.0–25.0 | 116.0 | 0.22–0.34 | 1.6–3.5/20–38 |
| *Dunadiella primolecta* | 23.1 | — | 0.09 | 14 |
| *Dunadiella tertiolecta* | 16.7–71.0 | — | 0.12 | — |
| *Dunadiella* sp. | 17.5–67.0 | 33.5 | — | — |
| *Ellipsoidion* sp. | 17.5–67.0 | 33.5 | — | — |
| *Euglena gracialis* | 14.0–20.0 | — | 7.70 | — |
| *Haematococcus pluvialis* | 25.0 | — | 0.05–0.06 | 10.2–36.4 |
| *Isochrysis galbana* | 7.0–40.0 | — | 0.32–1.6 | — |
| *Isochrysis* sp. | 7.1–33 | 37.8 | 0.08–0.17 | — |
| *Monodus subterraneous* | 16.0 | 30.4 | 0.19 | — |
| *Monallanthus salina* | 20.0–22.0 | — | 0.08 | 12.0 |
| *Nannochloris* sp. | 20.0–56.0 | 60.9–76.5 | 0.17–0.51 | — |
| *Nannocholoropsis oculata* | 22.7–29.7 | 84.0–142.0 | 0.37–0.48 | — |
| *Nannocholoropsis* sp. | 12.0–53.0 | 37.6–90.0 | 0.17–1.43 | 1.9–5.3 |
| *Neochloris oleoabundans* | 29.0–65.0 | 90.0–134.0 | — | — |
| *Nitzschia* sp. | 16.0–47.0 | — | — | 8.8–21.6 |
| *Ooscitis pusilla* | 10.5 | — | — | 40.6–45.8 |
| *Pavlova salina* | 30.9 | 49.4 | 0.16 | — |
| *Pavlova lutheri* | 35.5 | 40.2 | 0.14 | — |
| *Phaeodactylum tricornutum* | 18.0–57.0 | 44.8 | 0.003–1.9 | 2.4–21 |

**TABLE 3.5 (continued)**

Lipid Content and Productivities of Different Microalgal Species

| Marine and Freshwater Microalgae Species | Lipid Content (% Dry Weight Biomass) | Lipid Productivity (mg/L/day) | Volumetric Productivity of Biomass (g/L/day) | Areal Productivity of Biomass (g/m²/day) |
|---|---|---|---|---|
| *Porphyridium cruentum* | 9.0–18.8/60.7 | 34.8 | 0.36–1.5 | 25 |
| *Scenedesmus obliquus* | 11.0–55.0 | — | 0.004–0.74 | — |
| *Scenedesmus quadricauda* | 1.9–18.4 | 35.1 | 0.19 | — |
| *Scenedesmus* sp. | 19.6–21.1 | 40.8–53.9 | 0.03–0.26 | 2.43–13.52 |
| *Skeletonema* sp. | 13.3–31.8 | 27.3 | 0.09 | — |
| *Skeletonema costatum* | 13.5–51.3 | 17.4 | 0.08 | — |
| *Spirulina platensis* | 4.0–16.6 | — | 0.06–4.3 | 1.5–14.5/24–51 |
| *Spirulina maxima* | 4.0–9.0 | — | 0.21–0.25 | 25 |
| *Thalassiosira pseudonana* | 20.6 | 17.4 | 0.08 | — |
| *Tetraselmis suecica* | 8.5–23.0 | 27.0–36.4 | 0.12–0.32 | 19 |
| *Tetraselmis* sp. | 12.6–14.7 | 43.4 | 0.30 | — |

*Source:* Adapted from Microalgae for biodiesel production and other application: A review material, *A Renewable and Sustainable Energy Reviews*, 14(1), January 2010, 217–232.

The utilisation of microalgae for biofuel production offers the following advantages over higher plants:

- Microalgae synthesise and accumulate large quantities of neutral lipids (20–50% dry weight of biomass) and grow at high rates.
- Microalgae are capable of all-year-round production; therefore, oil yield per area of microalgae cultures could greatly exceed the yield of best oilseed crops.
- Microalgae need less water than terrestrial crops, therefore reducing the load on freshwater sources.
- Microalgae cultivation does not require the application of herbicides or pesticides.
- Microalgae consume $CO_2$ from flue gases emitted from fossil fuel-fired power plants and other sources, thereby reducing emissions of a major greenhouse gas (1 kg of dry algal biomass utilises about 1.83 kg of $CO_2$).
- It undertakes wastewater bioremediation by removal of $NH_4^+$, $NO_3^-$ and $PO_4^{3-}$ from a variety of wastewater sources (e.g., agricultural run-off, concentrated animal feed operations, and industrial and municipal wastewaters).

- Combined with their ability to grow under harsher conditions and their reduced needs for nutrients, microalgae can be cultivated in saline/brackish water/coastal seawater on non-arable land, and do not compete for resources with conventional agriculture.
- Depending on the microalgae species, other compounds may also be extracted, with valuable applications for different industrial sectors, including a large range of fine chemicals and bulk products, such as polyunsaturated fatty acids, natural dyes, polysaccharides, pigments, antioxidants, high-value bioactive compounds and proteins.

Following oil extraction, amylolytic enzymes are used to promote starch hydrolysis and formation of fermentable sugars. These sugars are fermented and distilled into bio-ethanol using conventional ethanol distillation technology.

### 3.3.1.2.2 Production of Biodiesel from Algae

Algae cultivation:

1. Raceway ponds: Chisti [19] has well described the structure of raceway ponds, which is in the form of a closed-loop recirculation channel operated at water depths of 15–20 cm. The mixing and circulation of algae are done with the help of paddles provided, which prevent sedimentation, whereas flow is controlled and assisted by bends of baffles present along the flow channel. They are built in concrete or compacted earth; therefore, it is less expensive to build. The microalgal biomass may be harvested by flocculation or centrifugation [26]. As it is an example of open culture, water is lost via evaporation and is very susceptible to contamination by unwanted species. They are easy to operate and are more durable than closed systems. Moreover, they require an extensive area for construction, and culture conditions such as light and temperature are harder to control in the open raceway ponds. They have low productivity of biomass than closed systems due to poor mixing. The largest raceway-based biomass production facility occupies an area of 440,000 m² as shown in Figure 3.27 [27,28].

2. Closed photobioreactors (PBRs): All the drawbacks that are associated with raceaway ponds are corrected in PBRs such as in terms of avoiding contamination, yielding higher culture densities and providing closer control over physico-chemical conditions. It essentially permits a single-species culture of microalgae for prolonged durations [29].

There are many types of photobioreactors:

a. *Tubular photobioreactors*: It is the most widely used photobioreactor. It has a tubing arrangement that may be straight, coiled or looped so

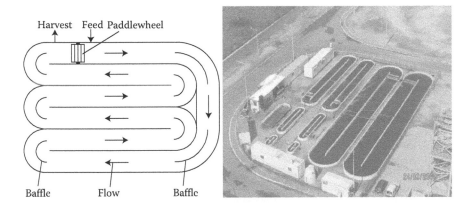

**FIGURE 3.27**
Image of large-scale seambiotic *Nannochloropsis* sp. culture ponds. (Image courtesy of Nature Beta Technologies Ltd, Eilat, Israel, subsidiary of Nikken Sohonsha Co., Gifu, Japan. Greenwell et al., *Journal of Royal Society Interface* 2010; 7(46):703–726.)

that maximum light can be captured. There is the arrangement of airlift systems, which help in the mixing and aeration of chambers. The tubes are mostly made up of plastic or glass. The diameter of the tube should be less than 0.1 m because light does not penetrate too deeply into the dense culture broth that is necessary for ensuring a high biomass productivity of the photobioreactor [19]. There are certain problems associated with these PBRs such as poor mass transfer, difficulty in controlling temperature and expensiveness. For example, several kilometres of tubes are necessary to produce commercial amounts of oil.

b. *Flat-plate photobioreactors*: Milner [31] first used the flat-plate PBR for algal cultivation as it provided a very large surface area for illumination. Following this work, Samson and Leduy [32] developed a flat reactor equipped with fluorescence lamps. These are also made up of transparent materials. It is relatively cheaper than the tubular one. Accumulation of dissolved oxygen concentrations in flat-plate photobioreactors is relatively low compared to horizontal tubular photobioreactors.

c. *Vertical column photobioreactors*: Vertical column photobioreactors are compact, low-cost and easy to operate monoseptically. There are certain benefits in using such PBRs such as high mass transfer, good mixing with low shear stress, low energy consumption, high potentials for scalability, easy to sterilise, readily tempered, good for immobilisation of algae, reduced photo-inhibition and photo-oxidation. However, it provides only a small illumination surface area and its construction requires sophisticated materials.

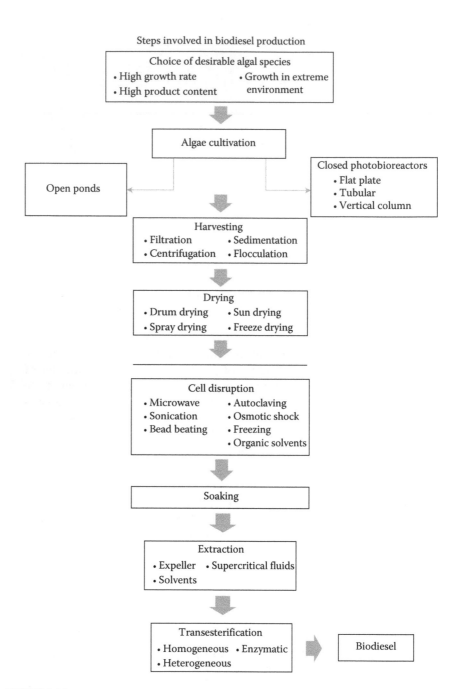

**FIGURE 3.28**
Different steps involved in biodiesel production.

Figure 3.28 shows the different steps in biodiesel production. The steps are explained as given below.

*Harvesting*: It refers to separating the algae effectively from the medium and then drying to process the desired material from algae. Harvesting is one of the major steps in biodiesel formation as it contributes about 20–30% of the total biomass production cost. It may involve one or more steps, but one needs to choose a method that is energy efficient and inexpensive. The most common methods are sedimentation, centrifugation, filtration and ultra-filtration, sometimes with an additional flocculation step or with a combination of flocculation–flotation. The method to be employed depends on the size and the property of algae. Dewatering is also an essential harvesting step. Harvesting is considered as a difficult step because of the small size of algae (Figure 3.29) [30,33].

1. *Filtration*: When the algae to be harvested are of low density, this method is employed, but it is limited to a small volume only or else it will clog the filter. To overcome this, a filter press is used, which works under pressure or a vacuum. This method is quite slow. Filtration is better suited for large microalgae such as *Coelastrum proboscideum* and *Spirulina platensis* but it cannot recover organisms with smaller dimensions such as *Scenedesmus*, *Dunaliella* or *Chlorella*. Membrane filtration and ultra-filtration are also used for fragile and small-scale production. However, this is quite expensive due to pumping and membrane replacement. Sometimes, they may be collected very well with a microstrainer. When a microstrainer is used to collect algae, the original suspension may be faintly green, which could be further concentrated (Figure 3.30) [10].

2. *Centrifugation*: Centrifugation is a method of separating algae from the medium by using a centrifuge to cause the algae to settle to the bottom

**FIGURE 3.29**
Harvesting algae from municipal waste.

**FIGURE 3.30**
Microalgal biomass recovered from the culture broth by filtration moves along a conveyor belt at Cyanotech Corporation (www.cyanotech.com), Hawaii, USA. (Photograph by Terry Luke. Courtesy of Honolulu Star-Bulletin.)

of a flask or tank. The centrifuge works using the sedimentation principle, where the centripetal acceleration is used to evenly distribute substances (usually present in a solution for small-scale applications) of greater and lesser density. This method is reasonably efficient, but sensitive algal cells may be damaged by pelleting against the rotor wall and the method is essentially unselective; all particles with a sedimentation rate above some limiting value will be collected [10].

3. *Flotation–flocculation*: Flocculation mainly aggregates the algal cells, which increases the effective particle size and hence eases sedimentation, centrifugal recovery and filtration. In this method, algae are removed in the form of scum. While adjusting the pH of the medium and the airflow rate, algae can be separated in the form of froth, which is called froth flotation. Flocculation involves the mixing of chemicals in the medium and thus forcing algae to form lumps. However, this method is uneconomical [30].

*Drying*: Once the harvesting is done, the next step to be followed is the dehydration of the biomass, which increases the shelf life of not only the biomass but also of the end product. Several methods have been employed to dry microalgae such as *Chlorella*, *Scenedesmus* and *Spirulina*, where the most common include spray-drying, drum-drying, freeze-drying and sun-drying. Owing to the high water content of the algal biomass, sun-drying is not a very effective method for algal powder production. Spray-drying is not economically feasible for low-value products such as biofuel or protein. Freeze-drying is too expensive for use in low-value products such as biodiesel. Sun-drying is most likely the least expensive option, and spray-drying is the method of choice in most installations. Drying and extraction of algal oils by conventional methods are both energy- and cost-intensive [30,33].

*Cell disruption*: Biodiesel production requires the previous release of lipids from their intracellular location. This step decides the overall yield, which thus depends on the method and the device employed. To obtain the metabolite of interest, cell disruption is followed by drying. If freeze-drying is used to dry the cells, this will also serve the purpose of disrupting the cell walls; in addition to this, mechanical type disruption could also be used for this purpose [30]. Many methods are employed for disruption, depending upon the properties of particular algae, such as the nature of the cell wall and the nature of the product to be obtained. Microwaves usually shatter the cells because of the high-frequency waves [34,35]. This is the most efficient and simplest method in use nowadays for large-scale production. Another method is sonication in which cells are disrupted using the cavitation effect, thus creating a cavity in the cell membrane [36,37]. Bead beating is another method in which the mechanical action of high-speed spinning fine beads is used to damage the cells [37,38]. Autoclaving is also sometimes done using high temperature and pressure and the use of a 10% NaCl solution to break the cell wall by osmotic pressure. Some non-mechanical methods are also in use such as freezing, use of organic solvents and osmotic shock and acid, base and enzyme reactions [39].

*Extraction*: There are a number of methods that are used for the extraction of oil from the feedstock, which includes use of a press or expeller, and the use of solvents (hexane) and supercritical fluid extraction. The use of press is a simple and easy technique, but the use of a solvent is the most common method followed, which is done with the help of a Soxhlet apparatus. Many solvents such as benzene, cyclo-hexane, hexane, acetone and chloroform can be used. The solvent destroys the algal cell wall and extracts oil from the aqueous medium because of their higher solubility in organic solvents than in water. Hexane is the most popular solvent as it is inexpensive. Fajardo et al. have reported an improved lipid extraction method by using a two-step process. In the first step, ethanol is used for the extraction of lipids, and in the second step, hexane is used for purifying the extracted lipids. The supercritical method helps in obtaining a larger yield but is comparatively expensive. Supercritical fluids are selective, thus providing high purity and product concentrations [33,40].

A supercritical fluid has the ability to extract almost 100% oil from the feedstock. In the supercritical fluid $CO_2$ extraction, $CO_2$ is liquefied under pressure and heated to the point where it has the properties of both a liquid and a gas. This liquefied fluid then acts as solvent in extracting the oil. The dry extraction procedure according to [41] as a modification of the wet extraction method by [42] was used to extract the lipid in microalgal cells.

Ultrasound is also one of the techniques employed for extraction. This method involves the exposure of algae to a high-intensity ultrasonic wave, which creates tiny cavitation bubbles around cells. The collapse of bubbles

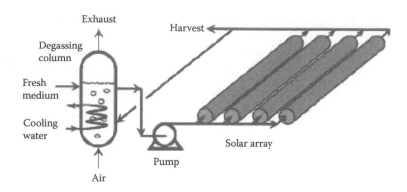

**FIGURE 3.31**
A tubular photobioreactor with parallel run horizontal tubes. (Adapted from A. E. Hokanson and R. M. Rowell, 1977. *Methanol from Wood Waste: A Technical and Economic Study. USDA Forest Service General Technical Report FPL 12*. Department of Agriculture, Forest Products Laboratory, Madison, Wisconsin, USA.)

emits shockwaves, shattering the cell wall and releasing the desired compounds into the solution.

*Biodiesel production*: The direct use of algal oils in the compression engines is restricted due to the high viscosity, which results in poor fuel atomisation, incomplete combustion and carbon deposition on the injector and the valve seats causing serious engine fouling. Other disadvantages of direct usage are low volatility and having a polyunsaturated character. Thus, to correct these constraints, the transesterification process is used, which has been discussed in the earlier section of this chapter.

Producing microalgal biomass is generally more expensive than growing crops. Photosynthetic growth requires light, carbon dioxide, water and inorganic salts. Temperature must generally remain within 20–30°C. To minimise expense, biodiesel production must rely on freely available sunlight, despite daily and seasonal variations in light levels.

A tubular photobioreactor [7] consists of an array of straight transparent tubes that are usually made of plastic or glass. This tubular array, or the solar collector, is where the sunlight is captured (Figure 3.31). The solar collector tubes are generally 0.1 m or less in diameter. Tube diameter is limited because light does not penetrate too deeply into the dense culture broth that is necessary for ensuring a high biomass productivity of the photobioreactor. Microalgal broth is circulated from a reservoir (i.e., the degassing column in Figure 3.31) to the solar collector and back to the reservoir. A continuous culture operation is used, as explained above.

### 3.3.2 Gas to Liquid Fuel

*Resources:* Natural gas (the final liquid fuel being GTL), coal (CTL), and residual biomass (BTL)

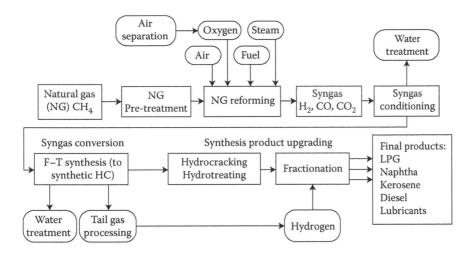

**FIGURE 3.32**
Generic gas to liquid (GTL) process.

As the percentage of raw material cost is about 50–75%, the petroleum fuel cost increases steeply with a crude oil price increase, which leads to unstable economic development. If it uses cheaper raw materials such as agro-waste, refinery effluents, petroleum coke, coal, naphtha, residual oil and natural gas, the end product price will be lower, resulting in energy suitability.

The GTL production technology comprises two chemical processes such as gasification and synthesis. First, the cheaper raw material is converted into carbon monoxide (CO) and hydrogen, known as synthesis gas, through a gasification process. Syngas is converted into valuable end products such as FTS diesel fuel, dimethyl ether, methanol and varied kinds of valuable liquid fuels using the synthesis process. The simple layout of the GTL process is shown in Figure 3.32.

The fundamental reactions of synthesis gas chemistry are methanol synthesis, FTS, oxo synthesis and methane synthesis. For example, FTS produces hydrocarbons of different lengths from a gas mixture of $H_2$ and CO (syngas) from biomass gasification called as bio-syngas, whereas oxo synthesis (hydroformylation) is an important industrial process for the production of aldehydes from alkenes. The properties of GTL fuel are shown in Table 3.6.

### 3.3.3 Fischer–Tropsch Diesel

FTS was discovered in the 1920s by German chemists F. Fischer and H. Tropsch. It was briefly used by Germany before and during World War II to produce fuels, and has generated varying levels of interest worldwide since that time. Today, it is used commercially to produce transportation fuels and chemicals at several sites in South Africa, from both coal and

**TABLE 3.6**

GTL Blends and Properties

| Properties | Units | EU Diesel (Ref) | GTL Diesel | 50:50 EU GTL Blend | 80:20 EU GTL Blend |
|---|---|---|---|---|---|
| Density at 15°C | Kg/m³ | 836 | 768 | 802 | 821 |
| Density at 20°C | Kg/m³ | 834 | 765 | 798 | 817 |
| Cetane number | — | 54 | 74 | 62 | 58 |
| Total sulphur | mg/kg | 7 | <1 | 4 | 6 |
| ASTM D86 distillation 10% | °C | 221 | 187 | 201 | 212 |
| 50% | °C | 277 | 251 | 264 | 270 |
| 95% | °C | 354 | 321 | 337 | 339 |
| FBP | °C | 360 | 329 | 346 | 350 |
| Flashpoint | °C | 82 | 59 | 66 | 76 |
| Kinematic viscosity at 40°C | mm² | 2.95 | 1.97 | 2054 | 2.79 |
| CFPP | °C | −17 | −19 | −18 | −17 |
| Cloud point | °C | −14 | −18 | −17 | −15 |
| Total aromatics | %m/m | 26.8 | 0.1 | 13.5 | 21.5 |
| Bi- and polycyclic aromatics | %m/m | 4.8 | 0.0 | 2.3 | 3.7 |
| Carbon content | %m/m | 86.5 | 85 | 85.7 | 86.2 |
| Hydrogen content | %m/m | 13.5 | 15 | 14.3 | 13.8 |
| C/H ratio (mass) | — | 6.41 | 5.67 | 6.02 | 6.25 |
| H/C ratio (molar) | — | 1.86 | 2.10 | 1.98 | 1.91 |
| Lower heating value | MJ/kg | 43.1 | 43.8 | 43.5 | 43.2 |
| HFRR scar diameter | | 394 | 370 | <400 | <400 |

natural gas, and at a site in Malaysia from natural gas. However, there is considerable interest in this technology for the conversion of stranded natural gas reserves into an easily transportable, liquid product.

The FT process converts a mixture of CO and $H_2$ (syngas) into a range of valuable hydrocarbons. Hence, it can be considered as an alternative to crude oil for the production of both liquid fuels (gasoline and diesel) and chemicals (in particular, 1-alkenes). The basic chemical reaction is indicated below.

GTL fuels are produced from natural gas, coal and biomass using the FTS process. The process involves air separation, gas processing, syngas production, conversion of syngas to syncrude and upgradation of syncrude by hydroprocessing to marketable products. Synthesis gas (CO and $H_2$) is produced from any hydrocarbon resource material as shown in Equation 3.20. It can be converted into valuable products as shown in Equations 3.21 through 3.23.

*3.3.3.1 Syngas Formation*

$$CH_n + O_2 = 0.5nH_2 + CO \qquad (3.20)$$

### 3.3.3.2 FTS Process

$$nCO + 2nH_2 \rightarrow (-C_nH_2-) + nH_2O \tag{3.21}$$

$$nCO + (2n + 1)H_2 \rightarrow C_nH_{2n}+1 + nH_2O \tag{3.22}$$

$$nCO + (n + m/2)H_2 \rightarrow C_nH_m + nH_2O \tag{3.23}$$

All reactions are exothermic and the product is a mixture of different hydrocarbons comprising paraffin and olefins as the major components. The products from FTS are mainly aliphatic straight-chain hydrocarbons ($C_xH_y$). $C_xH_y$ branched hydrocarbons, unsaturated hydrocarbons and primary alcohols are also formed in minor quantities. The distribution of the products depends on the catalyst and the process parameters such as temperature, pressure and residence time. The properties of F–T diesel are shown in Table 3.6 and its comparisons with other conventional fuels are given in Table 3.7.

The products include the following:

- Methane ($CH_4$)
- Ethane ($C_2H_6$)
- Ethene ($C_2H_4$)
- LPG ($C_3-C_4$)
- Gasoline ($C_5-C_{12}$)
- Diesel fuel ($C_{13}-C_{22}$)
- Light and waxes ($C_{23}-C_{33}$)

**TABLE 3.7**

Comparison of Properties of FT Diesel

| Property | Method | Diesel | Direct F–T | Petro SA COD |
|---|---|---|---|---|
| HHV MJ/kg | D240 | 43–48 | 45–48 | 45–48 |
| Density, 15°C | D4052 | 0.8464 | 0.7695–0.7905 | 0.8007 |
| Distillation, °C | D86 | | | |
| IBP | | 174 | 159–210 | 230 |
| 50% | | 253 | 244–300 | 254 |
| 90% | | 312 | 327–334 | 323 |
| FBP | | 344 | 338–358 | 361 |
| Cetane number | D613 | 44.9 | >74 | –50 |
| Sulphur, ppm | D5453 | 300 | <1 | <1 |
| Total aromatics | D5186 | –30 | 0.1–2 | –10 |
| Hydrogen, wt% | D5291 | 13–13.5 | –15 | –14.4 |
| Cloud point, °C | D2500 | –15 | 0 | –15 |
| Lubricity | | Good to poor | Poor | Poor |

Typical operating conditions for the FTS process are in the temperature and pressures range of 475–625 K and 15–40 atmosphere, respectively. The production layout of F–T diesel is shown in Figure 3.33. The kind and quantity of the liquid product obtained is determined by the reaction temperature, pressure and residence time, the type of reactor and the catalyst used. A cobalt (Co) catalyst yields a higher paraffin and has a higher conversion rate and a longer life. An iron (Fe) catalyst can also be used for higher olefins and oxygenates as it has a higher tolerance for sulphur and aromatic hydrocarbons. Since the FT reactions are highly exothermic, it is important to rapidly remove the heat.

*FT production: Coal-based designs*: The designs considered in Options 1 to 4 are all variations of the block flow diagram shown in Figure 3.34. Coal with conventional product upgrading (maximum distillate production) is given below:

### 3.3.3.3 Syngas Generation Area

*Coal receiving and storage*: Receives washed coal from a mine-mouth coal washing plant, stores the coal in piles, reclaims the coal from storage and delivers coal to the coal preparation plant.

*Coal preparation*: Dries and grinds the coal for use in coal gasifiers.

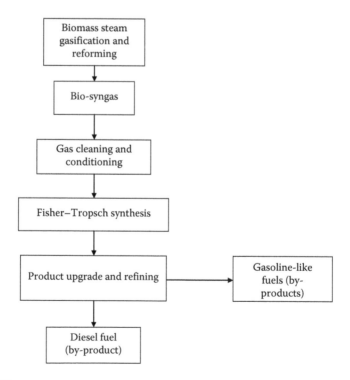

**FIGURE 3.33**
Production of F–T diesel using Fisher–Tropsch synthesis.

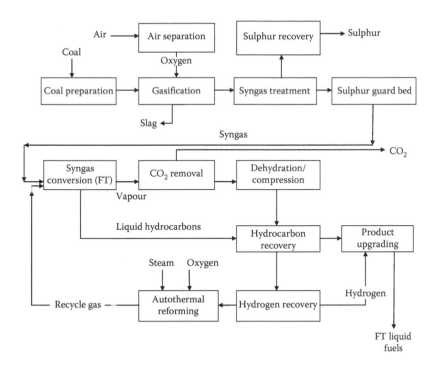

**FIGURE 3.34**
Block flow diagram of the coal liquefaction process. (Adapted from J. Wu, Y. Fang and Y. Wang, *Production of Syngas by Methane and Coal Co-Conversion in Fluidized Bed Reactor,* Institute of Coal Chemistry, Chinese Academy of Sciences, Taiyuan, People's Republic of China.)

*Air separation*: Provides high-purity (99.5%) oxygen, using cryogenic air separation, for gasification and ATR of recycle gas.

*Gasification*: Pressurises and feeds prepared coal to Shell gasifiers and gasifies coal; includes gas quench, high-temperature gas cooling, slag handling, fly-slag removal and handling and solid waste handling. $CO_2$ is used as the carrier gas for the feed coal.

*Syngas treatment* includes the following three plants:

*Syngas wet scrubbing*: Removes trace amounts of fine particles and humidifies the syngas.

*COS hydrolysis and gas cooling*: Converts COS to $H_2S$, HCN to $NH_3$ and cools the syngas.

*Acid gas removal*: Selectively removes $H_2S$ from the syngas using an amine solvent; solvent is regenerated and the $H_2S$-rich gas is sent to sulphur recovery.

*Sulphur guard bed*: Removes trace amounts of sulphur compounds, including $H_2S$, COS and $CS_2$, using ZnO beds, prior to the syngas entering the FT reactors.

*Sulphur recovery*: Receives sour ($H_2S$-rich) gas streams and converts $H_2S$ to elemental sulphur and any $NH_3$ to $N_2$ in a three-stage Claus unit. Tail gas is treated in a SCOT unit prior to discharge through a catalytic incinerator to the stack.

*Sour water stripping*: Strips the water used for syngas wet scrubbing. Wastewater is sent for wastewater treatment and the stripped gas to the sulphur plant.

*FT diesel production from biomass*: The production process of FT liquid fuels from biomass is shown in the block flow diagram in Figure 3.35. This design is for a much smaller plant having only a single gasification train and only

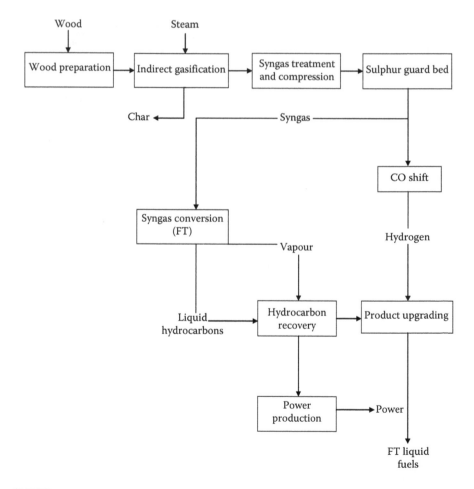

**FIGURE 3.35**

Block flow diagram of the biomass liquefaction process (Adapted from J. Wu, Y. Fang and Y. Wang, *Production of Syngas by Methane and Coal Co-Conversion in Fluidized Bed Reactor*, Institute of Coal Chemistry, Chinese Academy of Sciences, Taiyuan, People's Republic of China.)

producing 1156 bpd of FT liquid products versus the roughly 50,000 bpd produced in the previous designs. A breakdown of the various process plants appearing in the biomass design is given below.

*Wood receiving and storage*: Replaces coal receiving and storage.

*Wood preparation*: Replaces coal preparation; dries wood chips prior to gasification.

*Indirect gasification*: Feeds dried wood chips to a low-pressure, indirectly heated gasifier for gasification; includes a char combustor and a sand recirculation loop.

*Syngas treatment and compression*: Washes and cools syngas in a spray column before compressing syngas up to the pressures required for FT synthesis and power generation.

*CO shift*: Produces and purifies hydrogen from treated syngas used for FT product upgrading.

The sulphur guard bed is still required to remove trace amounts of sulphur compounds from the syngas (small amounts of sulphur are present in the biomass feed). Air separation, syngas wet scrubbing, COS hydrolysis and gas cooling, acid gas removal, sulphur recovery and sour water stripping of CO are not required.

### 3.3.3.4 FT Conversion Area

*Syngas conversion*: FT reactors and catalyst systems remain unchanged.

*Hydrocarbon recovery*: Cryogenic design has been replaced with a non-cryogenic system, which recovers only C5+ hydrocarbons and fractionates hydrocarbon liquids into naphtha, distillate and wax streams. Lighter hydrocarbons are used as fuel gas.

$CO_2$ removal, dehydration and compression, hydrogen recovery and ATR are not required.

*FT product upgrading*: Naphtha hydro-treating, distillate hydro-treating, wax hydrocracking, C5/C6 isomerisation and catalytic reforming are still included for product upgrading. C4 isomerisation, C3/C4/C5 alkylation and saturate gas plant are not required since light hydrocarbons are used as fuel in this design.

### 3.3.4 Dimethyl Ether

DME is the simplest ether. It is considered a leading alternative to petroleum-based fuels and liquefied natural gas. Its physical properties are similar to liquefied petroleum gas (LPG) and can be stored and delivered using existing infrastructures with minor modifications. DME is considered as a substitute for diesel fuel because it has a cetane number lying between 55 and 60.

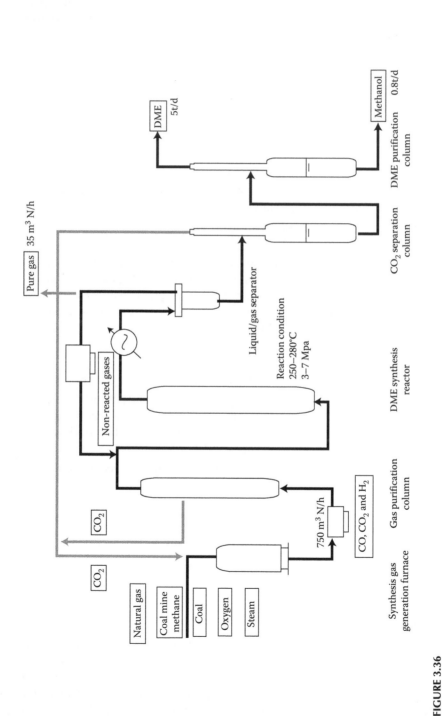

**FIGURE 3.36**
DME synthesis from a coal or natural gas plant. (Adapted from T. R. Tsao and P. Rao, 1987, *The Development of Liquid Phase Methanol Process: An Update,* Air Products and Chemicals, Inc. Proceedings: Eleventh Annual EPRI Contractors' Conference on Clean Liquid and Solid Fuels, pp. 3.1–3.46.)

DME is primarily produced by converting hydrocarbons sourced from natural gas or coal via gasification to syngas. Syngas is then converted into methanol in the presence of a catalyst (usually copper-based), with subsequent methanol dehydration in the presence of a different catalyst (for example, silica-alumina) resulting in the production of DME. As described, this is a two-step (indirect synthesis) process that starts with methanol synthesis and ends with DME synthesis (methanol dehydration). The same process can be conducted using organic waste or biomass. Approximately 50,000 tons were produced in the year 1985 in western Europe by using the methanol dehydration process.

DME can be produced through the GTL route as explained in the earlier section. The advantages of the GTL route is that fuels such as methanol and DME can both be produced through this route. DME can be easily produced from the feedstock of methanol. It may be noted that methanol is a good fuel for a spark ignition engine, whereas DME is a better fuel for a compression ignition engine. Methanol and DME can be a sustainable fuel for the transport sector. DME fuel production is shown in Figure 3.36. The feedstock such as natural gas, coal or any cheaper biomass is fed into the gasifier under the specified operating conditions. The feedstock is converted into synthesis gas (CO and $H_2$) and it is converted to methanol through the methanol synthesis process. Finally, DME is converted from methanol.

*DME production*: The process of producing DME (Figure 3.37) has four main sections:

- Reforming
- $CO_2$ absorption and recycling

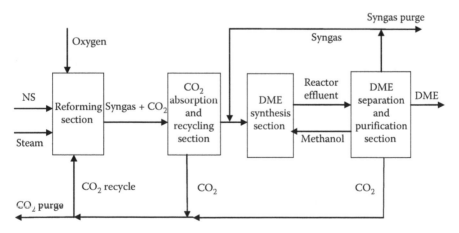

**FIGURE 3.37**
Simple block diagram of dimethyl ether production. (Adapted from A. E. Hokanson and R. M. Rowell, 1977. *Methanol from Wood Waste: A Technical and Economic Study. USDA Forest Service General Technical Report FPL* 12. Department of Agriculture, Forest Products Laboratory, Madison, Wisconsin, USA.)

- DME synthesis
- DME separation and purification

Within each of these sections, various technologies might be used and design decisions made.

In this process, natural gas and $CO_2$ are combined with a recycled carbon dioxide stream and heated in a fired heater. The recycled $CO_2$ comes from two locations: the first point is after the clean-up of the syngas and the second point is after the DME synthesis reaction. In the same fired heater, steam is heated in a separate convection coil. The natural gas and recycled $CO_2$ stream combines with the steam and flows to the burner of the tri-reformer. Oxygen is heated in a separate heater that uses high-pressure steam for the heating function. The temperatures are maintained such that reaction occurs instantaneously and a significant flame front is established.

The syngas exits the tri-reformer at temperatures of around 1000°C and pressures of about 3000 kPa. This hot gas is cooled down and compressed to 6000 kPa and routed through the $CO_2$ absorber, which is designed to extract $CO_2$ out of this stream using cold methanol.

The decarboxylated syngas, along with the unconverted reaction gas, is reheated up to temperatures of around 230°C and directed into the DME synthesis reactors. The gases leaving the DME synthesis reactors are cooled down in a series of sequential heat exchangers. The final heat exchanger is designed to cool the syngas down to about −40°C. Most of the DME is condensed along with all of the methanol and water exiting the DME reactor. The uncondensed gases, primarily comprising CO and hydrogen, are recycled back to the DME synthesis reactors. A small amount of purge is maintained to control the inert levels in the recycle loop.

The liquid DME/$CO_2$/methanol/water stream is then depressurised down to 3500 kPa and flows into the $CO_2$ column. All the $CO_2$ is rejected in the overheads. The bottoms are depressurised down to 1800 kPa and directed into a DME column. The product DME is recovered in the overheads and directed to the storage tanks. The bottoms are a small stream of methanol with water. This is combined with a portion of methanol solvent from the $CO_2$ stripper and directed into the methanol dehydration columns where the methanol is recovered. A portion of the recovered methanol is recycled back to the $CO_2$ absorber and the rest is directed into the methanol dehydration reactor for converting methanol into DME. A large portion of the $CO_2$ stream from the $CO_2$ absorption and DME purification section is recycled back to the reforming section and the rest is vented.

### 3.3.5 DEE

Diethyl ether has a high cetane number. It has a higher energy density than ethanol and is non-corrosive in nature. It can be mixed in any proportion

**TABLE 3.8**

Properties of Diethyl Ether

| Chemical formula | $C_4H_{10}O$ |
|---|---|
| Boiling point | 34.4°C |
| Reid vapour pressure | 1.1 bar at 37.8°C |
| Cetane number | >125 |
| Auto-ignition temperature | 160°C |
| Stoichiometric air/fuel ratio | 11.1 wt/wt |
| Flammability limits (vol%) | 9.5–36.0 (rich), 1.9 (lean) |
| Heating value | 33.897 MJ/kg |
| Viscosity | 0.23 centipoise at 20°C |
| Specific gravity | 0.714 |
| Density | 713 kg/m³ |

with diesel fuel. Conventional fuel injection systems can be used without any modification. Ethanol can be easily converted through a dehydration process to produce diethyl ether. The vapour pressure of diethyl ether is also lower than that of DME. Hence, this fuel will lead to less vapour lock problems in the fuel pump. In this work, diethyl ether was blended with diesel fuel.

*Diethyl ether* ($CH_3CH_2$–O–$CH_2CH_3$) is a great solvent for many things, but it is extremely flammable. Professional chemists will be well apprised of the hazards presented in using ether, but the layperson is less likely to be aware of these dangers. Diethyl ether vapours 'hug' the ground, and in dry air, explosive peroxides can form. In other words, even in a spark/flame-free environment, explosions can still happen when ether vapour is encountered. For this reason, it is probably a good idea to have some way of removing vapours from the vicinity (a fume hood would be a fine example) and one should not use ether on days with extremely low humidity. Because diethyl ether is so flammable and prone to ignition, this procedure should be carried out using a hot plate/stirrer designed for use in flammable environments. Such a heater/stirrer does not produce a contact spark when the hot plate is turned on, and generally employs a brushless AC motor for the stirrer, because DC motors with brushes generally produce small sparks that could ignite any stray vapours.

The important properties of diethyl ether are given in Table 3.8.

Figure 3.38 shows the production process flow diagram of diethyl ether by using ethanol as raw material. The chemical synthesis process is being carried out in a catalytic reactor at 250°C by passing through high pressure steam as shown in the figure. Liquid/liquid phases of water–ethanol/ethanol–DEE could be separated in a simple decant separator.

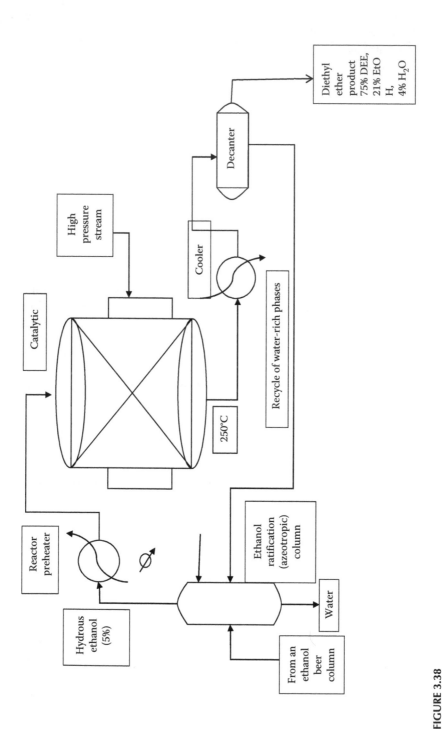

**FIGURE 3.38**

Production of diethyl ether from biomass ethanol. (Adapted from T. R. Tsao and P. Rao, 1987, *The Development of Liquid Phase Methanol Process: An Update,* Air Products and Chemicals, Inc. Proceedings: Eleventh Annual EPRI Contractors' Conference on Clean Liquid and Solid Fuels, pp. 3.1–3.46.)

## 3.4 Conclusion

Alternative fuels such as ethanol, methanol, butanol, biodiesel and F–T diesel could be utilised in a vehicle fleet using the existing infrastructure of crude oil and its derivative fuels. However, other alternative fuels need a major modification of the infrastructure for storage, transportation and distribution.

## References

1. J. Toyir et al., 2009. Sustainable process for the production of methanol from $CO_2$ and $H_2$ using Cu/Zno based multicomponent catalyst, *Physics Procedia*, 2(3), 1075–1079.
2. R. Feachem, D. J. Bradley, H. Garelick and D. D. Mora, 1983. *Sanitation and Disease: Health Aspects of Excreta and Waste Water Management.* J. Wiley and Sons, Chichester, UK.
3. Minowa et al., 1995. Oil production from algal oil of *Dunahilla tortioleta* by desert thermochemical liquification, *Fuel*, 74(12), 1735–1738.
4. F. O. Licht, cited in Renewable Fuels Association, Ethanol Industry Outlook 2008, 2009, and 2010, p. 16, 29, and 22. Available at www.ethanolrfa.org.
5. DOE Biomass Program, www1.eere.energy.gov.
6. http://www.afdc.energy.gov/afdc/ethanol/production_cellulosic.html.
7. Y. D. Park, J. J. Park and J. H. Lim, 1979. Research Reports of the Office of Rural Development, Suweon, Korea.
8. T. M. Mata, A. A. Martins and N. S. Caetano, 2010. Microalgae for biodiesel production and other application: A review, *Renewable and Sustainable Energy Reviews*, 14(1), 217–232.
9. Sridhar et al., 2001. Producer gas as a reciprocating engine fuel—An experimental analysis, *Biomass and Bioenergy*, 21(1), 61–72.
10. Ben-Zion Ginzburg, 1993. Liquid fuel from halophilic algae, A renewable source of non polluting energy, *Renewable Energy*, 3(2–3), 249–252.
11. T. Bond et al., 2011. History and future of domestic biogas plants in the developing world, *Energy for Sustainable Development*, 15(4), 347–354.
12. R. Chandra et al., 2011. Performance evaluation of constant speed IC engine on CNG, methane, enriched biogas and biogas, *Applied Energy*, 88(11), 3969–3977.
13. T.-Y. Mun, P.-G. Seon and J.-S. Kim, 2010. Production of a producer gas from woody waste via air gasification using activated carbon and a two-stage gasifier and characterization of tar, *Fuel* 89, 3226–3234.
14. S. Banga and P. K. Varshney, 2010. Effects of impurities on performance of biodiesel: A review, *Journal of Scientific and Industrial Research*, 69, 575–579.
15. D. Gosling, 1980. *Renewable Energy Resources in Thailand and the Philippines.* Department of Theology, University of Hull, UK.
16. R. E. Tate, K. C. Watts , C. A. W. Allen and K. I. Wilkie, 2006. The viscosities of three biodiesel fuels at temperatures up to 300 degree C, *Fuel*, 85, 1010–1015.

17. R. E. Tate, K. C. Watts, C. A. W. Allen and K. I. Wilkie, 2006. The densities of three biodiesel fuels at temperatures up to 300 degree C, *Fuel*, 85, 1004–1009.
18. T. Fujii, P. Khuwijitjaru, Y. Kimura and S. Adachi, 2006. Decomposition kinetics of monoacyl glycerol and fatty acid in subcritical water under temperature-programmed heating conditions, *Food Chemistry*, 94, 341–347.
19. Y. Chisti, 2007. Biodiesel from microalgae, *Biotechnology Advances*, 25, 294–306.
20. M. Wu, 2006. High quality biodiesel production from microalgae *Chlorella prothotudus* by heterotrophic growth in fermenters, *Journal of Biotechnology*, 126(4), 499–507.
21. www.doe.com.
22. X. Miaoa and Q. Wua, 2006. Biodiesel production from heterotrophic microalgal oil, *Bioresource Technology*, 96(6), 841–846.
23. U. Marchaim, 1983a. *3rd International Symposium on Anaerobic Digestion*, Boston, MA, pp. 343–355.
24. U. Azimov et al., 2011. Effect of syngas composition on combustion and exhaust emission characteristics in a pilot ignited dual fuel engine operated in premier combustion mode, *International Journal of Hydrogen Energy*, 36(18), 11985–11996.
25. D. Liu, 2008. Perspectives of microbial oils for biodiesel production, *Applied microbiology and Biotechnology*, 80(5), 749–756.
26. Hydrogen Solar Production Company, www.hydrogensolar.com.
27. P. Spolaore, C. Joannis-Cassan, E. Duran and A. 2006. Isambert, Commercial applications of microalgae, *Journal of Bioscience and Bioengineering*, 101, 87–96.
28. D. Campo et al., 2007. Outdoor cultivation of microalgae for carotenoid production, Current state and perspectives, *Applied Microbiology and Biotechnology*, 74(6), 1163–1174.
29. T. M. Mata et al., 2010. Microalgae for biodiesel production and other applications: A review, *Renewable and Sustainable Energy Reviews*, 14(1), 217–232.
30. H. C. Greenwell et al., 2010. Placing microalgae on the biofuels priority list: A review of the technological challenges, *Journal of Royal Society Interface*, 7(46), 703–726.
31. H. W. Milner, 1953. The chemical composition of algae. In: J. S. Burlew (Ed.), *Algal Culture: From Laboratory to Pilot Plant*. Carnegie Institute, Washington, pp. 285–302.
32. R. Samson and A. Leduy, 1985. Multistage continuous cultivation of bluegreen alga Spirulina maxima in the flat tank photobioreactors, *Canadian Journal of Chemical Engineering*, 63, 105–112.
33. H. M. Amaro et al., 2011. Advances and perspectives in using microalgae to produce biodiesel, *Applied Energy*, 88(10), 3402–3410.
34. G. Cravotto, L. Boffa, S. Mantegna et al., 2008. Improved extraction of vegetable oils under high-intensity ultrasound and/or microwaves, *Ultrasonics Sonochemistry* 15(5), 898–902.
35. M. Virot, V. Tomao, C. Ginies, F. Visinoni and F. Chemat, 2008. Microwave-integrated extraction of total fats and oils, *Journal of Chromatography* A, 1196–1197, 57–64.
36. C. R. Engler, 1985. Disruption of microbial cells. In: Moo-Yoong, M. (Ed.), *Comprehensive Biotechnology*, second ed. Pergamon Press, Oxford, pp. 305–324.
37. S. J Lee, B.D. Yoon and H. M. Oh, 1998. Rapid method for the determination of lipid from the green alga Botryococcus braunii, *Biotechnology Techniques*, 12, 553–556.

38. J. Geciova, D. Bury and P. Jelen, 2002. Methods for disruption of microbial cells for potential use in the dairy industry — a review, *International Dairy Journal*, 12, 541–553.
39. K. Arisoya, 2008. Oxidative and thermal instability of biodiesel, *Energy Sources, Part A: Recovery, Utilization, and Environmental Effects*, 30(16), 1516–1522.
40. P. F. M. Paul and W. S. Wise, 1971. *The Principles of Gas Extraction*, Mills and Boon, London.
41. M. Zhu et al., 2002. Extraction of lipids from Mortierella alpina and enrichment of arachidonic acid from the fungal lipids, *Bioresource Technology*, 84(1), 93–95.
42. E. G. Bligh and W. J. Dyer, 1959. A rapid method for total lipid extraction and purification, *Canadian Journal of Biochemistry and Physiology*, 37, 911–917.
43. A. E. Hokanson and R. M. Rowell, 1977. *Methanol from Wood Waste: A Technical and Economic Study. USDA Forest Service General Technical Report FPL 12*. Department of Agriculture, Forest Products Laboratory, Madison, Wisconsin, USA.
44. J. Wu, Y. Fang and Y. Wang, *Production of Syngas by Methane and Coal Co-Conversion in Fluidized Bed Reactor*, Institute of Coal Chemistry, Chinese Academy of Sciences, Taiyuan, People's Republic of China.
45. T. R. Tsao and P. Rao, 1987. *The Development of Liquid Phase Methanol Process: An Update*, Air Products and Chemicals, Inc. Proceedings: Eleventh Annual EPRI Contractors' Conference on Clean Liquid and Solid Fuels, pp. 3.1–3.46.
46. Y. Dote et al., 1994. Recovery of liquid fuel from hydrocarbon rich microalgae by thermochemical liquification, *Fuel*, 73(12), 1855–1857.

# 4

## Aviation Fuels

### 4.1 Introduction

Air traffic is predicted to grow by 5% per year and fuel demand by about 3% on a yearly basis. More than two billion people travelled by air in 2006. The current revenue passenger kilometre (RPK) is predicted using historical RPK, which is based on past data. Airbus and Boeing forecast that the air traffic will be increased from 5 trillion RPK in 2011 to 10 trillion RPK in 2025 as shown in Figure 4.1. The amount of goods transported by air mode increased significantly from 6.1 million tonnes per year to 37.7 million tonnes per year. Every tonne is transported on an average of 3780 km [1].

### 4.2 Fuel Quality Requirements of Aircraft Engines (Aircrafts and Helicopters)

The fundamental requirements for an aviation jet fuel are that it should have the following characteristics:

1. Low weight per unit heat of combustion to increase the payload such as more people/goods
2. Low volume per unit heat of combustion to allow fuel storage without compromising the aircraft size, weight and performance

Aviation fuel should ensure the following quality for high fuel efficiency with full safety of aircraft operation:

1. *Optimum distillation/volatile property:* It is important for the mixture formation of air with fuel at a wide range of operating temperature conditions unlike ground vehicles.
2. *High ignition quality:* Octane number for the aero piston engine and smoke point for the gas turbine aero engine.

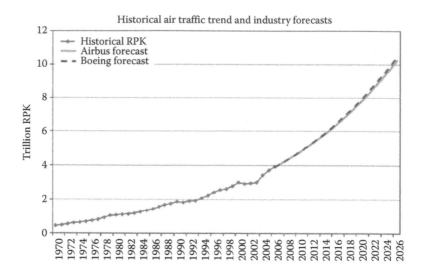

**FIGURE 4.1**
Historical data of RPK and Boeing predicted growth. (Adapted from E. Nygren, K. Aleklett and M. Hook, *Energy Policy*, 37, 4003–4010, 2009.)

3. *High cold flow characteristics:* Freezing point, cloud point and pour point should be very low.
4. *High energy content:* It reduces the volume and weight of the airbus.
5. *Corrosive resistance:* It should have corrosive resistance and material compatibility with aero engine parts.

Aviation fuels can be broadly classified into two groups as given below:

1. Aviation gasoline for reciprocating piston engines (mostly helicopters)
2. Aviation turbine fuels for use in turbo-propeller and turbo-jet engines (aircrafts)

## 4.3 Aviation Gasoline Fuel

Aviation gasoline fuel is used only for aero piston engines that are used in helicopters and low-power-output airbuses. The engine is a multi-cylinder spark ignition engine, which is chosen due to its lesser weight as compared to the compression ignition engine. The important fuel quality requirement of aviation gasoline for an aero piston engine is similar to a land vehicle as given below:

1. *Octane number:* It indicates anti-knocking properties. It is important for higher engine thermal efficiency by enhancing the compression ratio and smoother engine running.
2. *Distillation properties:* It is important for the mixture preparation of air and fuel.
3. *Density and viscosity:* It influences the atomisation of spray characteristics.
4. *No sulphur:* It eliminates corrosion problems.
5. *Energy density:* It reduces the system's weight and volume.

The knock rating of aviation gasoline is important for smooth combustion and enhancement of the thermal efficiency by the higher compression ratio of the engine (Table 4.1). The energy density relates to the power output of an aero engine and size of the fuel tank. The distillation characteristic of 10% volume of fuel evaporation indicates mixture preparation even at 41°C during the start-ability of an engine, whereas 50% and 90% show engine performance at engine cruising speed and deposit formation on the combustion chamber, respectively. The freezing point of the fuel is very important as the aircraft travels in a region where a large variation in temperature is present.

**TABLE 4.1**

Specification of Aviation Gasoline

| S. No. | Physico-Chemical Property | Aviation Gasoline | Petrol |
|--------|---------------------------|-------------------|--------|
| 1. | Knock rating | | 91 (octane number) |
| | Lean mixture, aviation rating | 105 | |
| | Rich mixture, supercharge method | 133 | |
| 2. | Tetraethyl lead gPb/L | 0.52 | — |
| 3. | Energy density (MJ/kg) | 43.73 | 46.529 |
| 4. | Density at 15°C (kg/m³) | 715 | 720–775 |
| 5. | Distillation | | |
| | Initial boiling point, °C | 41 | 70 |
| | 10% vol. evaporated at °C | 58 | 100 |
| | 40% vol. evaporated at °C | 92 | 150 |
| | 50% vol. evaporated at °C | 101 | 210 |
| | 90% vol. evaporated at °C | 128 | |
| | Final boiling point, °C | 156 | |
| | Sum of 10% and 50% evaporated temperatures, °C | 164 | |
| 6. | Vapour pressure, kPa | 41.6 | 55.158 |
| 7. | Freezing point, °C | Below −60 | −40 |
| 8. | Sulphur, % mass | 0.0003 | 50 (mg/kg) |
| 9. | Copper-strip corrosion (2 h at 100°C) | 1 | |

*Source:* Adapted from Air BP, *Handbook of Products*, www.bp.com/liveassets/bp_internet/ aviation/air_bp/STAGING/local_assets/downloads_pdfs/a/air_bp_products_hand book_04004_1.pdf, 2000.

The fuel should meet the desired specifications. If any property fails, the performance of the engine will deteriorate, which results in stalling of the engine or engine failure, or increase in the repair and maintenance cost of the system. The specifications of aviation fuels are given in Table 4.1.

## 4.4 Aviation Kerosene

Aviation kerosene-type jet fuel is used by civil gas turbine-based engine aircraft, and aviation gasoline is not used for this engine. The fuel is derived from feedstock of crude oil. The different grades of jet fuels such as Jet A-1, Jet Ts-1, Jet RJ-1, Jet TH and Jet Fuel No. 3 are used in different countries. The fuel grades are classified mainly based on its fuel quality such as sulphur, flash point, freezing point, energy content and smoke point as shown in Table 4.2.

*Aromatics content*: Benzene and toluene have an aromatic structure. Naphthalene and anthracene are known as polycyclic aromatic hydrocarbon (PAHC). The aromatic content is higher in crude oil, which helps to increase the octane number. However, it is responsible for soot/smoke/PAHC emission formation during combustion. The soot deposits on turbine blades may lead to durability problems.

*Sulphur*: Sulphur in fuel emits harmful emissions such as $SO_x$, sulphuric acid and sulphate, which enhance the corrosion problem on blades and other components.

*IBP and T10*: It indicates the mixture formation during the starting of the engine. The lesser value is preferable.

*Flash point*: It is the maximum temperature at which a liquid fuel can be stored and handled without any fire hazard. An inflammable mixture can

**TABLE 4.2**

Comparison of Different Jet Fuel Properties

| S. No. | Jet Fuel Properties | Jet A-1 | Jet Ts-1 | Jet RJ-1 | Jet TH | Jet Fuel No. 3 | Diesel |
|--------|---------------------|---------|----------|----------|--------|----------------|--------|
| 1. | Aromatics, % vol | 19.5 | 15.2 | 19.5 | 16.5 | 16 | |
| 2. | Sulphur, % mass | 0.02 | 0.04 | 0.02 | 0.01 | 0.02 | 50 (mg/kg) |
| 3. | Initial boiling point, °C | 156 | 138 | 140 | 142 | 153 | |
| 4. | T10, °C | 167 | 160 | 154 | 155 | 168 | |
| 5. | Flash point, °C | 42 | 31 | 35 | 40 | 39 | 35 |
| 6. | Freezing point, °C | −50 | −64 | −68 | −53 | −52 | −40 to −34.4 |
| 7. | Energy content, MJ/kg | 43.15 | 43.2 | — | 43.24 | 43.35 | 45.759 |
| 8. | Smoke point, mm | 25 | 28 | 26 | 22 | 25 | |

*Source:* Adapted from Air BP, *Handbook of Products*, www.bp.com/liveassets/bp_internet/ aviation/air_bp/STAGING/local_assets/downloads_pdfs/a/air_bp_products_hand book_04004_1.pdf, 2000.

form by chance using any localised ignition source, which is equal to its flash point temperature. It is important for transportation, storage and the fuel handling system, and for aircrafts to avoid fire accidents.

*Freezing point*: Fuel will start freezing at low temperature. It limits the altitude at which the airbus can fly since the surrounding air temperature decreases with an increase in altitude.

*Energy content*: It relates directly to the power output of the engine and the size of the fuel tank.

*Smoke point*: It measures the tendency of a liquid fuel to form soot. The higher the smoke point, the lesser the soot formation, resulting in a better flame stability of the gas turbine combustor.

## 4.5 Compressed Natural Gas

CNG is a cheaper fuel for a power-generating stationary gas turbine system. However, it is not suitable for aircrafts as CNG needs a relatively larger storage space as compared to aviation gasoline and kerosene. It needs a specialised fuel storage and handling system infrastructure. It is also a fossil fuel that has limited reserves and does not have any advantages as compared to aviation kerosene or aviation gasoline fuel. Hence, it is not preferable for aircrafts.

## 4.6 Liquefied Natural Gas

It is slightly better than CNG in terms of the smaller size of the fuel tank required to hold it. But it does not have any advantage as compared to aviation kerosene or other liquid fuel. Hence, liquid fuel is a better choice for the airbus as it has higher energy density and requires a smaller-sized fuel tank.

## 4.7 Biodiesel–Diesel Blend

Biodiesel can be produced by SVO through the transesterification process. The feedstocks for biodiesel production are plant/crop material such as *Jatropha curcas*, *Pongamia pinnata*, soya bean and sunflower. It has a number of advantages as a jet fuel such as a high flash point (>120°C), biodegradability and high ignition quality. However, it has some disadvantages such as poor cold flow characteristics (higher freezing point, cloud point), lower density (38 MJ/kg) and high deposit formation on the engine component and injector coking due to higher unsaturated compounds. The fuel quality of biodiesel needs to be improved for aviation application. It has a long way to go before it can compete with jet fuel.

## 4.8 Fischer–Tropsch Diesel

It is produced through the combined gasification and F–T synthesis process with the feedstock of cheaper raw materials such as coal, biomass, agro waste and bitumen. It has a higher heat content of about 43.9 MJ/kg, which is slightly higher than that of petroleum fuel. It also has good cold flow characteristics such as a freezing point of about −40°C. It has negligible sulphur, olefin and aromatic components, which result in less harmful emissions. It is one of the alternative fuels that could be considered for an aero engine if it is produced from coal or natural gas and it is a renewable fuel if it is produced from biomass or agro waste.

## 4.9 Carbon-Free Fuel: Compressed and Liquid Hydrogen

Hydrogen is an energy carrier which can be produced from a variety of renewable energy sources such as solar, biomass and wind. It can also be produced by the water electrolysis process, which is powered by a nuclear power plant. A hydrogen-fuelled engine does not emit any carbon-based emission such as CO, HC, soot, smoke, PM and $CO_2$. Even though hydrogen has the highest energy content (120 MJ/kg), it has poor volumetric heat content (9.6 MJ/Nm$^3$). It has the problem of high volume and weight per unit combustion, which is similar to CNG/LNG fuel.

Liquid hydrogen is a viable option in view of its higher energy density and volume. A liquid hydrogen-powered aeroplane is shown in Figure 4.2. It needs a larger tank, which reduces the fuel efficiency of short-range aircraft [3]. It needs a new infrastructure for fuel storage and handling. Metal hydride is an option to store hydrogen, however, it is in a research stage.

## 4.10 Carbon-Neutral and Sustainable Fuel

Alcohol fuels such as methanol, ethanol and butanol, and biodiesel are carbon-neutral fuels as the carbon emitted from combustion engines is recycled through crop/plant. This means there is no new addition of $CO_2$

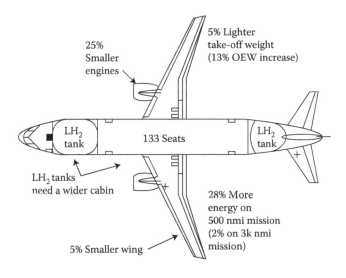

**FIGURE 4.2**
Liquid hydrogen-powered aeroplanes. (Adapted from D. Daggett et al., Alternative fuels and their potential impact on aviation, Paper presented at the 25th Congress of the International Council of the Aeronautical Sciences (ICAS) hosted by the German Society for Aeronautics and Astronautics, Hamburg, Germany, September 3–8, 2006, NASATM2006-214365, http://gltrs.grc. nasa. gov/reports/2006/TM-2006-214365.pdf, European Economic and Social Committee, 2009.)

emission into the atmosphere. But alcohol fuels have the problem of lower energy density, which requires a larger wing and engines, resulting in the reduction of the aircraft's fuel efficiency (Figure 4.3). In case of biodiesel, it has poor cold flow properties and high unsaturated compounds. If hydrogen is produced from renewable resources, it is a sustainable fuel. Energy density, safety and storage are the major problems. F–T diesel is only an alternative fuel that has a good potential for an aero engine. This sustainable fuel needs to be explored for aircrafts. Crude oil-based aviation kerosene fuel will play a crucial role for at least another two decades. A summary of the qualitative comparison of alternative fuels for aircraft with base aviation jet fuel is presented in Table 4.3.

## 4.11 Production Technology of Aviation Fuel

Crude oil is fed to the distillation column where straight-run light and heavy gasoline, kerosene and diesel are separated at atmospheric pressure. The bottoms from the atmospheric column are vacuum distilled to obtain gas oils for FCC or hydro-cracker feed. The gas oils may be hydro-treated to reduce sulphur and nitrogen to levels that will improve the performance of the FCC

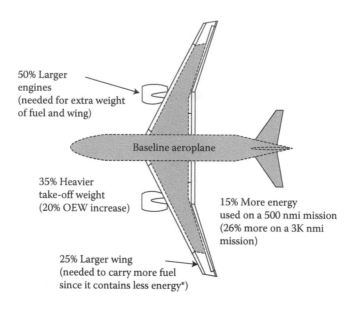

**FIGURE 4.3**

An ethanol-fuelled aeroplane. (Note: *Ethanol requires 64 percent more storage volume than kerosene fuel for the same amount of energy requirement.) (Adapted from D. Daggett et al., Alternative fuels and their potential impact on aviation, Paper presented at the 25th Congress of the International Council of the Aeronautical Sciences (ICAS) hosted by the German Society for Aeronautics and Astronautics, Hamburg, Germany, September 3–8, 2006, NASATM2006-214365, http://gltrs.grc. nasa. gov/reports/2006/TM-2006-214365.pdf, European Economic and Social Committee, 2009.)

**TABLE 4.3**

Summary of Key Properties of Alternative Fuels with a Comparison of Aviation Kerosene

| Fuel Quality | CNG | Compressed Hydrogen | Liquid Hydrogen | Biodiesel | Fisher– Tropsch Diesel | Alcohol (Methanol, Ethanol) |
|---|---|---|---|---|---|---|
| Volumetric energy content (MJ/kg) | Very poor | Very poor | Moderate | Good | Excellent | Poor |
| Freezing point | Excellent | Excellent | Excellent | Poor | Excellent | Excellent |
| Infrastructure | To be built | To be built | To be built | Good for existing infrastructure | Good for existing infrastructure | Slight modification needed for existing infrastructure |

process. A schematic layout of the production technology of the aviation fuel is shown in Figure 4.4.

The jet fuel produced by a refinery may be either a straight-run or hydro-processed product or it may be a blend of a straight-run, hydro-processed and/or hydro-cracked product. Small amounts of heavy gasoline components may

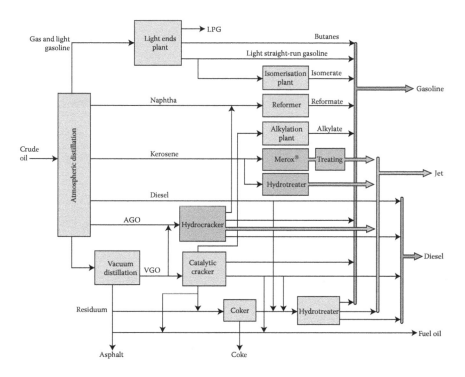

**FIGURE 4.4**
Production technology of aviation fuel. (Adapted from Chevron Global Aviation, http://www.cgabusinessdesk.com/document/aviation_tech_review.pdf.)

also be added. Straight-run kerosene from low-sulphur crude oil may meet all the jet fuel specifications. But straight-run kerosene is normally upgraded by Merox-treating, clay-treating or hydro-treating before it can be sold as jet fuel. The refinery must blend the available streams to meet all the performance, regulatory, economic and inventory requirements. Sophisticated computer programs have been developed to optimise all aspects of the refinery operation, including the final blending step. Refineries are optimised for overall performance, not just jet fuel production [4].

## 4.12 Volume and Weight of Aviation Fuel for Smaller- and Large-Sized Aircraft

Commercial aircrafts store fuel in their wings. Figure 4.5 shows the arrangement of fuel tanks in a Boeing 747-400. There are two main tanks and one reserve tank in each wing, along with a centre wing tank in the fuselage.

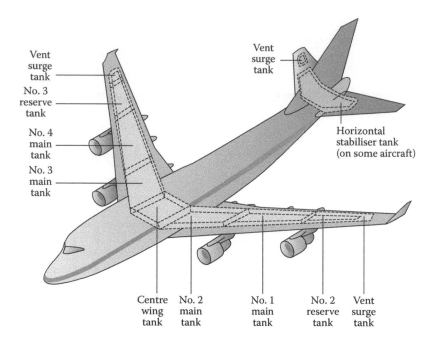

**FIGURE 4.5**
Boeing 747-400 fuel tank arrangement. (Adapted from Chevron Global Aviation, http://www. cgabusinessdesk.com/document/aviation_tech_review.pdf.)

Some 747-400s also have an additional fuel tank in the tail horizontal sta- biliser. Each main tank has a pump that supplies fuel to a manifold that feeds the engines. The Boeing 747-400 with its horizontal stabiliser tank fully loaded contains 216,389 L of 175,275 kg weight @ 0.810 g/mL. The Boeing 747-400 without its horizontal stabiliser tank fully loaded contains 203,897 L of 165,156 kg weight @ 0.810 g/mL. The Airbus A380 fully loaded contains 310,000 L of 251,100 kg weight @ 0.810 g/mL [4].

## 4.13 Emission and Effects by Aviation Fuel

Aircraft engine emissions have not received as much attention in recent years as emissions from other energy sources. This is because aviation only contributes a small proportion of emissions compared to ground vehicles and stationary sources. There are two main sources of aircraft emissions: the jet engines and the auxiliary power unit (APU). Most jet fuel is burnt in flight, so most of the emissions occur at an altitude, not at the ground

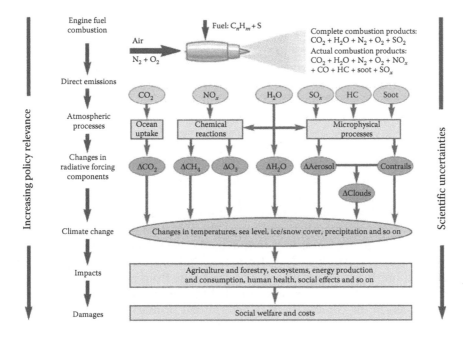

**FIGURE 4.6**
Schematic representation of aircraft emission and their effects. (Adapted from D. Wuebbles et al., *American Geophysical Union*, 88(14), 156–168, 2007.)

level [6]. A schematic representation of effects caused by aircraft emission is shown in Figure 4.6.

When hydrocarbons are completely combusted, the products are carbon dioxide and water. However, when jet fuel is burnt, other emissions, including sulphur oxides, nitrogen oxides, unburnt hydrocarbons and particulates (soot), are formed. This is due to the presence of both trace amounts of sulphur and nitrogen in the fuel and engine design and operating conditions.

The International Civil Aviation Organisation (ICAO) has established limits for emissions of nitrogen oxides, carbon monoxide, unburnt hydrocarbons and smoke from commercial jet engines. These limits were established for a defined landing and take-off cycle (LTO) to limit emissions near the ground level, but they also indirectly limit emissions at an altitude.

*Carbon dioxide* is a greenhouse gas that is implicated in climate change. Aviation contributes a relatively small proportion of anthropogenic carbon dioxide. In 1992, aviation emissions were about 2% of the total, and projections for 2050 state that aviation will contribute about 3% of the total. Carbon dioxide emissions are directly related to the amount of fuel consumed and can be reduced by increasing the efficiency of the engines and airframe.

*Water vapour* is the other major product of combustion. Emissions at the ground level are not an issue but water vapour emissions at cruise altitude can lead to the formation of contrails and aviation-induced cirrus clouds.

The effect of these contrails and clouds is uncertain but they are thought to contribute to climate change. This is another area of ongoing research.

*Sulphur oxides* are the result of the combustion of sulphur-containing compounds in the fuel, and thus are proportional to the fuel sulphur content. $SO_x$ emissions are thought to contribute to aerosols and particulate formation, although more research is needed to definitively answer this question. Over the past decade, there has been a worldwide trend to lower the sulphur content in motor gasoline and diesel fuel with some countries requiring near-zero sulphur today or in the near future. These limits have been mandated by government regulations driven by the need to reduce harmful emissions. A similar reduction has not occurred for jet fuel; the specifications continue to allow a maximum of 3000 ppm sulphur, although the worldwide average sulphur content in jet fuel appears to be between 500 and 1000 ppm.

*Nitrogen oxides* are mainly formed from the oxidation of atmospheric nitrogen at very high temperatures found in the combustor. Any fuel-bound nitrogen will also result in $NO_x$ formation. Nitrogen in the fuel is neither controlled nor typically measured, but can range from near zero to perhaps 20 ppm. Since $NO_x$ formation is controlled by the maximum local temperature reached in the combustor, it is determined primarily by engine design and operating conditions. Current engine design trends are intended to ensure more complete combustion at a rapid rate, resulting in more uniform combustion temperatures and thus lowering $NO_x$ emissions. $NO_x$ emissions are a matter of concern because they contribute to the formation of ozone near the ground level.

*Particulates* and unburnt hydrocarbons are the result of incomplete combustion. If present in a high enough concentration, these particulates will be visible as smoke or soot coming out of the engine. Modern jet engine combustors are designed to significantly reduce visible smoke and thus emit fewer particulates than older engines. Correlations to particulate emissions have been found among various fuel properties such as hydrogen content, hydrogen/carbon ratio, smoke point, aromatics and naphthalenes. However, the findings have not been consistent from one experiment to another. Overall, engine design and operating conditions play a greater role in particulate emissions than fuel properties. Particulates at the ground level can contribute to haze and smog formation and can be harmful if inhaled. The effect of particulates at cruise altitude is uncertain and is an area of ongoing research.

## 4.14 Conclusion

The aviation sector needs specific fuel quality requirements such as high energy content and good cold flow characteristics. All alternative fuels, except F–T diesel, have lower energy content and poor cold flow characteristics as compared to conventional fuel. Hence, the less emphasis may be given to alternative fuels (in the present scenario) in the aviation sector and more in the surface transportation sector.

## References

1. E. Nygren, K. Aleklett and M. Hook, Aviation fuel and future oil production scenarios, *Energy Policy*, 37, 4003–4010, 2009.
2. Air BP, *Handbook of Products*, www.bp.com/liveassets/bp_internet/aviation/ air_bp/STAGING/local_assets/downloads_pdfs/a/air_bp_products_hand-book_04004_1.pdf, 2000.
3. D. Daggett, R. Hadaller, R. Hendricks and R. Walther, Alternative fuels and their potential impact on aviation, Paper presented at the 25th Congress of the International Council of the Aeronautical Sciences (ICAS) hosted by the German Society for Aeronautics and Astronautics, Hamburg, Germany, September 3–8, 2006, NASATM2006-214365, http://gltrs.grc.nasa. gov/reports/2006/ TM-2006-214365.pdf, European Economic and Social Committee, 2009.
4. Chevron Global Aviation, http://www.cgabusinessdesk.com/document/avia-tion_tech_review.pdf.
5. Aviation and the Environment, GAO-03-252, United States General Accounting Office.
6. Okanagan University College in Canada, Department of Geography, University of Oxford, School of Geography; United States Environmental Protection Agency (EPA), Washington; Climate Change 1995. The Science of Climate Change, Contribution of Working Group 1 to the Second Assessment Report of the Intergovernmental Panel on Climate Change, UNEP and WMO, Cambridge University Press, 1996.
7. D. Wuebbles et al., Evaluating the impacts of aviation on climate change, EOS Transactions, *American Geophysical Union*, 88(14), 156–168, 2007.

# 5

# Utilisation of Alternative Fuels in Internal Combustion Engines/Vehicles

## 5.1 Spark Ignition Engines

### 5.1.1 General Introduction

Spark ignition (SI) engine-powered vehicles play a vital role in passenger transportation, which includes two-, three- and four-wheelers. In the SI engine, the ignition is initiated by an external aid (spark plug). People prefer SI engine-powered vehicles because they have a lower initial cost than CI engine-based vehicles. The SI engine operates with a lower compression ratio (CR: 6 to 11) than the CI engine (CR: 15:1 to 22:1), which leads to a lowering of the weight and complexity of the system, resulting in a reduced vehicle cost. The fuel economy of an SI engine-powered vehicle is on the lower side as compared to a CI engine-powered vehicle due to its reduced compression ratio. Hence, SI engine-powered vehicles are used for mass transportation over only short distances as compared to CI engine-powered vehicles. The details of passenger and mass transportation such as classification, vehicle type, maximum average distance travel, maximum number of hours of operation/day and reasons are given in Table 5.1. The values in the table are approximate and are only given for a better understanding of the topic. The values may change from passenger to passenger and goods to goods.

The selection of an SI or CI engine for mass or passenger transportation can be made using Equations 5.1a and 5.1b.

$$\text{Operating fuel cost (OC)/year} = \text{distance travel/year*} \quad (5.1a)$$
$$\text{(fuel cost/km)}$$

$$\text{Discounted cost (DC)/year} = \text{initial cost of vehicle*} \quad (5.1b)$$
$$\text{interest rate/year}$$

Operating cost includes the cost of fuel, oil and maintenance, whereas discounted cost includes the depreciation in the initial vehicle cost at a certain rate of interest. If the operating fuel cost is lesser than the discounted cost, the SI vehicle is a great choice for personal/mass transportation, whereas

**TABLE 5.1**

Details of Passenger and Mass Transportation

| S. No. | Classification | Vehicle: SI/CI and Fuel | Passenger/ Mass Transportation | Maximum Average Distance Travel (km) | Maximum Number of Hours Operation/ Day | Reason |
|---|---|---|---|---|---|---|
| 1 | Two-wheeler (bike, scooter, moped) | SI, Gasoline | Passenger | 50 | <1 | OC < DC |
| 2 | Three-wheeler (auto) | SI, Gasoline | Passenger | 50 | <1 | OC < DC |
| 3 | Three-wheeler (LCV) | CI, Diesel | Mass | 50 | >5 | OC > DC |
| 4 | Four-wheeler (car) | SI, Gasoline | Passenger | 200 | <2 | OC < DC |
| 5 | Four-wheeler (taxi and MUV: jeep, wagon) | CI, Diesel | Mass | 500 | >10 | OC > DC |
| 6 | Four-wheeler (bus) | CI, Diesel | Passenger | 1000 | >10 | OC > DC |
| 7 | Four-wheeler (lorry and truck) | CI, Diesel | Mass | | >10 | |
| 8 | Six-wheeler | CI, Diesel | Mass | 3000 | >10 | OC > DC |
| 9 | Above specialised six-wheeler | CI, Diesel | Mass | 3000 | >10 | OC > DC |

*Note:* OC, operating cost; DC, discounted cost.

the CI vehicle is preferable for longer-distance transportation. As 50–75% of the operating cost is for fuel, the choice of the vehicle with SI or CI will also depend on the fuel price. For example, natural gas reserve countries would obviously prefer SI vehicles for their vehicle fleet for both passenger and mass transportation as it would be more economical. The crude oil price increases steeply due to a variety of reasons, including geopolitical and fossil fuel depletion. The high price of crude oil directly affects the product price in almost all sectors, resulting in the unstable economic growth of any country. Alternative fuels have a major role to play to overcome this crisis.

### 5.1.2 Various Challenges with SI Engine

One of the remaining challenges in designing a more efficient and less polluting SI internal combustion (IC) engine is to increase the consistency of the process, leading from spark generation to a fully developed flame.

- Although the thermal efficiency is high, the combustion inside the cylinder is not 100% complete due to the improper mixing condition and the improper air–fuel ratio.
- Incomplete combustion always leads to unwanted emissions.
- Knocking is prominent due to improper combustion.

The combustion process is an important parameter in the operation of the SI engine; the modern-day research in the area of engine development is centred more towards the optimisation of combustion processes. This is an emerging area of work nowadays and also a complex one. The characterisation of combustion processes in the SI engine is one of the effective parameters that solves the problem of energy inefficiency of engines, and the combustion is characterised by means of flame characteristics such as the flame kernel growth rate, flame speed, flame development angle, rapid burning angle and overall burning angle drift velocity.

Drift velocity is also an important parameter to know whether the flame growth is uniform in all directions or offset with respect to the spark plug centre axis. If we know the actual value of drift velocity, we can optimise the combustion chamber design and the location of the spark plug. Simultaneously, it is a relatively newer field of study in the area of combustion. Very few research activities related to this idea have been seen in the literature and researchers have started to think over this important aspect, which is highly useful in understanding the finer details of some complex issues like knocking phenomena.

In addition, the heat release rate, combustion duration and ignition delay play an important part. The SI IC engine relies on the ability of the flame kernel to survive the high-strain-rate, unsteady environment of a turbulent flow field and successful transition into a fully developed flame to operate efficiently. Cyclic variations find their origin in the beginning of combustion and determine the working limits of the engine and the driving behaviour of the vehicle. These factors demonstrate the crucial importance of the knowledge of the flame kernel growth rate in SI engines.

The pollutant formation and energy conversion efficiency of homogeneous charge SI engines strongly depends on the way the flame develops, propagates and is finally quenched close to the combustion chamber walls. Several processes and parameters of interest like the lean burn limit, specific NO emissions, cyclic variability and knock tendency are clearly affected by the modes of interaction between thermochemistry and fluid mechanics during the combustion process. The study of the effect of various parameters during the process of combustion is one of the active areas of engine efficiency today and can be a major aspect of future engine development.

Some important points are as follows:

1. Under all conditions covered, the flame spreads in an approximate concentric pattern about the points of ignition. It follows that with a single ignition, the shortest combustion time will be obtained by placing the spark plug near the centre of the combustion chamber.

2. Still shorter combustion times can be achieved by using one or two spark plugs.

3. The burning velocity and the rate of pressure development increase nearly as fast as the engine speed, which explains why the engine can be operated at very high speeds with only a moderate increase in spark advance.

4. The increase in burning velocity with an increase in engine speed is believed to be due to an increase in small-scale random turbulence, which affects the structure and depth of the reaction zone and influences the rate at which the flame advances into the unburnt charge.

In general, the flame propagation in the combustion chamber is controlled by six factors:

1. Composition of fuel to be burnt
2. Temperature of unburnt charge
3. Degree of turbulence
4. Expansion of gases in the reaction zone
5. Piston movement
6. General swirl

### 5.1.3 Preliminary Studies Regarding Combustion Phenomenon

The combustion process begins with the formation of a laminar burning kernel. As such, a kernel grows and the flame surface departs from the ignition source; it begins to interact with the turbulent motion field of the cylinder charge. The flame surface is progressively corrugated and wrinkled by turbulence, and after transition, the flame becomes fully turbulent [1].

As the reacting surface is increased, the rate at which the fresh mixture is consumed is also increased, making the flame surface advancement in the unburnt region at a higher global speed—called turbulent flame speed—although locally the combustion is still taking place at the laminar flame speed, which is determined by the chemistry, pressure and temperature conditions (Figure 5.1) [1].

One of the most important intrinsic properties of any combustible mixture is its burning velocity. Burning velocity is defined as the relative velocity normal to the flame front, that is, the velocity with which unburnt gas moves into the flame front and is transformed. This property depends upon the mixture composition, the temperature of the unburnt gas, pressure and other parameters. Burning velocity is important because it is the property that influences the flame shape and important flame stability characteristics, such as blow-off and flashback. The laminar burning velocity is of fundamental importance for analysing and predicting the performance of a combustion engine. Therefore, knowledge of the burning velocity is important for improving the understanding of the fundamental combustion processes, and for direct practical application aimed at increasing the fuel efficiency and reducing the pollutants. The optimisation of combustion requires the tuning of the flame properties and its influence on how the flame kernel can

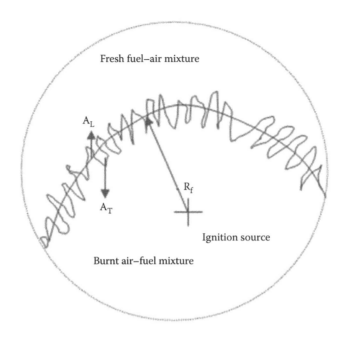

**FIGURE 5.1**
Flame development in combustion chamber. Rf, flame radius; AT, turbulent flame surface; AL, laminar flame surface.

spread and propagate through the mixture for best combustion timing and complete consumption of the fuel–air mixture.

## 5.1.4 Parameters Affecting Burning Velocity

1. *Air–fuel equivalence ratio (ø)*: At low turbulence levels, flame speed shows a peak at about stoichiometric air–fuel ratio. A decrease in burning velocity is observed with increasing lean mixtures. With the development of the flame, the effect of ø on burning velocity decreases as shown in Figure 5.2. Further, a low burning process results in the formation of peak pressure during the later part of the expansion process. For an optimum design, it is desirable that the peak pressure occurs within a few degrees after top dead center (TDC). Hence, a lean mixture operation requires an advancement of ignition timing. Laminar burning velocity is the function of pressure, temperature and equivalence as shown in Equation 5.2

$$Sl = Sl_0 \, (T_u/T_0)^{\alpha}(P_u/P_0)^{\beta} \qquad (5.2)$$

where $T_0 = 298$ K and $P_0 = 1$ bar are the reference temperature and pressure, and Slo, $\alpha$ and $\beta$ are constant specific for a given fuel–air equivalence ratio.

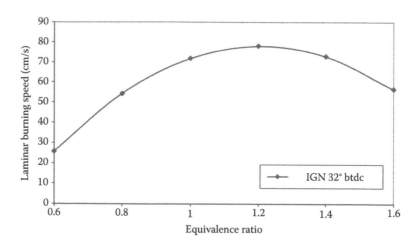

**FIGURE 5.2**
Laminar burning velocity versus equivalence ratio (ø) at 1600 rpm.

The values of constants are

$$\alpha = 2.18 - 0.8(ø - 1) \tag{5.2a}$$

$$\beta = -0.16 + 0.22(ø - 1) \tag{5.2b}$$

$$Slo = Bm + B_ø(ø - ø_m)^2 \tag{5.2c}$$

where ø is the equivalence ratio.

Laminar burning velocities at constant equivalence ratio (ø = 1) and for different speeds at each crank angle are shown in Figure 5.3. 0° CA shows the TDC position. At 1000 rpm, it is calculated that Tu is 627.8 K; at 1600 rpm, Tu is 538.36 K; at 2000 rpm, Tu is 488.988 K, and pressures are varying with respect to the crank angle. As the engine speed is increased, the maximum laminar velocity is decreased due to the different engine operating conditions in the cylinder at the time of ignition.

Maximum burning velocity ~85–90 cm/s (¢ = 1, 1000 rpm)

Maximum burning velocity ~65–70 cm/s (¢ = 1, 1600 rpm)

Maximum burning velocity ~55–60 cm/s (¢ = 1, 2000 rpm)

2. *Turbulence intensity*: The effect of high turbulence is to increase the flame propagation rate, which has a positive effect on the burning velocity. Burning velocity has an almost linear relationship with turbulent intensity. The turbulent burning velocity can be calculated

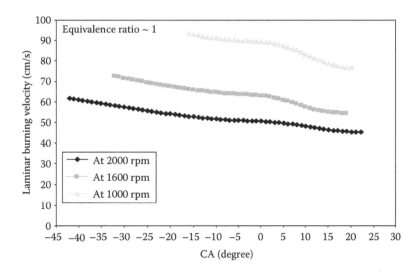

**FIGURE 5.3**
Laminar burning velocity versus CA.

using Equations 5.3 through 5.5 with inputs of laminar burning velocity and turbulent intensity.

$$St = Sl + Au' \tag{5.3}$$

where
   $St$ = terminal burning velocity
   $Sl$ = laminar burning velocity
   $A$ = constant value
   $u'$ = turbulent intensity and a function of the mean piston speed [2].

$$St = 3.9\, Sl + u' \tag{5.4}$$

$$St = Sl + \{1+ (2u'/Sl)\ 2\}1/2 \tag{5.5}$$

Similar data have been obtained for different values of engine speeds and the results are plotted. Turbulent burning velocities at a constant equivalence ratio ($\varnothing = 1$) and for different correlations at each crank angle are shown in Figure 5.4 and their values are also given in Table 5.2.

It is clearly seen from the figure that the absolute value of the predicted turbulent burning velocity is significantly different for each correlation. There is no change in turbulent burning velocity with Schelkin's correlation but turbulent burning velocity decreases with the crank angle with Kozachenko's correlation.

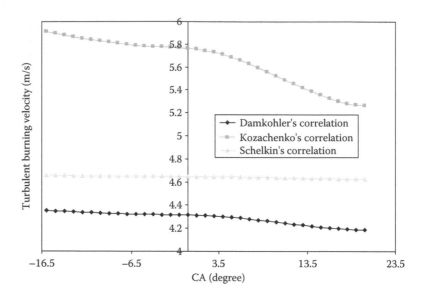

**FIGURE 5.4**
Turbulent burning velocity versus CA at 1000 rpm.

**TABLE 5.2**

Calculated Data for Turbulent Burning Velocity at 1000 rpm Engine Speed

| S. No. | C.A. | Sl (m/s) | u′ (m/s) | A | St (m/s) (Damkohler's) | St (m/s) (Kozachenko's) | St (m/s) (Schelkin's) |
|---|---|---|---|---|---|---|---|
| 1 | −16 | 0.93 | 2.28 | 1.5 | 4.35 | 5.90 | 4.65 |
| 2 | −12 | 0.92 | 2.28 | 1.5 | 4.33 | 5.85 | 4.65 |
| 3 | −8 | 0.90 | 2.28 | 1.5 | 4.32 | 5.80 | 4.64 |
| 4 | −4 | 0.89 | 2.28 | 1.5 | 4.31 | 5.77 | 4.64 |
| 5 | 0 (TDC) | 0.89 | 2.28 | 1.5 | 4.31 | 5.76 | 4.64 |
| 6 | 4 | 0.87 | 2.28 | 1.5 | 4.29 | 5.70 | 4.61 |
| 7 | 8 | 0.84 | 2.28 | 1.5 | 4.26 | 5.59 | 4.63 |
| 8 | 12 | 0.81 | 2.28 | 1.5 | 4.23 | 5.45 | 4.63 |
| 9 | 16 | 0.78 | 2.28 | 1.5 | 4.20 | 5.32 | 4.62 |
| 10 | 20 | 0.76 | 2.28 | 1.5 | 4.18 | 5.26 | 4.62 |

3. *Turbulence parameters*: Swirl and squish increase the burning velocity, the increase being greater with squish. The highest increase is with both swirl and squish.

4. *Type of fuel*: Flame photographs of propane and hydrogen–air flames show important differences in appearance, structure and behaviour. The hydrogen flame front is more spherical in outline than upstream side. Hydrogen burns at a much faster rate and has a much higher

burning velocity due to its mass rapid initial flame development and its having a faster, fully developed propagation phase.

5. *Engine speed*: The most significant effect shown is the fact that the variation in crank angle occupied by flame travel as the engine speed changes is very small. This implies that burning velocity increases nearly in proportion to the engine speed. The increase in burning velocity with increasing engine speed is due to the marked effect of turbulence, which has a tendency to increase the burning velocity. The travel time for the flame in terms of crank angle is more or less the same at different speeds. However, since the travel time for the piston changes with engine speed, the spark timing has to be changed at higher speeds to achieve peak pressures at the optimum point in the cycle.

6. *Cylinder size*: The ratio of burning velocity to piston speed is nearly the same for similar cylinders of different sizes. This implies the following:

   • At a given piston speed, burning time will be inversely proportional to the bore, and since speed is inversely proportional to the bore, the burning angle will be nearly independent of the bore.

   • At a given speed, average burning velocity will be nearly proportional to the bore, and an effective burning angle will again be independent of the bore.

## 5.1.5 Characterisation of Combustion Process

Combustion is characterised by parameters such as the flame development angle, rapid burning angle and overall burning angle. All these angles can be obtained from Figure 5.5.

*Flame development angle*: The crank angle interval between the spark discharge and the time when a small but significant fraction of the cylinder mass has burnt or fuel chemical energy has been released [3].

*Rapid burning angle*: The crank angle interval required to burn the bulk of the charge is defined as the interval between the end of the flame development stage and the end of the flame propagation process [3].

*Overall burning angle*: The duration of the overall burning process is the sum of flame development and rapid burning angle [3].

*Flame development angle* $\Delta\theta_d$—crank angle interval during which the flame kernel develops after spark ignition.

*Rapid burn angle* $\Delta\theta_b$—crank angle required for burning most of the mixture.

*Overall burning angle*—sum of flame development and rapid burning angles.

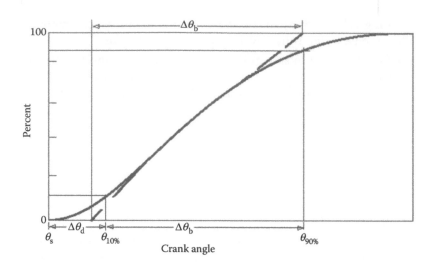

**FIGURE 5.5**
Parameters for characterising flame development.

In addition to heat release rate, combustion duration and ignition delay, mass fraction burnt rate is closely associated with the flame speed, area of the flame and unburnt charge density as given in Equation 5.6 [4]

$$\text{Mass burnt rate: } m_k = \rho_u \, A_k S_l \tag{5.6}$$

where
$S_l$ = laminar burning velocity (m/s)
$\rho_u$ = mean density of unburnt mixture (kg/m³)
$A_k$ = flame kernel surface (m²)

Mass burnt rate can be calculated by Equation 5.6. The mass fraction is characterised using the flame speed and area. Flame speed is a function of design parameters, operating variables and so on.

## 5.1.6 Study of Flame Kernel Growth Development

The birth of the flame kernel is a very complex process. Different mechanisms stimulate the growth of the kernel. An electrical spark will cause a breakdown in an isolating medium. A conductive plasma channel is formed with a high pressure and temperature. A shock wave and heat conduction and diffusion allow the kernel to expand before the combustion reactions take over and the combustion becomes self-sustainable. For the initial development of the kernel, three phases can be distinguished: the pre-breakdown phase (closed by breakdown), the plasma phase (expansion determined by shock wave and heat conduction) and the initial combustion phase (expansion determined by exothermic combustion reactions) [5].

### 5.1.6.1 Pre-Breakdown Phase

Initially, the gas between the electrodes is a perfect isolator. If voltage is applied between the electrodes, the electrons in the spark gap accelerate from the cathode and the anode. Collisions with gas molecules ionise these, and new electrons are produced. If the number of electrons increases sufficiently to make the discharge self-sustainable, a breakdown takes place and the pre-breakdown phase is closed. A very small conductive path is formed between the electrodes with a high pressure (around 20 MPa) and temperature (around 60,000 K) [5].

### 5.1.6.2 Plasma Phase

At the end of the pre-breakdown phase, the plasma channel is not in equilibrium. The very high pressure and temperature of the channel compared with the surrounding gas causes an immediate expansion of the channel, first by a shock wave and later by heat conduction. In SI engines, there is a definite purpose to study combustible mixtures. The question has to be posed if exothermic burning reactions do not have to be considered and if they cause expansion of the plasma. Indeed, combustion reactions (self-ignition) take place at temperatures lower than the inversion temperature only at the border of the plasma. In a very small channel with a high temperature, the expansion is dominated by thermal conduction and the combustion reactions can be neglected. If the temperature decreases, these reactions become more and more important. If the expansion by these combustion reactions prevails, the plasma phase is closed and the initial combustion phase starts (transition from self-ignition to flame propagation) [6].

### 5.1.6..3 Initial Combustion Phase

In the initial combustion phase, the combustion reactions dominate the expansion of the kernel. The kernel is no longer a plasma kernel but can be called a flame kernel. This kernel has to be sufficiently large to make it possible for the burning reactions to gain the opposite effects of the heat losses and flame stretch to ignite the mixture successfully. Ignition takes place if the expansion becomes self-sustainable. The burning reactions are then sufficiently powerful to lead the expansion of the kernel without supplementary energy supply [6].

*Effect of spark advance on the flame kernel growth rate*: The growth of the initial kernel is affected by the temperature and pressure of the combustion chamber when the spark is discharged. As the pressure increases, the breakdown energy is also increased, but the flame velocity is reduced. If the spark time is retarded, the pressure and temperature of the combustion chamber are increased simultaneously. Therefore, the velocity is affected by the parameters that contribute to the growth of the flame in opposite ways. As a result, kernel development is affected by spark advance. Figure 5.6 shows FKGR is maximum at minimum advance for best torque (MBT).

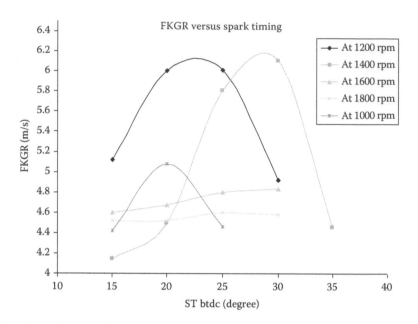

**FIGURE 5.6**
FKGR versus spark timing.

*Effect of engine speed on the flame kernel growth rate*: As the engine speed increases, the turbulence intensity in the cylinder and the fluid flow velocity are also increased. It provides a better condition for the initial flame kernel development. The heat transfer to the cylinder wall during the compression stroke is increased with the engine speed. However, its effect is small. Figure 5.7 shows variations of flame kernel growth rate with engine speed.

*Comparative analysis*: The flame kernel growth rate was predicted using mass fraction burnt at MBT spark timing and it compared with the measured values as shown in Figure 5.8.

Comparison shows that both the measured and predicted values satisfactorily agree with each other. However, this case is considered only for MBT conditions. As there is no significant difference between the curves at MBT conditions, it shows that the flame kernel growth rate can be calculated using mass fraction burnt data at MBT conditions within a reasonable error.

*Drift velocity*: It is an important parameter to know whether the flame growth is uniform in all directions or offset with respect to the spark plug centre axis. The drift velocity is used for the optimisation of spark plug orientation such as its position and angle in the cylinder head of the engine.

If flame growth is in the 360° direction as shown in Figure 5.9, it can be expressed in the following mathematical form:

$$\text{Drift velocity} = (v_1 + v_2 + \cdots + v_{180}) - (v_1' + v_2' + \cdots + v_{180}')  \qquad (5.7)$$

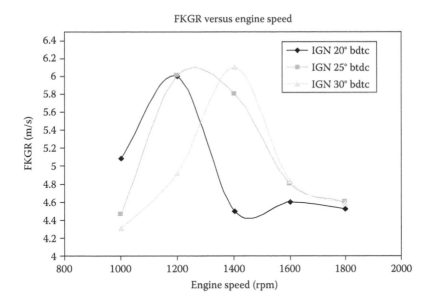

**FIGURE 5.7**
FKGR versus engine speed.

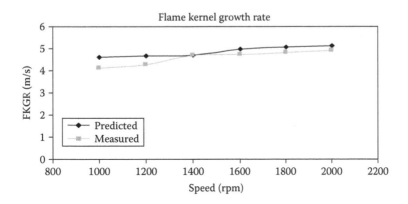

**FIGURE 5.8**
FKGR versus speed.

where

$v_1, v_2, \ldots$ = velocity in the forward direction
$v_1', v_2', \ldots$ = velocity in the backward direction

A top view of a typical combustion chamber is shown in Figure 5.10. The spark plug is located centrally, whereas the flame is progressively moving from the right to the left direction, and the velocity component in the right direction is very much higher as compared to the left direction. Owing to

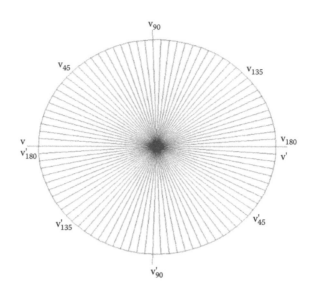

**FIGURE 5.9**
Mathematical presentation of drift velocity.

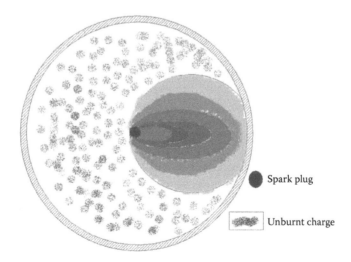

**FIGURE 5.10**
Visualisation of knocking tendency in a combustion chamber.

this effect, the accumulated charge on the left side of the chamber will not be able to burn completely. Hence, the possibility of knocking is more. The dotted black portion shows the unburnt charge.

The resultant of these components is defined as drift velocity. If drift velocity is minimum or approximately zero, then there is less chance of knocking as shown in Figure 5.11.

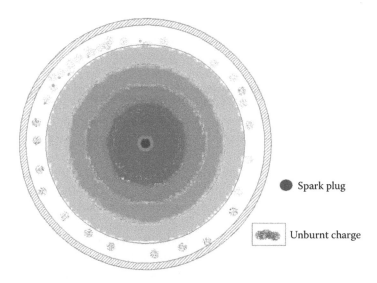

Spark plug

Unburnt charge

**FIGURE 5.11**
Best possible combustion with a minimum drift velocity.

It is clearly seen from the above analysis that the larger drift velocity may have the possibility of a knocking tendency. The drift velocity could be highly useful in the design of the combustion chamber and the location of the spark plug for any SI engine. Such studies should also be carried out for engines fuelled with alternative fuels.

For the observation of flame propagation and local self-ignition centres in an SI engine combustion, optical sensor elements and sensing techniques are used. In these systems, we integrate micro-optic elements into sensor units for the detection of flame radiation in the SI engine. The arrangement of sensor elements in a multi-channel configuration extends the simple flame radiation information into spatial information about flame properties such as the flame kernel growth and flame front propagation.

The AVL visiofem is an optical measuring system for the acquisition and visualisation of flame luminosity in gas and gasoline engines. It allows up to four light signals, decoupled from the combustion chamber by fibre-optic sensors, to be digitised with crank angle resolution.

The AVL visiofem output signals are connected to any indicating system. The AVL visiofem is controlled by the AVL indicom indicating software or by PC software.

The optical sensors that can be used with the AVL visiofem register the flame luminosity from the compression volume of the combustion chamber by means of micro-optic elements fitted in spark plugs.

The AVL visiofem sensor is a spark plug sensor that enables the user to investigate the formation of the flame kernel by means of eight cones of vision. Together, the two sensors permit the assessment of the curve of the flame luminosity in the combustion chamber.

Typical application areas for the AVL visiofem are:

*Idle stability*: Investigation of diffusion combustion that occurs during both injections into the intake manifold and direct injection into the combustion chamber due, for example, to fuel deposits on the inlet valves or piston surface as a result of poor-quality mixture preparation.

*Transient behaviour of the combustion changes during cold start*: The device (Figure 5.12) converts the light signals acquired by the sensor into analogue voltage signals for evaluation by an indicating device.

Light signals are recorded using photodiodes. The required spectral sensitivity is achieved by means of optical spectral filters. The measurement signals are launched into the photodiodes via fibre-optic cables. The photodiodes deliver current signals that are proportional to the luminous power and are converted into voltage signals by amplifiers. The analogue to digital converter then converts the voltage into digital signals.

*Optical connecting line*: The light signals that are acquired with a visiolution sensor are applied to the detectors (photodiodes) of the visiofem via glass-fibre connecting lines.

The AVL visiofem spark plug sensor (Figures 5.13 and 5.14) is used to observe the formation of the flame kernel shortly after ignition in the combustion chamber of gasoline engines. The sensor elements are integrated in a spark plug for the purpose.

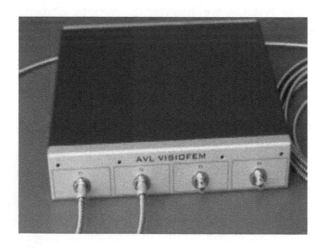

**FIGURE 5.12**
AVL visiofem.

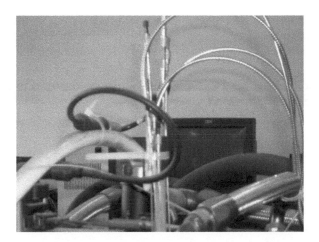

**FIGURE 5.13**
Sensor elements integrated in a spark plug.

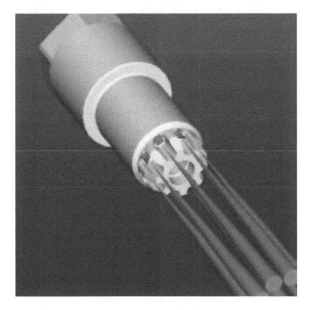

**FIGURE 5.14**
Spark plug sensor.

The visiofem sensor element consists of a micro-optic front-end element and an optic fibre connected to it. The optic fibres terminate in optic fibre connectors. The micro-optic front-end elements are 1.5 mm in diameter. The aperture angle of the cone of vision is ~14°.

When installing the AVL sensor-integrated spark plug in the engine, the data evaluation system needs to know the orientation that is determined based on the definition of the optical sensors' reference point:

1. The visiolution spark plug sensor can only be inserted in the spark plug in a certain position.
2. The reference point for the visiolution spark plug sensor with one ground electrode is the direction of the electrode.
3. The reference point for sensors with several ground electrodes is the direction of the one next to the shortest steel capillary tube.
4. The sensor that immediately follows the reference point in a clockwise direction is number 1 as identified by the shortest steel capillary tube.
5. The length of the steel capillary tube of a sensor group corresponds to the numbering, that is, number 1 has the shortest tube and number 8 the longest.

Figure 5.15 shows the drift velocity variation with respect to spark timing for different engine speeds.

It can be observed from Figure 5.15 that for all speeds at MBT conditions, the drift velocity is minimum, which interprets that at MBT condition, the tendency of knocking is lowest. It was also observed that if the engine speed is increased, the drift velocity increases, resulting in a knocking tendency. Drift velocity could be helpful to understand the knocking phenomenon for any fuel in an SI engine.

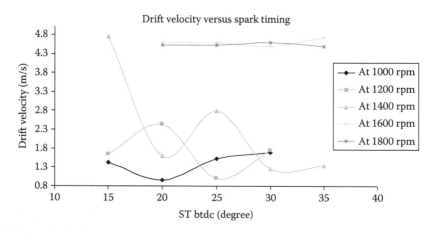

**FIGURE 5.15**
Drift velocity versus spark timing.

*Performance and emission parameters*: Performance and emission parameters were observed for the best and worst values of flame kernel growth. Results for different speeds are compared as shown in Figures 5.16 through 5.18.

As can be seen from the plots of CO and HC, the variations are negligible between the best and worst values of FKGR. This can be attributed to the fact that the values obtained for these emissions are from very early stages of flame development, and as such the net impact of an improved FKGR over the complete combustion range on the emissions is difficult to examine.

$NO_x$ is increased for some cases and decreased for other cases. $NO_x$ emissions are lower when combustion duration is high (lower value of FKGR). This is because it allows more time for the combustion to complete. Further, at a higher combustion duration, the peak temperature is low and therefore

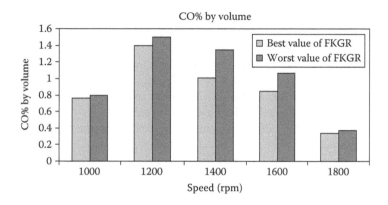

**FIGURE 5.16**
Effect of FKGR on the CO level.

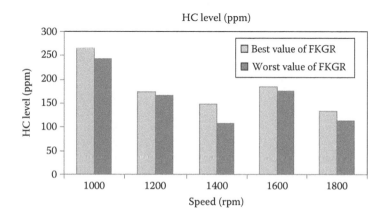

**FIGURE 5.17**
Effect of FKGR on HC level.

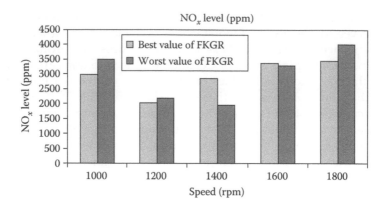

**FIGURE 5.18**
Effect of FKGR on $NO_x$ level.

the formation of $NO_x$ is reduced. On the other hand, decreasing the combustion duration (increasing the value of FKGR) beyond a certain limit reduces the concentration of $NO_x$ because of the lesser time for the exposure of products of combustion to the cylinder's peak temperature.

*Performance parameters*: Brake thermal efficiency calculated at the highest and lowest flame kernel growth speed is shown in Figure 5.19.

For all cases, brake thermal efficiency is increased for the best value of flame kernel growth speed. This is possibly due to the high flame kernel growth rate resulting in less combustion duration, which results in a near to constant volume combustion. This causes less heat losses during the expansion stroke, and thus most of the fuel energy is obtained as power rather than as losses.

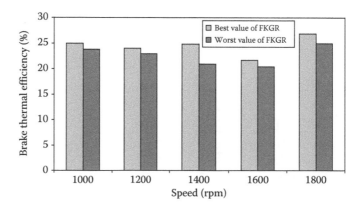

**FIGURE 5.19**
Effect of FKGR on brake thermal efficiency.

### 5.1.7 Carburettor, Manifold, Port and Direct In-Cylinder Injection

SI engines operate with four strokes, namely suction, compression, expansion and exhaust. Air and fuel are inducted during the suction stroke and the charge is compressed during the compression stroke. Spark is given near the end of the compression stroke for initiating ignition and combustion. The chemical energy of the fuel is converted into heat energy by combustion and it is then converted into mechanical shaft power by the reciprocating motion of the piston. The burnt products are expelled into the atmosphere during the exhaust stroke.

*Carburetion*: The mixing of fuel with air takes place in an intake manifold of the engine during the suction stroke using carburetion. A simple device called the carburettor does the mixing of fuel with air. However, it does not give the correct quantity of fuel with respect to load. This will result in higher emission levels and reduced fuel economy. In addition to this, some amount of air is replaced by fuel, which results in a reduction of the volumetric efficiency, which will further reduce the power output of the engine.

*Manifold injection*: The fuel is injected using the timed manifold injection system for accurate metered fuel induction with respect to the load. The fuel is only injected during the suction stroke. The closed-loop feedback control strategy is used for controlling the gasoline injector operation. The signals such as lambda, manifold absolute pressure (MAP), engine coolant temperature and lubricating oil temperature are given as input to the control system for supplying the correct amount of fuel injection.

*Port injection*: A solenoid injector is placed in the intake port and the direction of fuel spray is towards the rear surface of the intake valve. The fuel is injected through the port into the cylinder. Both the manifold injection and port injection improve the fuel economy of the vehicles and reduce emissions. There is still a problem of volumetric efficiency reduction.

*Direct injection*: The volumetric efficiency will be improved significantly using direct injection as it is like a CI engine. The injector is mounted on the engine cylinder head and the injection pressure should be higher than the in-cylinder pressure. The fuel is injected during the compression stroke so that volumetric efficiency improves significantly, which enhances the power, torque output and fuel economy of a vehicle. The maximum number of vehicles around the world uses either manifold injection or port injection, and the carburettor system is in the process of elimination due to its inability to meet the emission standards. The direct injection method is in the developmental stage, although a few of them have been commercially developed.

### 5.1.8 Different Methods That Can Be Adapted for Using CNG in SI Engines

Natural gas is mixed with air in the carburettor before allowing it to go inside the cylinder. Then the natural gas/air mixture is inducted into the cylinder, just like an air–petrol mixture in an SI engine, as shown in Figure 5.20. It is quite difficult to achieve a very accurate carburetion system that is dependent on the engine operating conditions. Despite all advancements, the carburetion system continues to fail to cater to the requirement under all operating conditions. Hence, injection techniques are gaining popularity and wider acceptance [7].

1. Manifold injection: The injector is located in the manifold to inject the fuel upstream of the air for better mixing of fuel and air, as shown in Figure 5.21. The manifold injection gives more time for the fuel to get mixed with the air properly before entering into the combustion chamber. The system is used when the fuel is very sensitive to heat changes such as backfiring and so on. The manifold injection is extremely relevant in view of the backfiring problem being encountered while using CNG. It promotes better mixing and the formation of appropriate mixture strength. However, in the case of a manifold injection, it causes a reduction in power as natural gas reduces the amount of air which leads to a lower power output. Further, an engine filled with the CNG fuel delivers lesser power output as compared to gasoline fuel due to the slow burning characteristic and low volumetric heat content of CNG fuel [8].

2. Normal induction

3. Direct injection: Natural gas is compressed to a high pressure of about 20 MPa and injected directly into the cylinder through a high-pressure injector during the start of the compression stroke. As shown in Figure 5.22, the direct injection of CNG into the cylinder of an SI engine after air is inducted and the intake valve is closed can

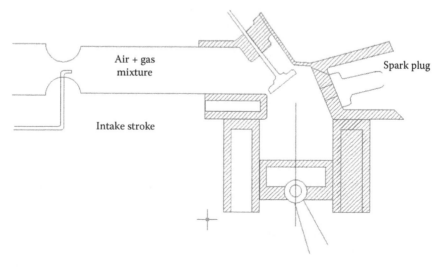

**FIGURE 5.20**
A typical diagram of a gas induction system in an SI engine.

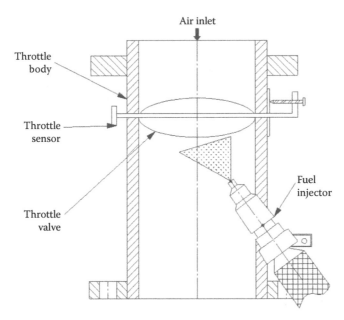

**FIGURE 5.21**
A schematic diagram of manifold injection.

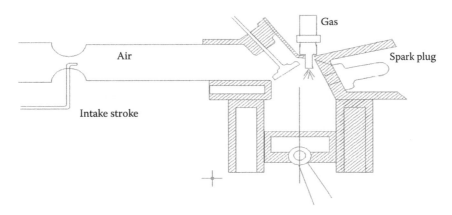

**FIGURE 5.22**
Schematic diagram of the direct injection system in an SI engine.

solve the problem of lower power output. However, such a system will have cost implications and the injector has to withstand high pressure and temperature in the cylinder. The use of the solenoid injector in direct injection creates considerable problems because the injector has to withstand the severity of the thermal environment and the force requirements are very high due to the strict controlled injection timing and the requirement for faster opening. Owing to working at high pressure, there can be a leakage problem as well. Further, such an injection system is in its preliminary stage of research and development (R&D). A few manufacturers have opted for it on a commercial scale. Large-scale commercial deployment may not be possible in the near future [9].

4. Port injection system: In the port injection system, the fuel is injected into the inlet valve port of each cylinder through an injector placed close to the upstream side of the intake valve. The main advantage of port injection is increased power and torque through improved volumetric efficiency. Other advantages include uniform fuel distribution, rapid response to changes in throttle position and precise control of the A/F ratio during cold start and engine warm-up. The fuel injection can be timed or continuous injection [6].

## 5.1.9  Continuous Injection

There are several ways of actuating the injectors, such as hydraulic, pneumatic and electromagnetic (solenoid operated). Solenoid-operated injectors are mainly being used. Hydraulic injectors have an advantage of low noise.

The solenoid-operated injectors can be based on static flow or dynamic flow. In static flow, the armature is kept attracted by the coil and the injector delivers the flow. This is related to the maximum power output of a particular engine. In dynamic flow, the injector is pulsed with a given pulse width, resulting in a flow, and it is important to determine the lowest load that can be applied to a particular engine. The accuracy of injectors is specified in terms of tolerances for static and dynamic flows to enhance the fast opening of the injector. An injector driver is a very sophisticated design allowing a very fast rise of current in the beginning to overcome the magnetic inertia. Subsequently, the current is reduced to minimise the heat build-up in the injector. In this way, it is possible to lower the pulse width, whereby the dynamic range decreases.

## 5.1.10  Timed Manifold Injection

Fuel is injected at regular intervals depending on the position of the piston. Injection takes place when the inlet valve opens and it ceases before the inlet valve closes. In a multi-cylinder engine, a separate injector is needed for each cylinder. Fuel is injected into each cylinder depending on the valve timing of that cylinder.

The performance and emission characteristics of SI engines fuelled by an alternative fuel are explained in the later part of this chapter. The IC engine parameters are optimised for base fuel such as gasoline or diesel. The majority of vehicles in the world have gasoline-fuelled SI engines and diesel-fuelled CI engines. Dedicated vehicles for the use of alternative fuels such as E15, E85, M15, M85 and B20 are very less. In many countries, utilisation of alternative fuels in IC engines is in the R&D stage.

The vehicle's design and operating parameters need to be optimised for better performance and emission characteristics of IC engines to utilise the alternative fuels more effectively. Flame speed also plays an important role.

## 5.1.11 Exhaust Gas Recirculation

The most effective way of reducing $NO_x$ emissions is to hold the combustion chamber temperature down. Although practical, this is a method in which the thermal efficiency of the engine gets reduced. We know that to obtain maximum engine thermal efficiency, it should be operated at the highest possible temperature. Probably the simplest and most practical method of reducing maximum flame temperature is to dilute the air–fuel mixture with a non-reacting parasite gas. This gas absorbs energy during combustion without contributing any energy input. The net result is a lower flame temperature. Any non-reacting gas would work as a diluent. Gases with larger specific heat absorb the most energy per unit mass and would therefore be required in the least amount.

Adding any non-reacting neutral gas to the inlet air–fuel mixture reduces the flame temperature and hence reduces $NO_x$ generation; exhaust gas recirculation (EGR) is one gas that is readily available for engine use. Exhaust gas recycling is done by ducting some of the exhaust flow back into the intake system. The amount of flow can be high as 30% of the total intake. EGR combines with the exhaust residual left in the cylinder from the previous cycle to effectively reduce the maximum combustion temperature. The flow rate of EGR is controlled by the engine management computer-controlled system or digital manometer or rotameter.

### 5.1.11.1 Oxides of Nitrogen

The exhaust gases of an engine can have up to 2000 ppm of oxides of nitrogen. Most of this will be nitrogen oxide (NO), with a small amount of nitrogen dioxide ($NO_2$). There will be other traces of other nitrogen–oxygen combinations. These are all grouped together as $NO_x$ with $x$ representing some suitable number. $NO_x$ is very undesirable; regulations to reduce $NO_x$ emission continue to become more and more stringent year by year. The released $NO_x$ reacts in the atmosphere to form ozone and is one of the major causes of photochemical smog.

$NO_x$ is formed mostly from nitrogen in the air. Nitrogen can also be found in fuel blends. Further, fuel may contain traces of $NH_3$, NC and HCN, but these would contribute only to a minor degree. There are a number of possible reactions that form NO. All the restrictions are probably occurring during the combustion process and immediately thereafter. These include but are not limited to

$$O + N_2 \rightarrow NO + N \tag{5.8}$$

$$N + O_2 \rightarrow NO + O \tag{5.9}$$

$$N + OH \rightarrow NO + H \tag{5.10}$$

NO, in turn, can further react to form $NO_2$ by various means, including

$$NO + H_2O \rightarrow NO_2 + H_2 \tag{5.11}$$

$$NO + O_2 \rightarrow NO_2 + O_2 \tag{5.12}$$

At low temperatures, atmospheric nitrogen exists as a stable diatomic molecule. Therefore, only very small trace amounts of nitrogen are found. However, at high temperature that occurs in the combustion chamber of the engine, some diatomic nitrogen ($N_2$) breaks down to monatomic nitrogen (N), which is reactive:

$$N_2 \rightarrow 2N \tag{5.13}$$

It may be noted that the chemical equilibrium constant for the above equation is highly dependent on temperature. A significant amount of nitrogen is generated in the temperature range of 2500–3000 K, which can exist in an engine. Other gases that are stable at low temperatures but become reactive and contribute to the formation of $NO_x$ at high temperatures, including oxygen and water vapour, and their breakdown are as follows:

$$O_2 \rightarrow 2O \tag{5.14}$$

$$H_2O \rightarrow OH + 1/2H_2 \tag{5.15}$$

If one goes a little deeper into combustion chemistry, it can be understood that chemical equations 5.8 through 5.15 all react much further to the right as high combustion chamber temperatures are reached. The higher the combustion reaction temperature, the more the diatomic nitrogen, $N_2$, will dissociate with monatomic nitrogen, N, and the more the $NO_x$ that will be formed. At a low temperature, very little $NO_x$ is created [10].

The performance and emission characteristics of SI engines for alternative fuels such as the methanol–gasoline blend, ethanol–gasoline blend, LPG, CNG and hydrogen are explained below.

### 5.1.12 Ethanol–Gasoline Blends

As ethanol is an octane number booster, it enhances the octane number of gasoline. A typical experimental set-up for the assessment of the performance, combustion and emission characteristics of an engine for different percentages of ethanol–gasoline blends is shown in Figure 5.23. An eddy current dynamometer was connected with an AVL research engine for loading the engine. A piezoelectric transducer was mounted on the cylinder head for the measurement of in-cylinder gas pressure. The charge of the cylinder pressure transducer was converted into an analogue signal using a charge amplifier and it was then converted to a digital signal using the data acquisition system for post-processing of combustion characteristics. A crank angle encoder was mounted on the engine shaft for the measurement of crank angle for every cycle. A solenoid injector was mounted in the intake manifold of the engine for injecting gasoline or ethanol–gasoline blend during the suction stroke. A schematic layout of an experimental test rig is shown in Figure 5.23. A view of an experimental set-up with an AVL research engine is shown in Figure 5.24. Ethanol–gasoline can be blended off-line (premixed with desired quantity of ethanol) or on-line as shown in Figure 5.25.

There is no significant variation in the torque with ethanol–gasoline blends of E10 and E30 as compared to base gasoline. However, the torque decreases significantly with E70 as shown in Figure 5.26a. It is known that torque is a main function of the brake mean effective pressure (BMEP) and swept volume, and thus torque decreases as indicated mean effective pressure (IMEP) decreases with E70. BMEP is a function of the brake thermal efficiency, volumetric efficiency, fuel–air ratio, density of air and calorific value. BSEC is the least for E70 but torque is less as shown in Figure 5.26b. The fall

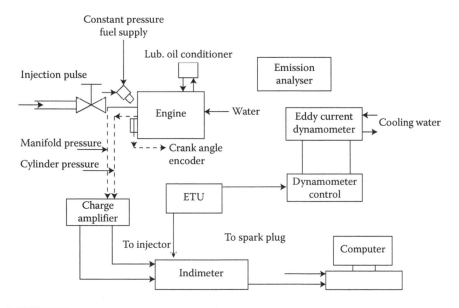

**FIGURE 5.23**
Schematic layout of an experimental test rig. (Adapted from A. Kumar, M. K. Gajendra Babu and D. S. Khatri, *Journal of SAE Special Publications*, SAE Paper number 2009-01-0137, SP-2241, 2009.)

**FIGURE 5.24**
Experimental set-up with an AVL research engine.

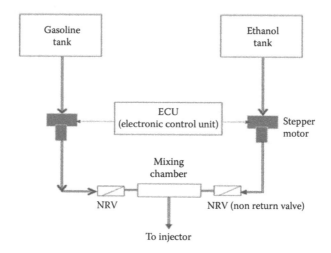

**FIGURE 5.25**
Schematic diagram of an on-line fuel blending system.

in BSEC at E70 is attributed to the improved combustion. The CO emission also decreases significantly to support the reduction in BSEC. IMEP is the highest at E30 as shown in Figure 5.27.

The engine torque improved to the level of base gasoline by optimising engine operating parameters such as the start of injection, duration of injection, injection pressure and spark timing, as illustrated in Figure 5.28 [7].

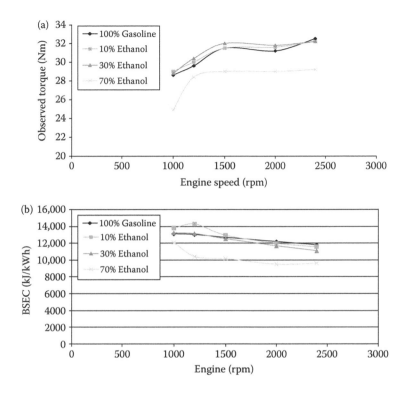

**FIGURE 5.26**
(a) Variation in torque with different ethanol–gasoline blends. (b) Brake-specific energy comparison among gasoline–ethanol blends. (Adapted from A. Kumar, M. K. Gajendra Babu and D. S. Khatri, *Journal of SAE Special Publications*, SAE Paper number 2009-01-0137, SP-2241, 2009.)

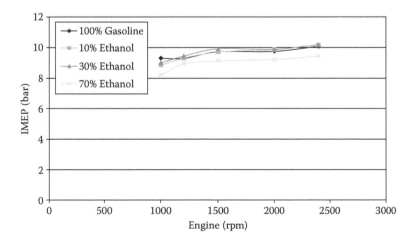

**FIGURE 5.27**
IMEP comparison among gasoline–ethanol blends.

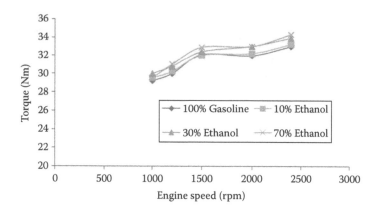

**FIGURE 5.28**
Comparison of WOT performance after optimisation.

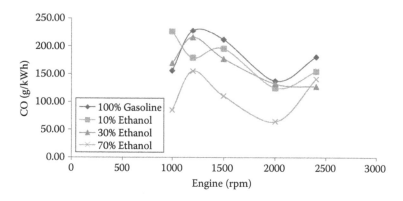

**FIGURE 5.29**
Comparison of WOT emission (CO) after optimisation.

CO emission falls appreciably with E70 as depicted in Figure 5.29, which is primarily due to the presence of oxygen in ethanol. This helps the oxidation of CO into $CO_2$ during combustion. There is no significant change in HC emission as shown in Figure 5.30.

Figure 5.31 shows the elevated level of $NO_x$ emission when the ethanol content is increased. $NO_x$ is a strong function of temperature, oxygen concentration and residence time. Air containing nitrogen reacts at high temperature, resulting in $NO_x$ formation. As ethanol containing oxygen leads to high $NO_x$ emission levels, they need to be reduced using a suitable technology such as EGR.

### 5.1.13 Methanol–Gasoline Blends

Methanol fuel has a greater resistance to knocking. As it is an oxygenated fuel, its emissions are less than that of neat gasoline. The performance can

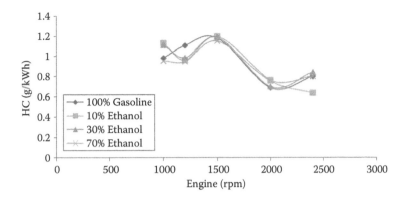

**FIGURE 5.30**
Comparison of WOT emissions (HC) after optimisation.

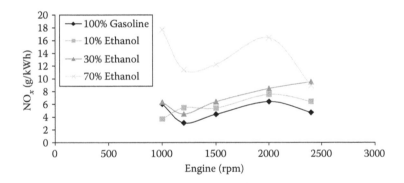

**FIGURE 5.31**
Comparison of WOT emissions ($NO_x$) after optimisation.

be increased with an increase in the compression ratio of an engine. The compression ratio could be increased up to 10:1 without knocking during combustion [11]. Figure 5.32 shows the effect of methanol fuels on brake thermal efficiency at several compression ratios. The value of maximum brake thermal efficiency obtained with gasoline is 19.2% at a CR of 6:1. The values of brake thermal efficiency obtained with methanol are about 22.1%, 27.3% and 30.3% at CRs of 6:1, 8:1 and 10:1, respectively.

Figure 5.33 shows the effect of methanol on HC emission at various compression ratios. HC emission rises with an increase in the compression ratio of an engine for methanol fuel. The methanol fuel operation reduces the cylinder gas temperature as the heat of vapourisation of methanol is higher (about 3.0 times) than that of gasoline, which results in the reduced oxidation rate of fuel. The increase in the surface area to the volume of the combustion chamber is also a reason for higher HC emission. The increase in the compression ratio increases the combustion chamber surface/volume ratio.

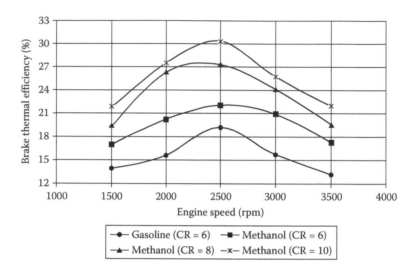

**FIGURE 5.32**
The effects of gasoline and neat methanol fuels on brake thermal efficiency at various compression ratios. (Adapted from M. Bahattin Celik, B. Ozdalyan and F. Alkan, *Fuel* 90, 1591–1598, 2011.)

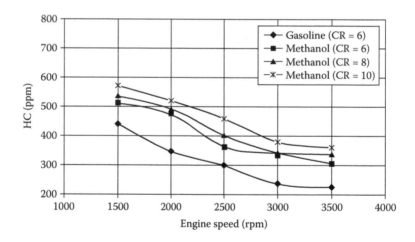

**FIGURE 5.33**
The effects of gasoline and methanol fuels on HC at various compression ratios.

The effect of methanol operation on $NO_x$ at various compression ratios is given in Figure 5.34. At all the compression ratios with methanol, $NO_x$ is lower than that of gasoline. Since methanol has a higher heat of vaporisation than gasoline, the mixture temperature at the end of the compression stroke decreases. Hence, the combustion temperature decreases, resulting in the reduced formation of $NO_x$ emission. As the compression ratio increases, the combustion temperature increases, leading to a higher $NO_x$ formation.

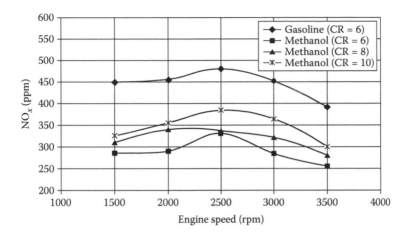

**FIGURE 5.34**
The effects of gasoline and methanol fuels on $NO_x$ at various compression ratios.

Figure 5.35 shows the effect of methanol fuel on CO emission at different compression ratios. CO emission shows a lower level with methanol than with base gasoline. It reduces with an increase in engine speed. The mass flow rate of fuel increases with an increase in engine speed. In cylinder gas, the temperature increases with load; this results in lower CO emissions.

The effect of methanol on $CO_2$ at various CRs is given in Figure 5.36. Carbon dioxide is non-toxic but it contributes to the greenhouse effect. At all compression ratios with methanol operation, the $CO_2$ levels are lower than the levels of base gasoline.

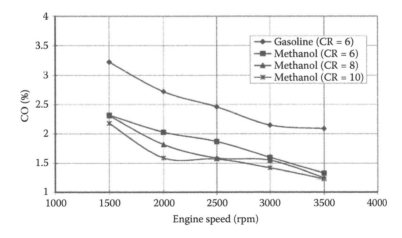

**FIGURE 5.35**
Effects of gasoline and methanol fuels on CO at various compression ratios.

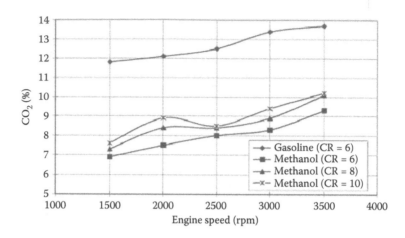

**FIGURE 5.36**
Effects of gasoline and methanol fuels on $CO_2$ at various compression ratios. (Adapted from J. B. Heywood, *Internal Combustion Engine Fundamentals*, International Edition, McGraw-Hill Book Company, 1989.)

### 5.1.14 Butanol–Gasoline Blends

Butanol is an oxygenated fuel that can be blended with gasoline in any proportion. It could be a supplementary fuel to gasoline. Figure 5.37 shows the effect of preheating inlet air on brake-specific fuel consumption (BSFC) at different fuel/air equivalence ratios. The results show that as the inlet air temperature increases, BSFC increases [12].

Figure 5.38 shows the influence of the equivalence ratio on different butanol–gasoline blends. Blends with 40% butanol or less show similar HC emissions than gasoline. However, it was possible to operate the engine with a slightly leaner mixture with B40, compared to the leanest gasoline mixture. For a stoichiometric mixture, B0, B20 and B40 yielded similar UHC emissions [13]. Further, increasing the butanol content brings the emission levels to above that of gasoline with an 18% increase in the case of B60, while B80 shows a 47% increase in UHC emissions compared to gasoline. If butanol is used in a conventional SI engine that is optimised for gasoline fuel, BSFC increases. However, it can be reduced by optimising the design and operating parameters. Increasing the compression ratio is also an effective method to decrease BSFC.

It is observed from Figure 5.39 that for a stoichiometric mixture, B20, B40 and B60 generate lower CO emissions than pure gasoline. B80, on the other hand, produces the same level of CO emissions as compared to gasoline for stoichiometric mixtures and it gives the highest emission levels at all other equivalence ratios. For lean mixtures, CO emissions generally increase as does the butanol concentration. The complete CO oxidation is more difficult with butanol fuel. However, it is also observed that for a given butanol concentration, CO emissions are relatively constant for lean mixtures. This

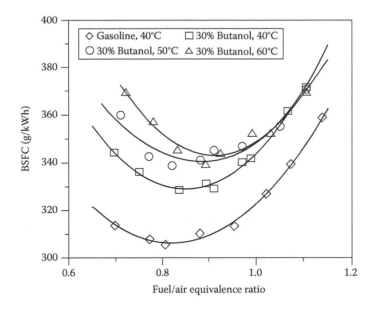

**FIGURE 5.37**
The effect of inlet air temperature on BSFC. (Adapted from F. N. Alasfour, *Applied Thermal Engineering* 18, 245–256, 1998.)

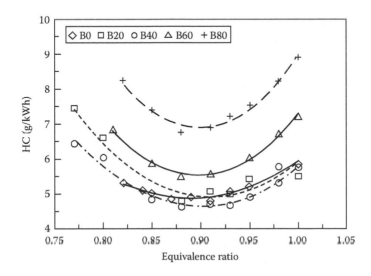

**FIGURE 5.38**
Influence of equivalence ratio on UHC for different gasoline–butanol blends. (Adapted from H. Willems and R. Sierens, *Journal of Engineering for Gas Turbines and Power, ASME Transactions*, 125(2), 479–484, 2003.)

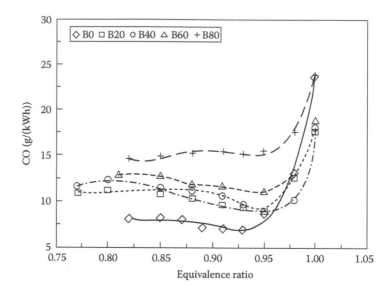

**FIGURE 5.39**

Carbon monoxide emissions as a function of equivalence ratio and blends. (Adapted from J. Dernotte et al., *Oil and Gas Science and Technology—Rev. IFP*, 65, 345–351, 2010.)

behaviour of CO is due to the fact that the equivalence ratio controls CO emissions until a lean mixture is reached.

Figure 5.40 shows the variation of $NO_x$ emission according to the butanol content of the gasoline. When the concentration is less than 60%, there is no significant change in $NO_x$ emissions for stoichiometric mixtures. A slight decrease of $NO_x$ emissions, however, can be observed when the mixture's equivalence ratio is less than or equal to 0.85 for blends with a butanol concentration higher than 40%. On the other hand, B80 offers a noticeable difference in $NO_x$ emissions for all equivalence ratios. This can be linked to the incomplete combustion as measured by the increase in UHC emissions presented in Figure 5.38. These higher UHC emissions for B80 have resulted in a lower heat release and, therefore, a lower gas temperature.

### 5.1.15 Hydrogen

As hydrogen is one of the potential fuels that does not have carbon, the formation of hydrocarbon, carbon monoxide and carbon dioxide during combustion can be completely avoided; however, traces of these compounds may be formed due to the partial burning of lubricating oil in the combustion chamber. When hydrogen is burnt, hydrogen combustion does not produce toxic products such as hydrocarbons, carbon monoxide, oxide of sulphur, organic acids or carbon dioxide. Instead, its main product is water. Hence, hydrogen can be considered as a potential alternative fuel.

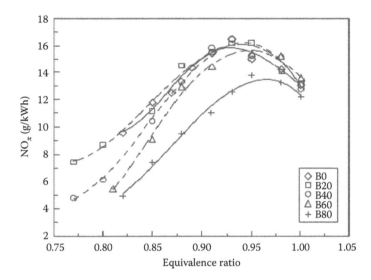

**FIGURE 5.40**
Influence of butanol addition and equivalence ratio on specific $NO_x$ emission.

Hydrogen has advantageous properties such as a high flame speed, short quenching distance, high heating value and high diffusivity. The burning velocity is so high that a very rapid combustion process can be achieved. The limits of flammability of hydrogen vary from an equivalence ratio ($\phi$) of 0.1 to 7.1, Hence, the engine can be operated over a wide range of air/fuel ratio. Hydrogen as a fuel for the IC engine is best suited to the SI mode due to its high self-ignition temperature.

Figure 5.41 shows that there is an increase in the brake thermal efficiency with hydrogen because of its rapid burning due to which the combustion takes place almost near constant volume conditions. Also, with hydrogen, the engine runs on a very lean mixture, which enhances the thermal efficiency. Also, it is seen that a variation in the hydrogen supply pressure has little effect on the brake thermal efficiency as expected. The part-load efficiency with hydrogen is higher. The lower throttling losses are expected to have played a significant role. However, the maximum power output gets enhanced with the higher injection pressure.

Figure 5.42 shows that BSFC with hydrogen is much lower than that with gasoline. At the optimal thermal efficiency point, it is around 0.32 kg/kWh for gasoline, and for hydrogen it is around 0.11 kg/kWh. This is attributed mainly to the increased thermal efficiency and higher calorific value of hydrogen.

Figure 5.43 shows the variation of HC emission with brake power for both gasoline and hydrogen through manifold injection. The hydrocarbon levels present in the exhaust gas of the gasoline were in the range of 2200–4200 ppm. The hydrocarbon level present in the exhaust gas with hydrogen was less than 200 ppm. Even though there is no carbon present in the fuel,

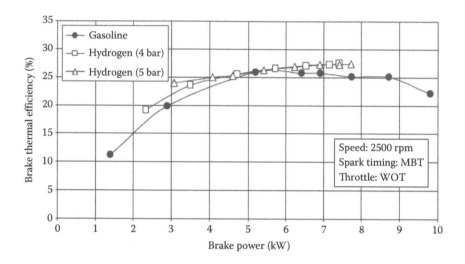

**FIGURE 5.41**
Variation of brake thermal efficiency with brake power for gasoline and hydrogen engine. (Adapted from R. Hari Ganesh et al., *Renewable Energy* 33, 1324–1333, 2008.)

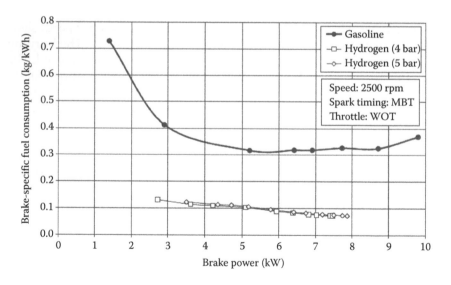

**FIGURE 5.42**
Variation of BSFC with brake power for gasoline and hydrogen engine. (Adapted from R. Hari Ganesh et al., *Renewable Energy* 33, 1324–1333, 2008.)

hydrocarbons were present in the exhaust gas due to the evaporation of the lubricating oil.

From Figure 5.44, it is observed that up to about 4.5kW (at equivalence ratio = 0.5), $NO_x$ emissions were negligible (almost zero). This is due to the fact that the amount of air present was high, which brings down the bulk

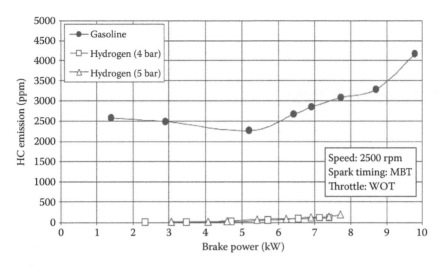

**FIGURE 5.43**
Variation of HC emission with brake power for gasoline and hydrogen engine. (Adapted from R. Hari Ganesh et al., *Renewable Energy* 33, 1324–1333, 2008.)

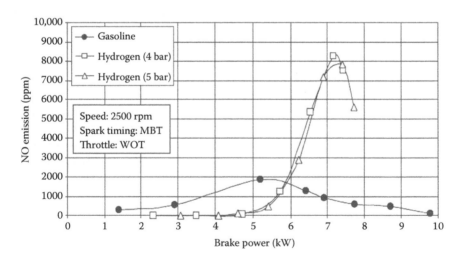

**FIGURE 5.44**
Variation of NO with brake power for gasoline and hydrogen engine.

temperature and, in turn, reduces the overall combustion temperature. When the equivalence ratio was increased, the temperature of the burnt gases went up. The combustion rate had also increased substantially, which led to an increase in $NO_x$. The availability of excess oxygen also supported the NO formation. The NO concentration reached a peak value of around 8000 ppm at a brake power of around 7kW (at equivalence ratio = 0.85). Further, an increase of brake power through the equivalence ratio reduces the $NO_x$ emission. This is due to the lack of oxygen supply with reduced air consumption, which can be verified with the extremely low exhaust oxygen concentration.

### 5.1.16 Compressed Natural Gas

Natural gas is a fossil fuel primarily comprising methane (87–94%). It is one of the cleanest burning alternative fuels. It can be used in the form of

1. Compressed natural gas
2. Liquefied natural gas

CNG is a natural gas under pressure that remains clear, odourless, non-poisonous and non-corrosive. CNG is more suitable for an SI engine due to its high octane number (ON > 120). Gaseous fuels have higher diffusivity resulting in a better homogeneous charge due to the better mixing of air and fuel.

#### 5.1.16.1 Difference in Performance and Power Output between CNG and Gasoline

It has been experienced that a gasoline engine will produce less power on CNG than on gasoline and motorists may feel that a car's performance on CNG is not satisfactory.

It is necessary to consider three separate aspects that can contribute to the performance and power difference:

1. *Fuel-related factors*: The space occupied in the engine cylinder by CNG is greater than that by gasoline, with the result that the fuel–air mixture inside the engine chamber contains 8–10% less air in the CNG mode. Less oxygen means less fuel burnt and hence less power produced. Since CNG burns more slowly than petrol, it produces less power and slower idle speed for the same ignition timing and throttle position. The difference in the burning rate can be partially compensated by advancing the ignition timing for CNG. Idling speed also needs to be increased in the CNG mode to prevent engine stalling. Natural gas also requires much higher ignition temperatures to burn in the combustion cylinder. The activation energy for initiating ignition for CNG fuel is higher than gasoline resulting to larger ignition lag. In addition to this, specific energy release during combustion would be lesser for CNG even with the optimized ignition timing and improved volumetric efficiency [14].
2. *Gasoline/engine-related factors*: The power output also depends on the individual engine design features, particularly in the engine combustion chamber, valve timing, inlet manifold heating and positioning of the air cleaner intake. In most gasoline engines, the air cleaner

draws hot air from immediately behind the radiator or in the vicinity of the exhaust manifold. But as the heating of the intake CNG charge will cause a reduction in the density of the charge, for maximum power, the entry of intake air should be nearer to the engine.

3. *CNG conversion system-related factors*: A major factor in the optimal performance of the conversion kit is the quality of installation. Optimisation of the gas mixture and ignition timing and spark strength during installation is necessary to obtain an acceptable power performance.

Figures 5.45 through 5.48 show variation of brake power with respect to speed as compared to that of the motor spirit [15]. It is observed from the graphs that the power developed by the engine using CNG operation on

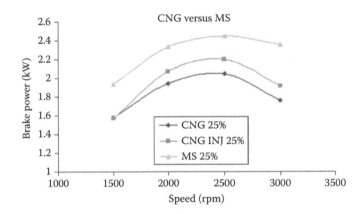

**FIGURE 5.45**
Speed versus brake power at 25% throttle condition.

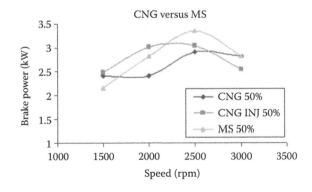

**FIGURE 5.46**
Speed versus brake power at 50% throttle condition.

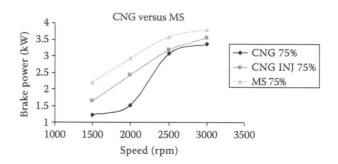

**FIGURE 5.47**
Speed versus brake power at 75% throttle condition.

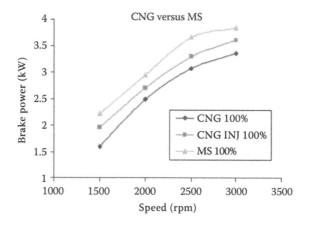

**FIGURE 5.48**
Speed versus brake power at 100% throttle condition.

carburettor mode is less as compared to that on gasoline carburettor. The power loss can be attributed to the displacement of air with CNG, and its slow burning characteristics as shown in graphs. However, when the manifold injection technique was adopted, the power developed improved by a value of 7% on an average and the emissions were much lower.

### 5.1.16.2 Fuel Consumption

Figures 5.49 through 5.51 show the variation of BSFC against speed, brake power and BMEP with respect to BSFC at 25% throttle positions. By using the CNG injection system, fuel consumption is low because fuel and air mixing improves considerably. The injector is controlled by an electronic control unit, which accurately meters the fuel injected and the timing. This leads to a considerable fuel saving, and hence the fuel consumption is reduced. At

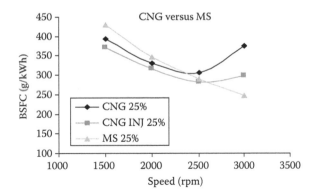

**FIGURE 5.49**
Speed versus brake-specific fuel consumption at 25% throttle condition.

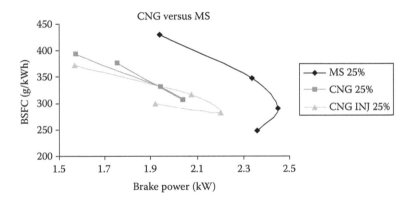

**FIGURE 5.50**
Brake power versus brake-specific fuel consumption at 25% throttle condition.

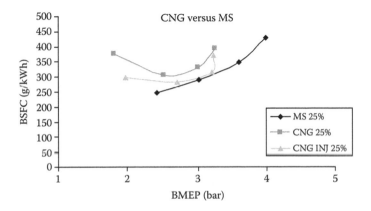

**FIGURE 5.51**
Brake mean effective pressure versus brake-specific fuel consumption at 25% throttle condition.

a speed of 1500 rpm, 1.6 kW brake power and 2.1 bar BMEP, 8.34% fuel was saved by adopting manifold injection.

### 5.1.16.3 Brake Thermal Efficiency

As shown in Figures 5.52 through 5.54, there is an increase in the efficiency of the CNG version as compared to gasoline operation. For the CNG carburettor mode, the thermal efficiency is more than that of gasoline at higher speeds or throttle position. At lower speeds, the CNG supply remains inadequate due to a lower suction pressure. The brake thermal efficiency under manifold injection mode was observed to be maximum. It is observed that under part-load condition, the CNG injection system performance is better than that under the carburettor mode.

At a speed of 2000 rpm, 2.1kW brake power and 2.5 bar BMEP, the brake thermal efficiency increased by 9.52% when manifold injection was adopted.

### 5.1.16.4 Effect of Speed on Emissions

The most remarkable feature using a CNG-fuelled engine is that it is a cleaner and greener fuel, as demonstrated by several investigations. As seen from

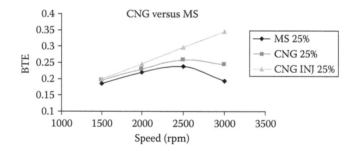

**FIGURE 5.52**
Speed versus brake thermal efficiency at 25% throttle condition.

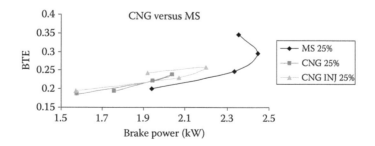

**FIGURE 5.53**
Brake power versus brake thermal efficiency at 25% throttle condition.

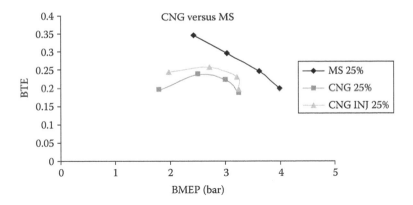

**FIGURE 5.54**
Brake mean effective pressure versus brake thermal efficiency at 25% throttle condition.

Figures 5.55 through 5.57, carbon monoxide, being a product of incomplete combustion, is therefore totally dependent on the air–fuel ratio. Owing to the gaseous nature of CNG, it easily mixes with air because of its diffusivity at high pressure. Hence, it gives lower CO emissions in the CNG carburettor mode and it is further reduced under the CNG injection mode.

At a speed of 1500 rpm, brake power of 1.6 kW and a BMEP of 1.8 bar, a CO reduction of 9% was observed when manifold injection was employed.

Similarly, Figures 5.58 through 5.60 show the variation of hydrocarbon emission with respect to speed, brake power and BMEP. The main source of hydrocarbons is due to the composition and incomplete combustion occurring due to the uneven mixture formation. It is observed from the graphs plotted that the hydrocarbons decrease with increase in speeds. It has been observed that HC emissions under the manifold injection mode are the lowest.

At a speed of 1500 rpm, 1.6 kW brake power and 1.9 bar BMEP, there was 10% reduction in hydrocarbons due to manifold injection.

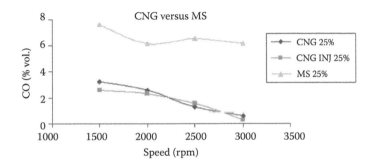

**FIGURE 5.55**
Speed versus carbon monoxide emissions at 25% throttle condition.

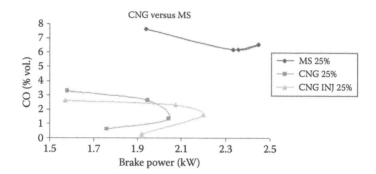

**FIGURE 5.56**
Brake power versus carbon monoxide emissions at 25% throttle condition.

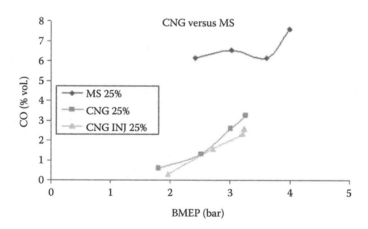

**FIGURE 5.57**
Brake mean effective pressure versus carbon monoxide emissions at 25% throttle condition.

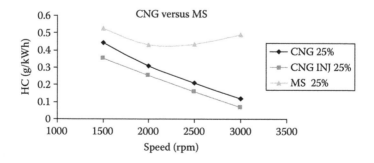

**FIGURE 5.58**
Speed versus hydrocarbon emissions at 25% throttle condition.

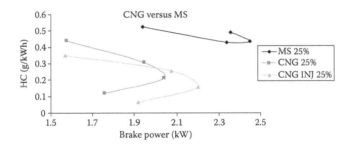

**FIGURE 5.59**
Brake power versus hydrocarbon emissions at 25% throttle condition.

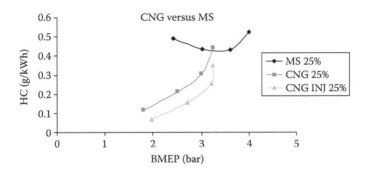

**FIGURE 5.60**
Brake mean effective pressure versus hydrocarbon emissions at 25% throttle condition.

The formation of nitrogen oxide is basically a function of reaction temperature and duration of reaction and availability of oxygen. It can be seen from Figures 5.61 through 5.63 that $NO_x$ formation increases with higher temperatures and greater availability of oxygen. Natural gas has a low flame temperature (2148 K) that tends to reduce the peak $NO_x$ formation.

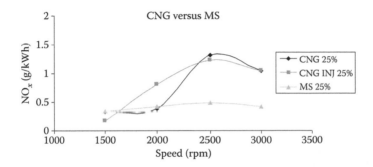

**FIGURE 5.61**
Speed versus nitrogen oxide emissions at 25% throttle condition.

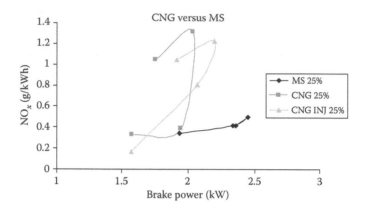

**FIGURE 5.62**
Brake power versus nitrogen oxide emissions at 25% throttle condition.

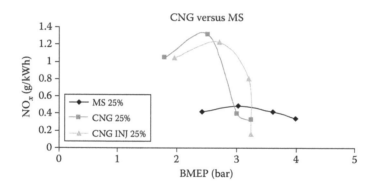

**FIGURE 5.63**
Brake mean effective pressure versus nitrogen oxide emissions at 25% throttle condition.

However, owing to the engine operation near about the stoichiometric value, $NO_x$ emissions increase. The graphs indicate a higher $NO_x$ emission in the case of CNG than gasoline, which is primarily due to the higher cylinder gas temperature.

Figures 5.64 and 5.65 show the variation of carbon dioxide with respect to speed, brake power and BMEP. The higher $CO_2$ under CNG injection mode was due to improved mixing and combustion as compared to the carburetted mode.

The following are the conclusions on the effects of injection characteristics on the performance and emission characteristics of the research engine:

- Power output and BMEP increase with an increase in the injection duration at all loads. Brake power and BMEP increased from 2.95 kW, 7.09 bar with the injection duration of 50° CA to 3.01 kW, 7.24 bar corresponding to an injection duration of 65° CA.

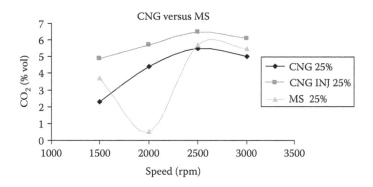

**FIGURE 5.64**
Speed versus carbon dioxide emissions at 25% throttle condition.

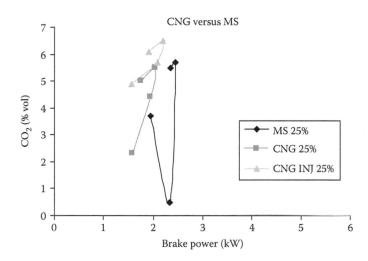

**FIGURE 5.65**
Brake power versus carbon dioxide emissions at 25% throttle condition.

- Reduction in CO and HC was 0.3% when the injection timing was increased from 72° CA to 117° CA, corresponding to the same increase in injection timing.
- Higher injection duration increases $NO_x$ emission at all loads.
- It is observed from the experimental results that there is no significant influence of start of injection on performance and emission characteristics of the engine.

Conclusions on the effects of the use of CNG fuel on performance and emission characteristics of the SI engine using timed manifold injection and the results are compared with base CNG induction and base gasoline in Table 5.3.

**TABLE 5.3**

Effects on the Use of CNG Fuel on Timed Manifold Injection SI Engine

| Parameters | CNG Induction (%) at Throttle Position and rpm | Effects |
|---|---|---|
| Brake thermal efficiency | 30.37% at 2500 rpm and 75% throttle position | 18.6% increase |
| Power | At 2500 rpm due to decrease in volumetric efficiency | Significant drop |
| CO | At 2000 rpm and 75% throttle position | Drastically decreases as compared to gasoline |
| HC | At 2000 rpm and 75% throttle position | Significant decrease by 55 ppm |
| $NO_x$ | At 2000 rpm and 75% throttle position | Significant increase in $NO_x$ emission by 1.7 g/kWh |

The usefulness of the TMI is demonstrated through this study in an SI engine. The TMI method gives significant benefits in performance improvement and emission reduction of any SI engine for the use of CNG as compared to the conventional induction method. However, power drop and an increase in $NO_x$ emission are a major problem in the use of CNG in an SI engine. Further investigations have to be carried out to overcome it.

A typical comparison of a set of experiments on a two-cylinder engine is shown in Figure 5.66. It shows the brake power output at a throttle position of 80%. The brake power of the CNG-operated engine is lower than that of the gasoline version throughout the speed range. Displacement of air by

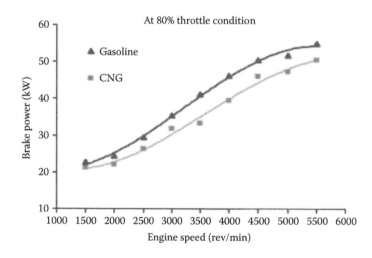

**FIGURE 5.66**

Brake power versus engine speed at 80% throttle condition. (Adapted from M. I. Jahirul et al., *Applied Thermal Engineering* 30, 2219–2226, 2010.)

natural gas and by the slower flame velocity of CNG were the main reasons for the lower brake power as compared to that by gasoline. As a result, the air volumetric efficiency and the charge energy density per injection into the engine cylinder reduced the CNG content. In the case of liquid fuels, it was considered that the fuel did not reduce the amount of air drawn into the cylinder. Hence, a gasoline-fuel-designed engine that was converted to CNG operation would produce a significantly lower peak power.

Figure 5.67 shows the variation of BSFC over the speed range of 1500–5500 rpm. For the engine running on CNG, BSFC was always lower than for that running on gasoline throughout the speed range. This was primarily due to the higher heating value (HHV) and slow burning characteristics of CNG as compared to gasoline. At a low throttle operation, SFC increased at higher engine speeds due to the rapid increase in friction power. SFC rapidly dropped in the lower speed range and nearly levelled off at medium speeds and finally spurted in the high speed range (Figure 5.67). At lower speeds, the heat lost to the combustion chamber walls was proportionately greater, resulting in higher fuel consumption for the power produced. At higher speeds, the frictional power was rapidly increasing, resulting in a slower increase in the brake power than the rate in fuel consumption, with a consequent increase in the BSFC.

Figure 5.68 shows that the HC emission of CNG was lower than that of gasoline throughout the speed range. Poor mixing of air and fuel, local rich regions and incomplete combustion produces CO. Figure 5.69 shows that the CO emission of the CNG-operated engine was significantly lower than that of the gasoline version throughout the speed range. Higher combustion temperature was

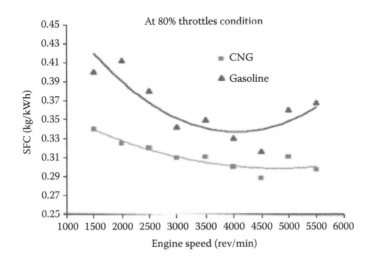

**FIGURE 5.67**
Specific fuel consumption versus engine speed at 80% throttles condition. (Adapted from M. I. Jahirul et al., *Applied Thermal Engineering* 30, 2219–2226, 2010.)

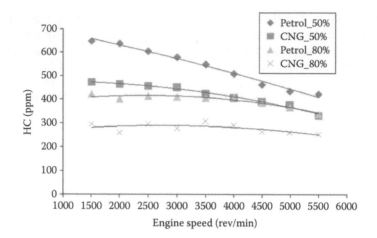

**FIGURE 5.68**
Hydrocarbon (HC) emission over a speed range at 50% and 80% throttle condition for gasoline and CNG. (Adapted from M. I. Jahirul et al., *Applied Thermal Engineering* 30, 2219–2226, 2010.)

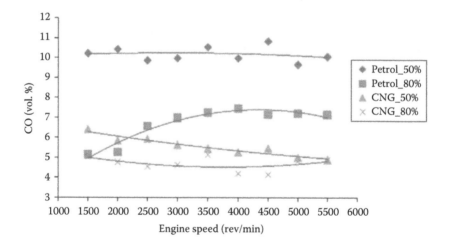

**FIGURE 5.69**
Carbon monoxide (CO) emission over a speed range at 50% and 80% throttle condition for gasoline and CNG. (Adapted from M. I. Jahirul et al., *Applied Thermal Engineering* 30, 2219–2226, 2010.)

another reason for the low CO emission of the CNG-fuelled engine. During the combustion process, CO converts to $CO_2$, especially at higher temperatures.

$CO_2$ emission of the CNG-fuelled engine was found to be lower than that of the gasoline version throughout the speed range as shown in Figure 5.70. If CO emission is higher, $CO_2$ emission will be lower and vice versa. However, it may be noted that $CO_2$ emission varies with respect to the thermal efficiency of any engine. CNG, which is composed mostly of methane

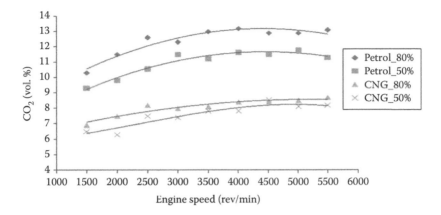

**FIGURE 5.70**
Carbon dioxide ($CO_2$) emission over a speed range at 50% and 80% throttle condition for gasoline and CNG.

($CH_4$) and the gasoline ($C_8H_{18}$) compound, is packed with less hydrogen per carbon (2.5). Thus, the percentage of carbon in the methane (CNG) was lower than that in the gasoline. This led to a lower emission of $CO_2$ for the CNG-operated engine than the gasoline fuel operation. $CO_2$ emission increased with the increase in engine speed for both the CNG and gasoline fuels, which is attributed to higher fuel conversion efficiency.

NO is produced more in the post-flame gases than in the flame front. The mixture that burnt early in the combustion process was being compressed to a higher temperature, thus increasing the NO formation rate; as the combustion proceeded, the cylinder pressure also increased. The comparative emissions of the oxides of nitrogen ($NO_x$) by the CNG and the gasoline fuels are shown in Figure 5.71. The $NO_x$ emission was strongly related to the lean fuel with a high cylinder temperature or high peak combustion temperature. A fuel with a high heat release rate in the premix or rapid combustion phase and a lower heat release rate during the mixed controlled combustion phase would produce the $NO_x$.

## 5.1.17 Liquefied Petroleum Gas

LPG is well known as a clean alternative fuel for vehicles because it contains less carbon molecules than gasoline or diesel. The major constituents of LPG are ethane ($C_2H_6$), propane ($C_3H_8$), butane ($C_4H_{10}$) and pentane ($C_5H_{12}$). Its higher ratio of carbon (C) to hydrogen (H) reduces the amount of $CO_2$ and other non-regulated emissions, such as formaldehyde and acetaldehydes. LPG also has many other advantages such as high octane number, high combustion value, less carbon accumulation, easy storage and low cost. LPG fuel is preferred as a clean alternative fuel for IC engines due to

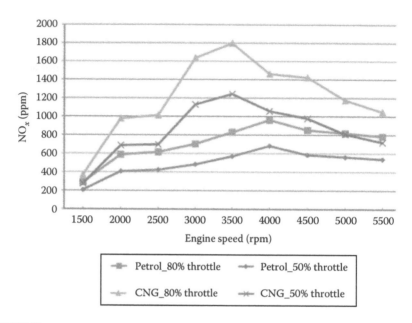

**FIGURE 5.71**
Nitrogen oxides ($NO_x$) emission at 50% and 80% throttle condition for gasoline and CNG.

its easy availability and storage, low cost, high octane number, high combustion efficiency and low exhaust emissions with respect to other fuels.

Brake thermal efficiency rises with an increase in BMEP and reduces with an increase in the LPG usage level as shown in Figure 5.72. The major reason for this decrease in brake thermal efficiency (BTE) with LPG usage may be ascribed to the decrease in the volumetric efficiency during LPG usage.

Figures 5.73 through 5.75 illustrate the effect of the relative air–fuel ratio on CO, $CO_2$ and HC emissions for the gasoline- and LPG-fuelled versions. It can be seen from these figures that as the relative air–fuel ratio increases, the CO emissions get lowered, and there is no serious change over the lean burn operating condition. $CO_2$ emission gets reduced with the lower carbon number of fuel (LPG) and a leaner mixture (Figure 5.74). $CO_2$ emission decreases significantly with an increase in the equivalence ratio beyond the stoichiometric ratio. The trend is due to incomplete combustion as the air quantity in charge is lesser. However, the CO emission will be higher. The HC emission also follows a similar trend.

### 5.1.17.1 Comparison among Gasoline, CNG, E10 and LPG Fuels

Figure 5.76 shows the torque comparison of an SI engine under high A/F ratios and CNG at $\lambda = 1$ up to 4000 rpm. The average torque drop in the CNG-fuelled version as compared to gasoline was observed to be 10%. The analysis of the results also indicates that the drop in torque is high at higher

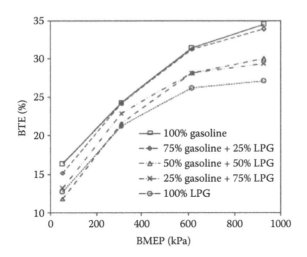

**FIGURE 5.72**
Variation of BTE with BMEP. (Adapted from M. Gumus, *Fuel Processing Technology* 92, 1862–1867, 2011.)

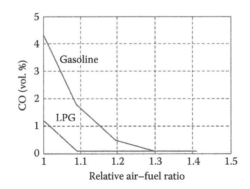

**FIGURE 5.73**
Variation of CO emissions with relative air–fuel ratio (100% LPG). (Adapted from M. A. Ceviz and F. Yuksel, *Renewable Energy* 31, 1950–1960, 2006.)

engine speed zones than at low-engine-speed zones. It was observed that in the CNG operation, there is no improvement in brake torque with increased richness levels. In fact, CNG is more resilient than LPG to torque increase with increasing richness levels.

In Figure 5.77, CNG and LPG showed a decrease in BSEC at all part-load operating points. E10 showed no major difference in BSEC as compared to neat gasoline. The improvement in part-load BSEC of gaseous fuels is attributed to the difference in the stoichiometric A/F ratio between gaseous fuels and gasoline. The results also indicate that BSEC of LPG is marginally better than that of CNG in most of the part-load operating points.

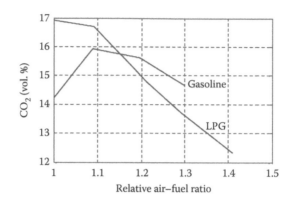

**FIGURE 5.74**
Variation of $CO_2$ emissions with relative air–fuel ratio. (Adapted from M. A. Ceviz and F. Yuksel, *Renewable Energy* 31, 1950–1960, 2006.)

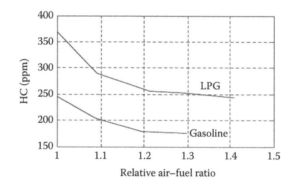

**FIGURE 5.75**
Variation of HC emissions with relative air–fuel ratio. (Adapted from M. A. Ceviz and F. Yuksel, *Renewable Energy* 31, 1950–1960, 2006.)

$CO_2$ emission with E10 follows the same trend as that of neat gasoline but with a marginal 2–3% reduction as shown in Figure 5.78. CNG, on the other hand, showed an average of 18% reduction in $CO_2$ up to 4500 rpm. At 5000 rpm, the $CO_2$ emission with neat gasoline drops drastically due to the high richness levels. This results in the shift in chemical equilibrium of $CO \leftrightarrow CO_2$ formation, which consequently results in more emissions of CO than $CO_2$. On the other hand, gaseous fuels operate at comparatively leaner λ values, resulting in high $CO_2$ levels at this engine speed. $NO_x$ emission is the highest with E10 due to ethanol containing oxygen (Figure 5.79).

A reduction in HC and CO emissions was observed with gaseous fuels at part load as compared to neat gasoline (Figures 5.80 and 5.81). This is attributed to the high H/C ratio, simpler hydrocarbon structure and

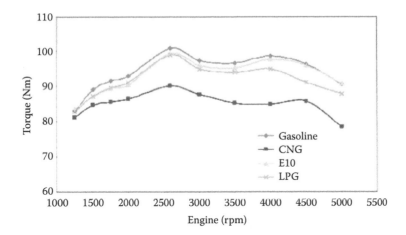

**FIGURE 5.76**
Torque comparison of CNG, LPG and E10 with gasoline. (Adapted from R. Muthu Shanmugam et al., *SAE International Journal of Fuels and Lubricants*, SAE no. 2010-01-0740, 3.)

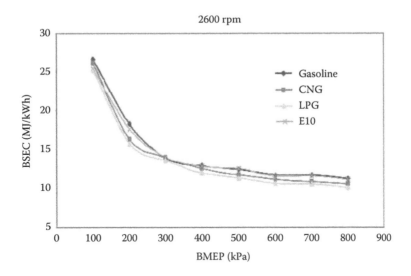

**FIGURE 5.77**
Brake-specific energy consumption (BSEC) comparison of CNG, LPG and E10 with gasoline. (Adapted from R. Muthu Shanmugam et al., *SAE International Journal of Fuels and Lubricants*, SAE no. 2010-01-0740, 3.)

improved combustion characteristics of gaseous fuels. Also, stoichiometric operation at WOT with gaseous fuels further reduces the CO and HC emissions. However, the same phenomenon is responsible for the increase in $NO_x$ emissions under WOT with gaseous fuels as already shown in Figure 5.79.

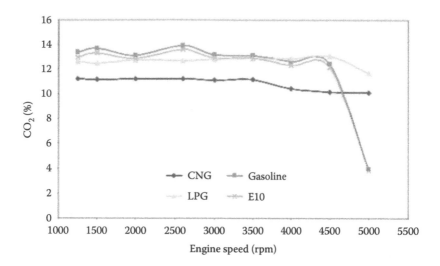

**FIGURE 5.78**
$CO_2$ comparison of CNG, LPG and E10 with gasoline. (Adapted from Ch. Beidl et al., *International Symposium on Internal Combustion Diagnostics*, AVL Deutscwand, pp. 6–25, June 2002.)

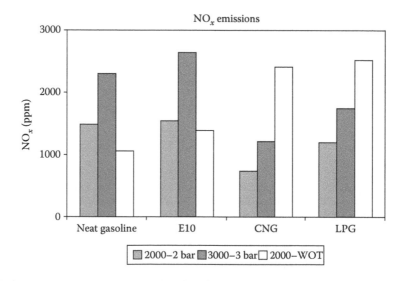

**FIGURE 5.79**
$NO_x$ comparison of CNG, LPG and E10 with gasoline. (Adapted from R. Muthu Shanmugam et al., *SAE International Journal of Fuels and Lubricants*, SAE no. 2010-01-0740, 3.)

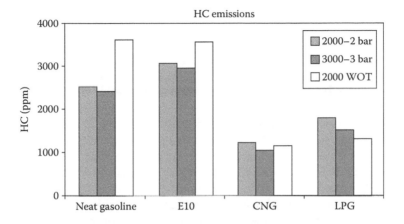

**FIGURE 5.80**
HC comparison of CNG, LPG and E10 with gasoline. (Adapted from R. Muthu Shanmugam et al., *SAE International Journal of Fuels and Lubricants*, SAE no. 2010-01-0740, 3.)

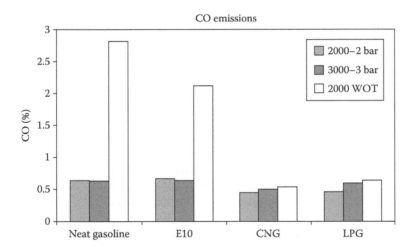

**FIGURE 5.81**
CO comparison of CNG, LPG and E10 with gasoline. (Adapted from R. Muthu Shanmugam et al., *SAE International Journal of Fuels and Lubricants*, SAE no. 2010-01-0740, 3.)

## 5.2 Compression Ignition Engines

A CI engine, also known as an diesel engine, is an IC engine that operates using the diesel cycle. Invented in 1892 by German engineer Rudolf Diesel, it was based on the hot bulb engine design and patented on 23 February 1893. There are two classes of diesel engines currently in use: two-stroke and four-stroke. The four-stroke type is the 'classic' version, tracing its lineage back to Rudolf Diesel's prototype. It is also the most commonly used form, being the preferred power source for many motor vehicles, especially buses and trucks. Much larger engines, such as those used for railroad locomotive application and marine propulsion, are often two-stroke units that offer a more favourable horsepower-to-weight ratio and a better fuel economy. The most powerful engines in the world are two-stroke cycle diesels of mammoth proportions. These so-called low-speed diesels are able to achieve thermal efficiencies approaching 50%. Diesel engines are more efficient than gasoline (petrol) engines of the same power, resulting in lower fuel consumption. However, such a comparison does not take into account that diesel fuel is denser and contains about 15% more energy by volume [20,21]. Also, since fuel is sold on a volumetric basis, more energy is available at a cheaper price. This fact has tilted the balance in favor of a common commuter towards diesel engines.

Diesel engines use compression ignition, a process by which fuel is injected after the air is compressed in the combustion chamber, causing the fuel to self-ignite by heat of compression. The compression ratios employed lie between 16:1 and 25:1. This extremely high level of compression causes the air temperature to increase to 700–900°C [21] at the end of the compression stroke.

Diesel engines employ high compression ratios to achieve the required temperatures to ignite the fuel, unlike the spark plug in a gasoline engine. To operate the engine at such high compression ratios, a sturdy structure is required. This essentially increases the overall weight of the engine and reduces the power-to-weight ratio. A larger displacement diesel engine is also required to produce the same power as a gasoline engine.

Diesel engines emit low levels of carbon monoxide as they burn the fuel in excess air even at full load, at which point the quantity of fuel injected per cycle is still about 50% lean of stoichiometric. However, they can produce black soot (or more specifically, diesel particulate matter) from their exhaust, which consists of unburnt hydrocarbon compounds. This is caused by local low temperatures where the fuel is not fully atomised. These local low temperatures occur at the cylinder walls and at the outside of big droplets of fuel. At these areas where it is relatively cold, the mixture is rich (contrary to the overall mixture that is lean). The rich mixture has less air to burn and some of the fuel turns into a carbon deposit. Liquid fuel burns partially, resulting in soot formation. Together with $NO_x$, the emissions of soot characterise the diesel combustion process. For present engines, a trade-off between these two emissions is observed, which poses a major challenge to reach future legislation for both emissions. The major advantages of the diesel engine as compared to the SI engine are the low pumping losses, due to the absence of a throttle, and a higher compression ratio, leading to a higher efficiency.

It was found in many literature surveys that the introduction of EGR generally influences diesel engine combustion in three ways: thermal, chemical and dilution. The thermal effect was related to the increase of inlet charge temperature that affects volumetric efficiency and the increase of charge specific heat capacity due to the presence of $CO_2$ and $H_2O$ and $N_2$. On the other hand, the chemical effect is related to the dissociation of species during combustion, while dilution is related to the reduction of $O_2$ availability. The process of EGR induction in the combustion chamber reduces the oxygen concentration and peak combustion temperature, which results in reduced $NO_x$. In the EGR technique, both hot and cold EGR can be applied. The reduction of oxygen concentration with hot EGR has an adverse effect on soot emissions. The combined effect of increased temperature and decreased oxygen concentration with the use of hot EGR results in a slight increase in $NO_x$. The $O_2$ reduction in the engine cylinder with the use of cold EGR is higher than that with the use of hot EGR. The combined effect of $O_2$ reduction and large specific heat of inert gases in cold EGR reduces $NO_x$ drastically compared to hot EGR.

### 5.2.1 Emissions

Despite being economical and efficient, diesel engines basically suffer from the main drawback of high amounts of particulate matter as well as oxides of nitrogen in the exhaust emissions [22]. Major research emphasis is given to their reduction due to their toxicity and damaging effects to both human health and the environment.

The $N_2O$ emission from the automobile sector occurs in combustion at low (<950°C) temperatures and they are affected by fuel type, operating conditions and excess air level, oxygen in fuel and catalytic activity. For example, one of the promising technologies such as EGR, is used to enhance the control of $NO_x$ emission by means of reducing the combustion temperature, which may result in high $N_2O$ emission. Most of the research in the world is focussing on reducing $NO_x$ emission by reducing the gas temperature, but it may be negatively resulting in a higher $N_2O$ emission.

$N_2O$ is reportedly increasing in the atmosphere at an estimated rate of 0.7 ppb per year. Since $N_2O$ is a stable compound, it is transported to the stratosphere where it is photochemically oxidised to nitric oxide (NO), a contributor to catalytic ozone depletion [23]. It is estimated that the global warming potential (GWP) of $N_2O$ is about 320 times more than its $CO_2$ equivalent [22,23].

Particulate matter or soot content from diesel exhaust is another matter of concern. They tend to react in the presence of their pollutants during their residence time in the atmosphere producing harmful by-products. The production of smog and respiratory diseases can be attributed to them.

Efforts have been made to solve these emission-related issues with techniques that improve fuel consumption. These involve efficient injection systems, modified injector designs and electronic controls. Although improvements in the injection and combustion systems have reduced both emissions considerably, it is unlikely that future emission legislations can be met. After-treatment systems will be necessary, but the systems introduced so far are quite complex and expensive and their durability is still an issue. The possible solution lies in an entirely different approach that combines the strong points of both SI and CI engines. This approach, termed homogeneous charged compression ignition (HCCI), prepares the fuel–air mixture under the SI engine mode and ignites it in the CI process, which will be discussed later.

## 5.2.2 Biodiesel

The utilisation of a renewable fuel, such as biodiesel in diesel engines in the form of blends, could reduce emissions such as the particulates, CO and HC. As biodiesel has a higher ignition quality (CN > 58) than diesel (CN = 50), it is more suitable for diesel engines. Biodiesel can be derived from vegetable oils and fats. Biodiesel has a higher flash point temperature, higher cetane number, lower sulphur content and lower aromatics than petroleum diesel fuel. It could also be expected to reduce exhaust gas emissions due to fuel containing oxygen. The main problems of the current Indian diesel fuel are the low flash point (35°C against the world average of 52°C) and high sulphur content, which can be improved by blending it with biodiesel. The emission of greenhouse gases ($CO_2$) can be reduced by the use of biodiesel in diesel engines as it will be recycled by the crop plant, resulting in no new addition into the atmosphere.

It is a carbon-neutral fuel as it adds no net carbon dioxide to the environment. Researchers worldwide are on a consensus that biodiesel, irrespective of the feedstock used, results in a decrease in the emissions of hydrocarbons (HC), carbon monoxide (CO), particulate matter (PM) emissions and sulphur dioxide ($SO_2$). Only $NO_x$ are reported to increase, which is due to the oxygen content in the biodiesel. Other authors have suggested various strategies to eliminate the $NO_x$ emission. Szybist et al. suggested changing the chemical composition of feedstock by increasing methyl oleate and the addition of cetane improvers. This reduces the iodine value. Boehman et al. and Kegl recommended retarding the injection timing, which could lead to a decrease in $NO_x$ emission.

Typical engine combustion reaction for biodiesel:

$$\text{Biodiesel} + \text{air} (N_2 + O_2) = CO_2 + CO + H_2O + N_2 + O_2 + (HC) + O_3 + NO_2$$

### 5.2.2.1 Comparison with Other Fuels

The major problem with most of the alternative fuels is that they are not directly compatible to complement the current engines. Ethanol is a slightly polar, hygroscopic and corrosive liquid in the case of a steel storage infrastructure. Gasoline, on the other hand, is non-polar, does not absorb water and is non-corrosive. Although a small amount of ethanol can be mixed with gasoline with no ill effect, a large fraction of ethanol requires a different distribution mechanism, different pumps and will suffer the problems outlined by Struben et al. in attempting to gain adoption [24].

This problem has also been seen by less controversial alternative fuels such as diesel and liquefied natural gas, which are relatively compatible with current pump technology, competitive with gasoline as a fuel and utilises technology that is inexpensive and readily available.

Biodiesel has a large advantage over other alternative fuels, such as ethanol, hydrogen and electricity, in that it is compatible chemically in most of the existing diesel fuel. Generally, blends of up to 20% biodiesel with regular diesel (B20) are compatible with existing diesel engines, while some minor engine modifications are required to run on up to 100% biodiesel (B100) to avoid maintenance and performance problems. Biodiesel is sometimes slightly hygroscopic due to incomplete esterification; however, overall, it is hydrophobic and should not absorb large amounts of water. The largest issue with biodiesel is that, owing to the esters in its make-up, it will more rapidly degrade natural rubber gaskets and hoses. However, natural rubber has not been commonly used in engines for about 15 years, and most modern diesel engines will be able to run on B100 without modification. The oxygen content of biodiesel improves the combustion process and decreases its oxidation potential. The structural oxygen content of a fuel improves the combustion efficiency due to the increase in the homogeneity of oxygen with the fuel during the combustion process. Owing to this, the combustion efficiency of biodiesel is higher than that of petrodiesel. Also, the combustion efficiency of methanol/ethanol is higher than that of gasoline. A visual inspection of the injector types would indicate no difference between the biodiesel fuels when tested on petrodiesel. The overall injector coking is considerably low. Biodiesel contains 11% oxygen by weight and no sulphur. The use of biodiesel can extend the life of diesel engines because it is more lubricating than petroleum diesel fuel. Biodiesel has got better lubricant properties than petrodiesel. As shown in Table 5.4, the HHVs of biodiesels are relatively high. The HHVs of biodiesels (39–41 MJ/kg) are slightly lower than those of gasoline (46 MJ/kg), petrodiesel (43 MJ/kg) or petroleum (42 MJ/kg), but higher than that of coal (32–37 MJ/kg). General characteristics of karanja biodiesel is given in Table 5.5.

**TABLE 5.4**

Comparison of Chemical Properties and Higher Heating Values of Biodiesel and Petrodiesel

| Chemical Property | Biodiesel (Methyl Ester) | Diesel |
| --- | --- | --- |
| Ash (wt%) | 0.002–0.036 | 0.006–0.010 |
| Sulphur (wt%) | 0.006–0.020 | 0.020–0.050 |
| Nitrogen (wt%) | 0.002–0.007 | 0.0001–0.003 |
| Aromatics (vol%) | 0 | 28–38 |
| Iodine number | 65–156 | 0 |
| HHV (MJ/kg) | 39.2–40.6 | 45.1–45.6 |

**TABLE 5.5**

Characteristics of Karanja Biodiesel (KOME)

| Characteristics | Value | Test Methods |
|---|---|---|
| Density | 891 kg/m³ | ASTM D 4052 |
| Viscosity at 40 | 4.08 cst | ASTM D 445 |
| Flash point C | 145 | ASTM D 93 |
| Cloud point C | 4 | ASTM D 2500 |
| Cetane number | 53.6 | ASTM D 613 |
| Ash content | 0.004 | ASTM D 482 |
| Carbon residue | 0.016 | ASTM D 4530 |
| Moisture | 0.005 | Karl–Fisher titrator |
| Pour point | 0 | ASTM D 2500 |
| Phosphorus | % ppm <10 | ASTM D 4951 |

### 5.2.2.1.1 *Effect of Biodiesel Addition on BSFC and BP*

Higher levels of blends are needed to produce the same amount of energy due to its higher specific gravity and lower calorific value as compared to diesel fuel (Figure 5.82). Greater BSFC with LOME (linseed oil methyl ester) was mainly due to the increased fuel flow rather than power. The general advantages of the use of biodiesel–diesel blends in a diesel engine are as given below:

- Biodiesel contains ~10% (in weight) oxygen, which helps in complete combustion.
- As calorific value of biodiesel is lesser than base diesel, the more amount of biodiesel fuel have relatively to be injected to the in-cylinder for maintaining same power output as compared to base

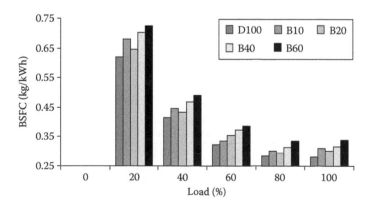

**FIGURE 5.82**
BSFC versus load.

diesel. However, the biodiesel has higher density than base diesel resulting to more mass of fuel which are to be injected automatically. So, the net effect on power drop has to be studied quantitatively in a diesel engine for biodiesel fuel.

- A more viscous blend means less internal leakage in the fuel pump.

### 5.2.2.2 Effect of Biodiesel Addition on CO and UBHC Emissions

*Increase in load*: CO emission increases at higher loads due to the fall in the A/F ratio. The blends have lower CO emissions due to the presence of more oxygen in the fuel. The higher oxygen content and cetane number of LOME lead to lower HC emissions as compared to diesel fuel. These trends can be seen in Figures 5.83 and 5.84.

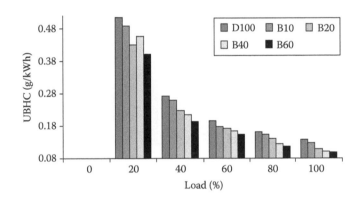

**FIGURE 5.83**
Variation in CO emission with loads.

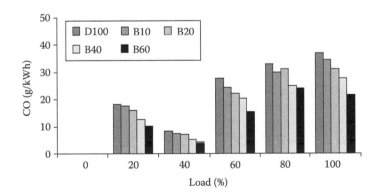

**FIGURE 5.84**
Variation in UBHC emission with loads.

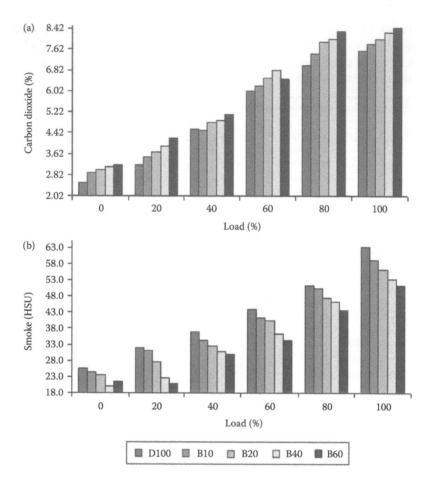

**FIGURE 5.85**
(a) Smoke versus load. (b) Carbon dioxide versus load.

### 5.2.2.3 Effect of Biodiesel Addition on Smoke and CO₂

Higher brake power and lower HC show improved combustion of fuel. Hence, lower smoke density with LOME blends as compared to diesel is observed. The $CO_2$ emissions with the blend were higher than those with the diesel fuel due to the increase in the mass of fuel injected and improved combustion of the fuel-borne oxygen. Higher $CO_2$ is an indication of the complete combustion of fuel. Hence, higher exhaust gas temperatures are observed. These trends can be seen in Figure 5.85a and b.

### 5.2.2.4 Effect of Biodiesel Addition on $T_{EG}$ and $NO_x$

The higher temperatures of combustion and the presence of fuel oxygen with the blend caused higher $NO_x$ emissions (Figure 5.86b). There are three factors

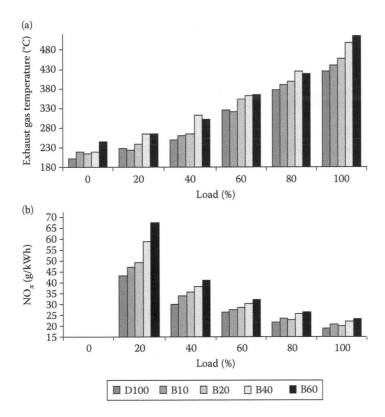

**FIGURE 5.86**
(a) Exhaust temperature versus load and (b) $NO_x$ versus load.

that affect the $NO_x$ emission formation. They are oxygen concentration, combustion gas temperature and reaction time. The increase in the exhaust gas temperature is due to the slow combustion of LOME as compared to diesel fuel as shown in Figure 5.86a.

### 5.2.2.5 Effect of Biodiesel Addition on the Combustion Process

The cylinder peak pressure of diesel fuel is higher than that of diesel–biodiesel blends as shown in Figure 5.87. This is due to the reduction in premixed combustion and poor volatility of LOME. With LOME, the ignition delay is longer than for diesel fuel, which leads to higher exhaust gas temperature due to the decomposition of LOME. This analysis suggests that injection timing should be advanced for ester fuels for complete combustion by providing more time. The rate of pressure rise is lesser with B40 due to biodiesel having a higher cetane number, as shown in Figure 5.88. The change in peak pressure with respect to crank angle for different biodiesel-diesel blend is shown in Figure 5.89.

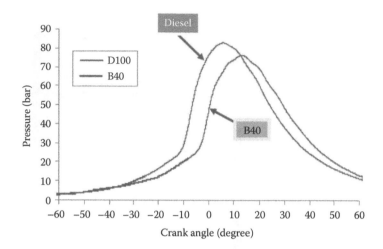

**FIGURE 5.87**
Pressure versus crank angle.

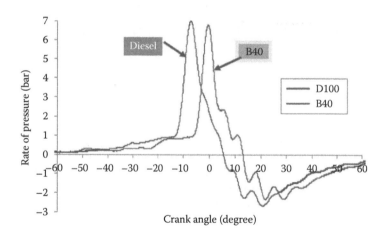

**FIGURE 5.88**
Rate of pressure versus crank angle.

### 5.2.2.6 Effect of Biodiesel Addition on Combustion Characteristics

Figure 5.91 shows variation in brake power and brake-specific fuel consumption with respect to fuel injection pressure. The brake power is lower with lowest and highest injection pressure whereas it is higher at moderate injection pressure.

This is due to the change in dynamic injection timing resulting to change in start of combustion. If start of combustion occurs nearby top dead centre, the power and SFC will be higher. If it is too advance or retard from top dead centre that affects power output and BSFC. Peak pressure depends upon the combustion rate in the initial stages, which is influenced by the fuel taking

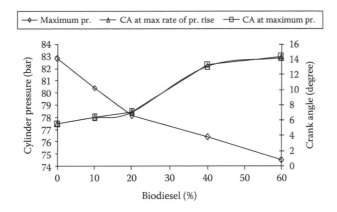

**FIGURE 5.89**
Pressure rise versus biodiesel blends.

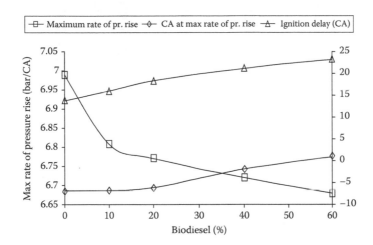

**FIGURE 5.90**
Cylinder pressure versus biodiesel blends.

part in the uncontrolled heat release phase. The high viscosity and low volatility of the biodiesel blends lead to poor mixture preparation with the air and atomisation during the ignition delay. The rate of pressure rise decreases with an increase in the biodiesel blends due to the higher cetane number of biodiesel (CN = 52–58) as shown in Figure 5.90.

### 5.2.2.7 Effect of Injection Pressure on Brake Power and Brake-Specific Fuel Consumption

An increase in injection pressure decreases the particle diameter and causes the diesel–biodiesel fuel spray to vaporise quickly (Figure 5.91a and b).

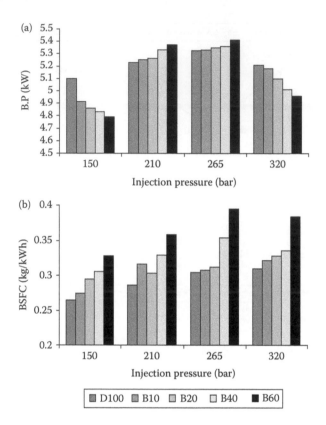

**FIGURE 5.91**
(a) Brake power and (b) brake-specific fuel consumption versus injection pressure.

However, initial combustion with the spray was restricted to a smaller region near the injector, and the flame spread around the chamber through slow propagation. The injection with high pressure would have more chance of fuel jet strike on the combustion chamber wall called as wall impingement. The chemical energy of the injected fuel could not be converted completely into heat energy due to affecting mixture formation process resulting to poor combustion. Hence, brake power is lower at a very high injection pressure. The BSFC values of the blends were higher than those for the diesel fuel due to the higher specific gravity, viscosity and lower heating value.

### 5.2.2.8 *Effect of Injection Pressure on Exhaust Gas Temperature and Nitrogen Oxides Emission*

The enhancement of injection pressure results in a higher spray velocity at the outlet of the nozzle, which produces smaller Sauter mean diameters of the fuel; hence, complete combustion results in higher combustion gas temperature, and consequently higher exhaust gas temperature, resulting in an increased $NO_x$ (Figure 5.92a and b). The biodiesel has oxygen content which enhances in-cylinder temperature and $NO_x$ emission due to proper combustion. But, at high injection pressure, $NO_x$ emission decreased due to reduction in ignition delay which decrease fuel accumulation during the period resulting to lesser premixed combustion phase than diffusive phase.

### 5.2.2.9 Effect of Injection Pressure on (a) CO and (b) UBHC

Atomisation and the mixing of fuel depend on the fuel properties (viscosity, surface tension, self-ignition temperature) and fuel injection system. Increasing the injection pressure from 150 to 320 bar causes lower CO and HC emissions, due to the good mixing of fuel and air, better atomisation and improved combustion of the smaller droplets (Figure 5.93a and b).

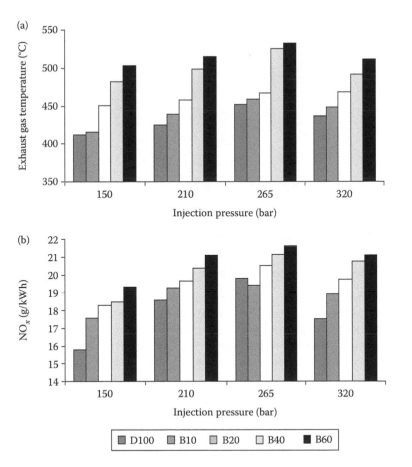

**FIGURE 5.92**
(a) Exhaust gas temperature versus injection pressure. (b) $NO_x$ versus injection pressure.

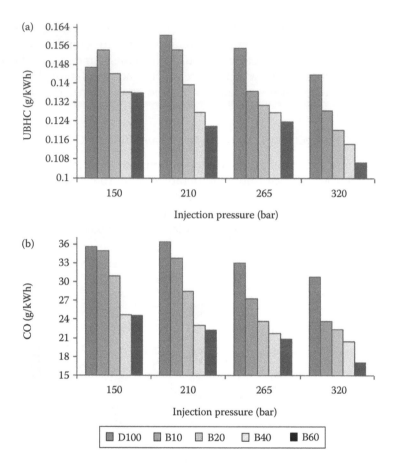

**FIGURE 5.93**
(a) CO and (b) UBHC versus injection pressure.

### 5.2.2.10 Effect of Injection Pressure on CO₂ and Smoke

It is well known that the amount of carbon dioxide is proportional to the amount of fuel burnt (Figure 5.94a and b). Therefore, as the injection pressure increases, the $CO_2$ emission will also increase due to the improvement in the combustion process. Higher power indicates a better and more complete combustion of fuel. Hence, lower smoke density values are achieved with biodiesel blends as compared to diesel. The smoke level is affected by injection pressure.

### 5.2.2.11 Effect of Injection Timing on Brake Power and Brake-Specific Fuel Consumption

BSFC is lowest at an injection timing of 26 BTDC as shown in Figure 5.97. As the injection timing is retarded, there is a decrease in brake power of the

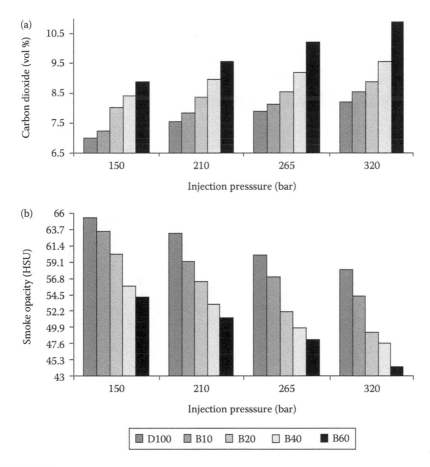

**FIGURE 5.94**
(a) Carbon dioxide and (b) smoke versus injection pressure.

engine. It may be due to the inadequate time available for mixture preparation, lower temperature for fuel evaporation and higher start of injection timing of biodiesel than diesel (Figure 5.95a and b). The data show that the advanced timing system incurs a penalty on fuel consumption as the governor injects more fuel than when running on standard time units.

### 5.2.2.12 Effect of Injection Timing on Exhaust Gas Temperature ($T_{EG}$) and $NO_x$

Operating with retarded injection timing shows a higher exhaust gas temperature than with advanced injection timing. The higher viscosity and low volatility of the fuel will retard the rate of combustion and hence cause a higher ignition delay. Slow burning can result in a higher exhaust gas temperature. Advancing the injection timing will give more time for the ester fuel to evaporate Hence, smoother combustion with lower ignition delay

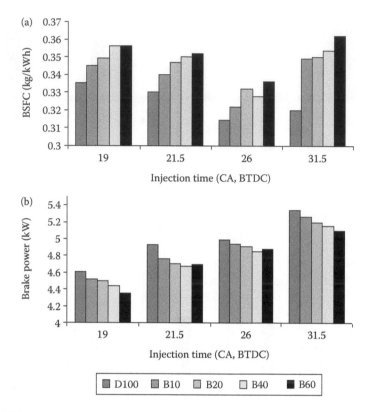

**FIGURE 5.95**
(a) Brake power and (b) BSFC versus injection time.

is presently leading to lower exhaust gas temperature than the retardation case. $NO_x$ emission is higher with biodiesel blend as compared to base diesel, which is depicted in Figure 5.96a and b.

### 5.2.2.13 Effect of Injection Timing on CO and UBHC Emissions

By increasing the injection advance, the CO and HC emissions are reduced due to improved combustion as compared to the retarded timing as shown in Figure 5.97a and b. Generally, retarding the delivery fuel timing can result in higher CO and HC emissions due to delayed combustion.

### 5.2.2.14 Effect of Injection Timing on CO₂ and Smoke

Advancement of injection timing results in an increase in the $CO_2$ emission, which is an indication of improved combustion of the fuel due to the higher time availability and oxygen presence in the ester fuel. Retarding the ignition timing will decrease the combustion rate and hence lead to lower

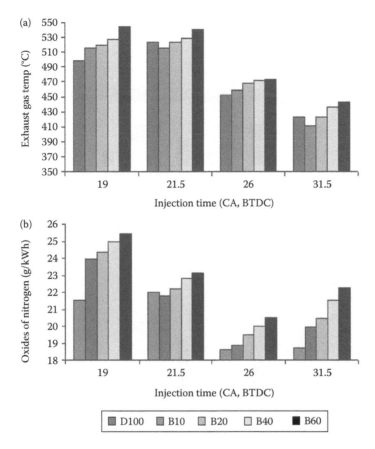

**FIGURE 5.96**
(a) Exhaust temperature and (b) NO$_x$ versus injection time.

CO$_2$ (Figure 5.98a). When the ignition timing is advanced, the smoke level is reduced due to the availability of more time for the fuel to evaporate and mix with the air and complete the combustion (Figure 5.98b). Results get reversed under retarded conditions of injection timing.

## 5.2.3 Experimental Setup for EGR

A constant-speed (1500 rpm), single-cylinder, four-stroke, air-cooled diesel engine was used for this study. The schematic diagram of the EGR set-up is shown in Figures 5.99 through 5.102. Engine systems were equipped with several experimental sub-systems and were instrumented at the appropriate location to evaluate the performance parameters such as the A/F ratio, BSFC and BTE. A dynamometer was used to load the engine. A digital manometer was used to measure the inlet airflow and a rotameter to measure the exhaust the gas into the inlet manifold. Thermocouples were

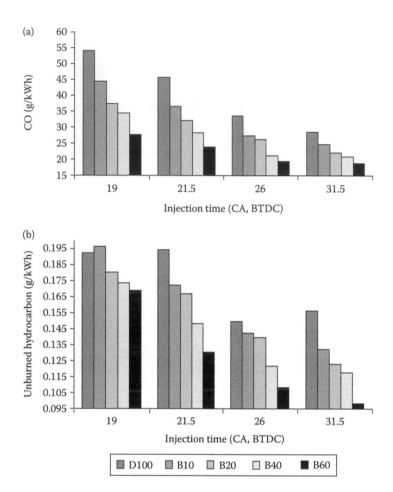

**FIGURE 5.97**
(a) CO and (b) unburned hydrocarbon versus injection time.

installed to monitor the gas temperature at the inlet and exhaust ducts as well as the cylinder wall temperatures. Fuel was fed to the injector pump under gravity and the volumetric flow rate was monitored. The speed was calibrated on an AVL digital monitor. Exhaust gas emission was measured by an exhaust gas analyser (AVL-437). It measured HC, CO and $CO_2$ with the infrared analyser (IP), and $O_2$ and $NO_x$ with the electrochemical method. A smoke opacity meter was used to measure the exhaust gas opacity. The quantity of EGR was regulated by controller valves that were installed in the EGR system.

A surge tank was used to dampen the fluctuations of the recirculated exhaust. A rotameter was installed in the EGR line after the surge tank to measure the flow rate of the recirculated exhaust gas.

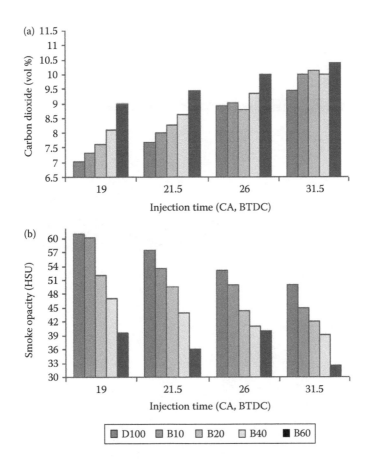

**FIGURE 5.98**
(a) Carbon dioxide and (b) smoke opacity versus injection time.

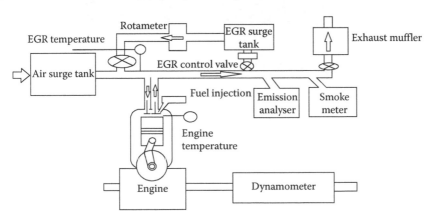

**FIGURE 5.99**
Line diagram of a typical diesel engine with EGR setup.

**FIGURE 5.100**
A photographic view of diesel engine.

**FIGURE 5.101**
A photographic view of diesel engine with EGR setup.

To evaluate the performance and emission characteristics, the load on the engine was varied from 0% to 100%. The engine was run with diesel and biodiesel without EGR to generate the baseline data. The exhaust gas was cooled with the water circulated inside the surge tank and a hollow cylindrical section was fabricated with copper as the material. The temperature was observed to be in the range of 46–57°C, which varied with the load and EGR percentage.

**FIGURE 5.102**
A photographic view of inlet manifold with EGR inlet and air inlet.

### 5.2.3.1 Design of Intercooler

A intercooler works by passing hot EGR to be cooled through a network with a cooler metal surface, which is exposed to a coolant. The EGR passages were designed to expose the hot gases to the maximum allowable cooled surface area. Water was used as the cooling medium, and copper and aluminium were used for the design of the intercooler. Copper has good thermal conductivity and the temperature obtained was approximately 43–49°C.

In the design part of the intercooler as shown in Figures 5.103 through 5.105, two surge tanks of cylindrical shape were fabricated, one with a capacity of 5 L and the other with a 7 L capacity. The material for the surge tank was made from aluminium because it is light in weight and has good thermal conductivity. EGR gas was passed through the surge tank and appropriate flow was calibrated through a rotameter. It was attached in line with the system and through a valve adjustment. Hot EGR passes through the copper tube and for cooling, it passes through the surge tank and is adjusted by the valve to obtain the desired flow through the rotameter.

The temperature reduction that was achieved using the EGR system was from 550°C to 46°C. In this system, the logarithmic mean temperature difference (LMTD) method was used to specify the characteristic dimension of the shell and tube counter-flow heat exchanger (intercooler).

For sensible heat transfer, the heat transfer rate was estimated by

$$Q = m_h C_{p,h}(T_{h,i} - T_{h,o}) = m_c C_{p,c}(T_{c,o} - T_{c,i}) \qquad (5.16)$$

$$A = Q/(U \times F \times \text{LMTD}) \qquad (5.17)$$

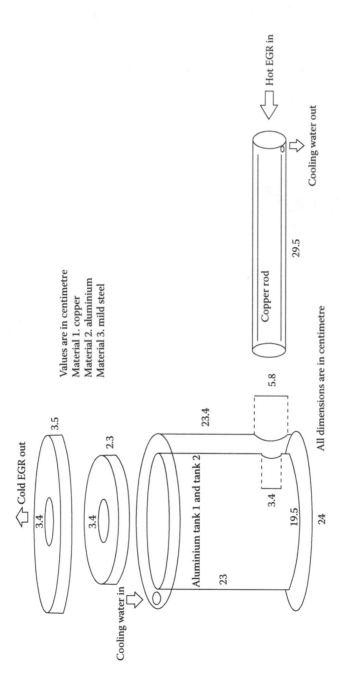

**FIGURE 5.103**
Intercooler parts and assembly.

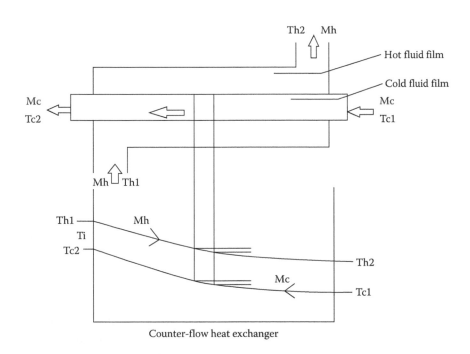

**FIGURE 5.104**
Counter-flow intercooler/heat exchanger.

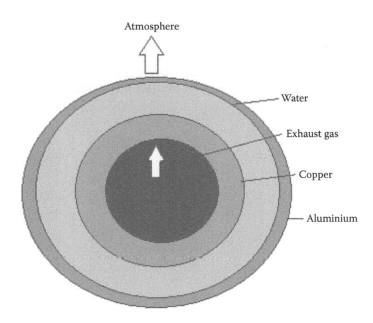

**FIGURE 5.105**
Cross-sectional view of intercooler.

One of the important measures of performance of the intercooler is its effectiveness. This is designed as the ratio of the actual heat transfer to the theoretical maximum heat transfer. The actual heat transfer computed from either the energy lost by the hot fluid or the energy gained by the cold fluid is given by Equation 5.18.

$$E = (mc_p)_h(T_{hi} - T_{ho})/C_{min}(T_{hi} - T_{ci}) = (mc_p)_c(T_{ci} - T_{co})/C_{min}(T_{hi} - T_{ci}) \quad (5.18)$$

The intercooler effectiveness defines the ability of the device to reduce the EGR temperature for the given operating condition (i.e., mass flow rate of the EGR and coolant and the coolant inlet temperature). For the counter-flow heat exchanger, the effectiveness is given as

$$E = \left[1 - exp^{-NTU(1-C_{min}/C_{max})}\right] / \left[1 - C_{min}/C_{max}(e)^{-NTU(1-C_{min}/C_{max})}\right] \quad (5.19)$$

### 5.2.3.2 Effect of Using EGR with Biodiesel Blends

The performance and emission characteristics of a diesel engine fuelled with karanja biodiesel (B20) with different percentages of EGR are discussed below. The exhaust gas was cooled from 500°C to a temperature range between 34°C and 44°C at full load using a developed intercooler.

### 5.2.3.3 Brake Thermal Efficiency

As shown in Figure 5.106, the BTE of biodiesel is greater than that of diesel throughout the measured load range.

Figure 5.107 indicates the variation of BTE at different levels of EGR from 0% to 16%. The brake thermal efficiency decreases marginally with EGR at lower engine loads. The reason behind this is that the higher inlet

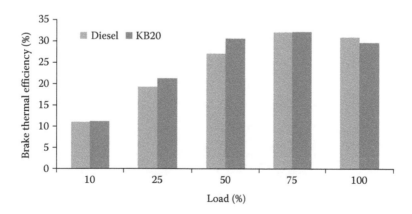

**FIGURE 5.106**
Comparison of BTE for diesel and biodiesel without EGR.

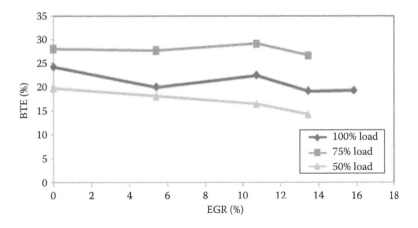

**FIGURE 5.107**
Variation in BTE for biodiesel with different percentage of EGR at different loads.

temperature of air helps to complete combustion. In addition to this, being at a slightly higher pressure than atmospheric pressure, EGR might also have reduced pumping losses. The chemical effect associated with the dissociation of carbon monoxide to form a free radical can also improve the efficiency.

### 5.2.3.4 Brake-Specific Fuel Consumption

BSFC decreases with an increase of load, varying from 10% load to 100% load (Figure 5.108).

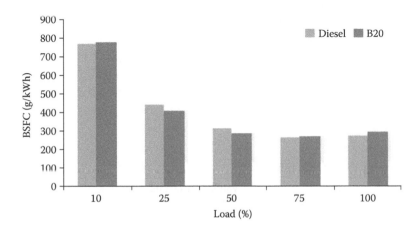

**FIGURE 5.108**
Comparison of BSFC for diesel and biodiesel.

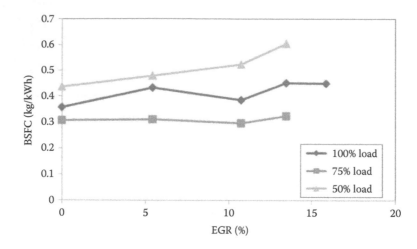

**FIGURE 5.109**
Variation in BSFC for biodiesel with different percentage of EGR at different loads.

The variation of EGR with BSFC with B20 fuel at full load is shown in Figure 5.109. The main reason for the lower BSFC for 75% load above 5% EGR may be explained by the higher calorific value of B20. Furthermore, B20 as an oxygenated fuel has a beneficial effect on combustion, especially in fuel-rich zones.

### 5.2.3.5 Calculating Volumetric Efficiency

Volume flow rate of air $Va = Cd \times A \times \sqrt{2} \times g \times h \times$ (density water/density air) $= 0.0583$ m³/s

Swept volume $= \pi/4 \times D2 \times L \times N/2 \times 60 = 0.0826$ m³/s

Volumetric efficiency $= 70.58111\%$

The volumetric efficiency with EGR addition for different biodiesel blends is calculated and shown in Table 5.6. As shown in Figure 5.110, the volumetric efficiency decreases with respect to EGR%. The air flow decreases because the air is replaced by cooled EGR.

**TABLE 5.6**

Calculation for Volumetric Efficiency

| S. No. | $m_{air}$ | EGR (%) | Volume Flow Rate of EGR (m³/s) | $m_{EGR}$ (kg/s) | $\eta_{vol}$ |
|---|---|---|---|---|---|
| 1 | 0.0583 | 0 | 0.0583 | 0.050258621 | 70.58111 |
| 2 | 0.0583 | 0.06 | 0.0443 | 0.000356444 | 60.41426 |
| 3 | 0.0583 | 0.1 | 0.0423 | 0.003207997 | 56.96201 |

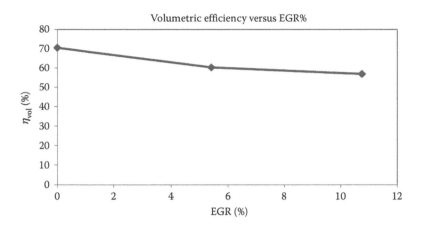

**FIGURE 5.110**
Variation in volumetric efficiency with increase in percentage of EGR.

### 5.2.3.6 *Engine Emission for Biodiesel with Different Percentage of Exhaust Gas Recirculation*

#### 5.2.3.6.1 *Carbon Monoxide*

Figures 5.111 and 5.112 [25] indicate CO variation with various EGR levels and engine loadings. The graph shows that the CO emissions increase with EGR and load. The possible reason could be attributed to a reduction in the availability of excess oxygen for combustion. Oxygen concentration resulted in rich air–fuel mixtures at different locations inside the combustion chamber. This heterogeneous mixture did not burn properly and might have resulted in higher CO emissions. Higher values of CO were observed at full load. At 50% load, CO emission also increased with an increase in EGR.

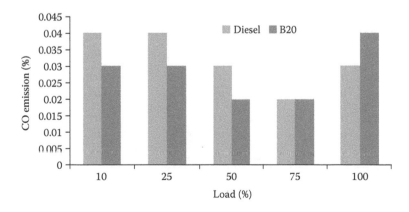

**FIGURE 5.111**
Comparison of CO emission for diesel and biodiesel.

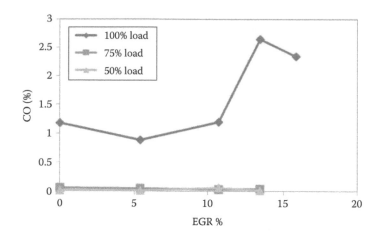

**FIGURE 5.112**
Variation in CO emission for biodiesel with different percentage of EGR at different engine loads.

### 5.2.3.6.2 Hydrocarbon

Figures 5.113 and 5.114 show the variation of HC emissions for different values of EGR under different load conditions. HC increases sharply from 420 ppm without EGR to 535 ppm with 15.68% EGR at full load. This high increase in emission is due to the reduced availability of oxygen to the heterogeneous mixture inside the cylinder. This heterogeneous mixture may not burn properly, resulting in a higher value of hydrocarbon. Beyond 13.46% EGR, HC emissions increase sharply at all the loads.

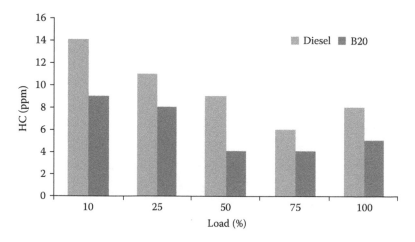

**FIGURE 5.113**
Comparison of HC emission for diesel and biodiesel.

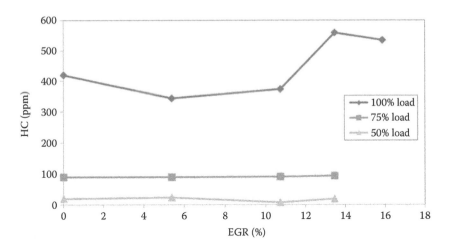

**FIGURE 5.114**
Variation in HC emission for biodiesel with different percentage of EGR at different loads.

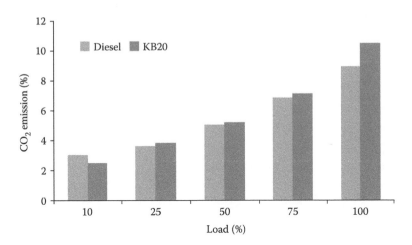

**FIGURE 5.115**
Comparison of $CO_2$ emission for diesel and biodiesel.

### 5.2.3.6.3 Carbon Dioxide

Figures 5.115 and 5.116 [25] show the variation of $CO_2$ in the exhaust of the biodiesel-fuelled diesel engine at different percentages of EGR. $CO_2$ emissions were lower at all loads. The reduction in the availability of $O_2$ due to EGR could have led to incomplete combustion, resulting in the fall of $CO_2$.

### 5.2.3.6.4 Oxide of Nitrogen

Figures 5.117 and 5.118 show the variation of $NO_x$ emissions with the EGR rate for different loads. EGR reduces the oxygen concentration in the charge.

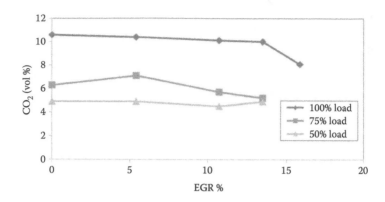

**FIGURE 5.116**
Variation in $CO_2$ emission for biodiesel with different percentage of EGR at different loads.

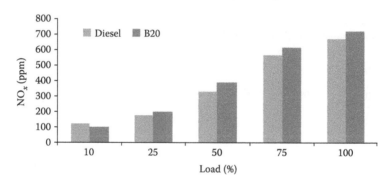

**FIGURE 5.117**
Comparison of $NO_x$ emission for diesel and biodiesel.

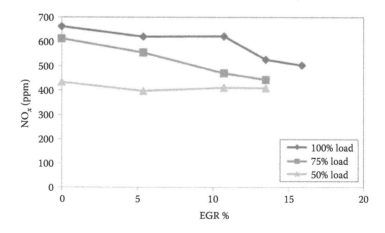

**FIGURE 5.118**
Variation in $NO_x$ emission for biodiesel with different percentage of EGR at different loads.

Consequently, the combustion pressure and temperature decreased. The temperature of the exhaust gas for the use of biodiesel with EGR was found to be lower, which indicates a lower cylinder gas temperature. As the cylinder gas temperature decreases due to the high specific heat of EGR, the formation of $NO_x$ also decreases. A high rate of EGR is found to be ineffective at lower loads as inert gases are low in concentration and oxygen rate is higher in exhaust. At full load, $NO_x$ decreased from 662 ppm without EGR to 503 ppm, corresponding to an EGR of 15.86%. At lower loads (50% and 75%) with 15.86%, $NO_x$ emission for biodiesel reduced from 435 to 409 and from 613 to 443, respectively.

### 5.2.3.6.5 Smoke Opacity

Figures 5.119 and 5.120 show the smoke opacity of different diesel and biodiesel blends and at different EGR rates. Smoke opacity increases with an increasing EGR rate and an increasing engine load. EGR reduces the availability of oxygen for the complete combustion of fuel, which results in a relatively incomplete combustion, leading to an increase in smoke emission.

### 5.2.3.7 Engine Combustion Characteristics for Biodiesel with a Different Percentage of Exhaust Gas Recirculation

The ignition delay period is the so-called preparatory phase during which some fuel has already been admitted but not yet ignited. This period is counted from the start of injection to the point where the pressure–time curve separated from the motoring curve indicated as the start of combustion. Ignition delay is the highest for 15% EGR (Figure 5.121). Ignition

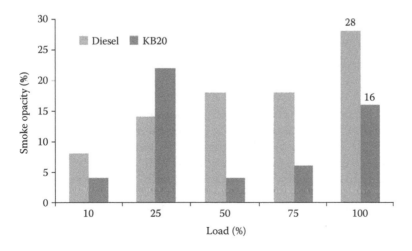

**FIGURE 5.119**
Comparison of smoke opacity for diesel and biodiesel.

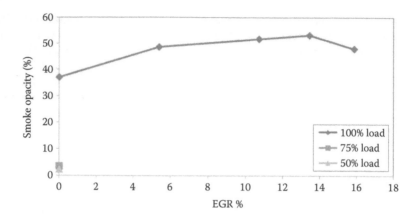

**FIGURE 5.120**
Variation in smoke opacity for biodiesel with different percentage of EGR at different loads.

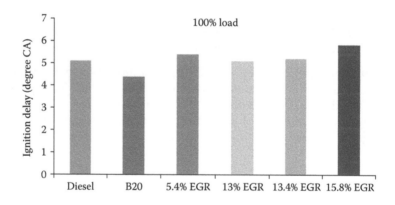

**FIGURE 5.121**
Variation in ignition delay with diesel B20 and different EGR percentages.

delay has an important effect on $NO_x$ emission and knocking. If ignition delay is higher, the fuel accumulation during the period is higher, resulting in higher $NO_x$ emission due to higher temperature, and knocking would occur. In case of EGR, even though ignition delay is higher, $NO_x$ emission is lower due to enhancement of the specific heat of charge by the inert gas induction.

The peak pressure for 13% EGR is about 68 bar at 10° crank angle. A typical pressure–time diagram collected from the engine is illustrated in Figure 5.122. The pressure–crank angle data are used to calculate the pressure rise rate or slope of the pressure–time curve at each data point. It can be seen from this figure that the slope of the pressure–crank angle curve increases

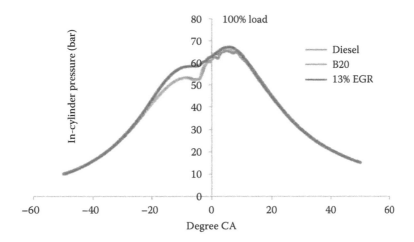

**FIGURE 5.122**
Variation in cylinder gas pressure with biodiesel diesel and 13% EGR at 100% load.

during the compression and combustion period until it reaches the highest value at a certain crank angle, and then the slope starts to decrease, while the pressure is still increasing, until the maximum pressure point is reached. The peak pressures for both the fuels are almost the same.

In Figure 5.123, the maximum heat release rate for 13% EGR is 35 kJ/m³ at 4° BTDC. This is lower than that for diesel and B20 and decreases with the crank angle.

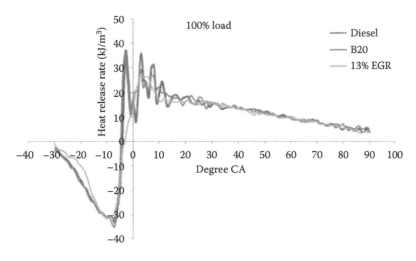

**FIGURE 5.123**
Heat release rate for diesel, B20 and 13% EGR at 100% load.

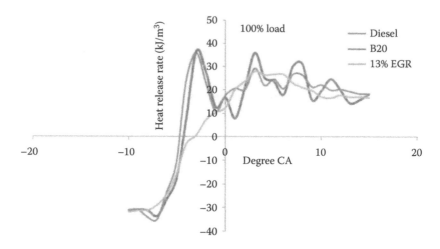

**FIGURE 5.124**
Heat release rate for diesel, B20 and 13% EGR at 100% load.

The heat release rate for 13% EGR is the lowest among all due to the charge dilution and reduction in oxygen concentration resulting in slow combustion (Figure 5.124).

### 5.2.3.8 Conclusions

The following conclusions are drawn based on the experimental results relating to the biodiesel:

*Performance and emission characteristics of biodiesel–diesel blend B20 as compared to base diesel*

- CO and HC emission decreases from 0.04%, 9 ppm with base diesel to 0.03%, 8 ppm with B20, respectively.
- Smoke decreases drastically from 28% opacity with base diesel to 16% with B20.
- $NO_x$ emission increased due to biodiesel having oxygen concentration and automatic advance injection timing due to a higher bulk modulus. $NO_x$ emission increased from 665 ppm with base diesel to 717 ppm with B20.

*Performance and emission characteristics of biodiesel–diesel blend B20 with different EGR% as compared to base biodiesel blend B20*

- HC increases sharply from 420 ppm without EGR to 535 ppm with 15.68% EGR at full load. Beyond 13.46% EGR, HC emissions increase sharply at all loads.

- CO emissions decrease at 50% EGR and its value at 100% load is 2.9 and it decreases after 12% of EGR. Higher values of CO were observed at full load. At 50% load, CO emission also increases with an increase in EGR.
- $CO_2$ emissions for biodiesel blends with EGR are lower at all loads.
- Brake thermal efficiency decreases with an increase in EGR% at all loads.
- The volumetric efficiency decreases with an increase of EGR as exhaust gas replaces the inducted air.
- The temperature of the exhaust gas was found to be lower with EGR.
- Smoke opacity increases with an increasing EGR rate and engine load. EGR reduces the availability of oxygen for the complete combustion of fuel, which results in a relatively incomplete combustion, leading to an increase in smoke emission.
- The ignition delay is maximum for 15% EGR and minimum for B20.
- At full load, $NO_x$ decreases from 662 to 503 ppm with the introduction of 15.86% EGR. At a load of 50%, it is reduced from 435 to 409 ppm and at 75% load, it falls from 613 to 443 ppm.
- CO and HC emission decreases from 0.04%, 9 ppm with base diesel to 0.03%, 8 ppm with B20, respectively.

It can be concluded that the use of biodiesel–diesel blend (B20) in a diesel engine could reduce CO, HC and smoke drastically. However, $NO_x$ emission increases significantly. $NO_x$ emission could be reduced for a biodiesel-fuelled diesel engine using cooled EGR and it could be reduced to the level below base B20 and even below base diesel. $NO_x$ emission decreased from 717 ppm with B20 to 619, 622 and 503 ppm for B20 + 5%EGR, B20 + 10.7%EGR and B20 + 15.86% EGR, respectively. On the whole, the engine could be operated for the use of biodiesel with even 5% EGR for the benefits of $NO_x$ emission reduction without an adverse effect of other emissions such as CO, HC and smoke/PM.

### 5.2.4 Fischer–Tropsch Diesel

Fischer–Tropsch (F–T) diesel fuel quality is characterised by a high cetane number, a near zero sulphur content and a very low aromatic level. F–T diesel fuel can be produced from synthesis gas (CO and $H_2$) through the F–T synthetic processes, using either natural gas or coal as the feedstock. The properties of different kinds of F–T diesel fuel can vary depending on the processing and refining technologies. Table 5.7 shows the properties of F–T diesel and No. 0 diesel fuel.

The engine tests were carried out on a water-cooled, single-cylinder, four-stroke, naturally aspirated direct injection diesel engine, and the performance and emission characteristics are illustrated below.

**TABLE 5.7**

Main properties of F–T and No. 0 Diesel

| Properties | F–T Diesel | No. 0 Diesel |
|---|---|---|
| Liquid density (g/cm³) at 20°C | 0.76 | 0.8312 |
| Kinematic viscosity (mm²/s) at 20°C | 3.276 | 3.0–8.0 |
| Flash point (°C) | 74 | >65 |
| Freezing point (°C) | –2 | <0 |
| Cold filter plugging point (°C) | 2 | <4 |
| Initial boiling point (°C) | 150 | 180 |
| Final boiling point (°C) | 350 | 370 |
| Sulphur (wt.%) | <0.00005 | <0.2 |
| Aromatic (vol.%) | 0.1 | 34.68 |
| Lower heating value (MJ/kg) | 43.9 | 42.6 |
| Cetane number | 74.8 | >45 |
| Carbon content (wt.%) | 85 | 86 |
| Hydrogen content (wt.%) | 15 | 14 |

It can be seen from Figure 5.125 that the peak combustion pressure with F–T diesel fuel is slightly lower and occurs at a later crank angle as compared to No. 0 diesel. The maximum rate of pressure rise for F–T diesel fuel operation is far lower and occurs at an earlier crank angle as compared to No. 0 diesel. This is because the higher cetane number of F–T diesel fuel can considerably shorten the ignition delay and thus reduce the premixed combustion phase, resulting in a reduction in the maximum combustion pressure and rate of pressure rise in the cylinder. The lower maximum pressure and rate of pressure rise with F–T diesel fuel leads to a smoother combustion (spread over a slightly longer period), resulting in a reduction in the mechanical load and combustion noise.

Figure 5.126 compares the BSFC and the brake fuel conversion efficiency of the engine fuelled with F–T and No. 0 diesel fuels. BSFC with F–T diesel is reduced and brake fuel conversion efficiency is increased. The reason for

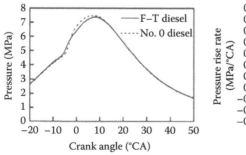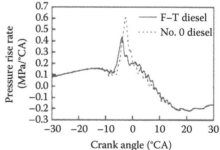

**FIGURE 5.125**

Comparison of in-cylinder pressure and rate of pressure rise. (Adapted from H. Yongcheng et al., *Journal of Automobile Engineering*, 220, 827–835, 2006.)

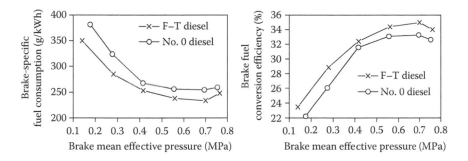

**FIGURE 5.126**
Brake-specific fuel consumption and brake fuel conversion efficiency for F–T and No. 0 diesel fuels. (Adapted from H. Yongcheng et al., *Journal of Automobile Engineering*, 220, 827–835, 2006.)

the improvement in fuel economy is that the engine fuelled with F–T diesel fuel has a lower peak value of in-cylinder pressure and a much lower rate of pressure rise, which reduce the mechanical load and realise smooth combustion (over a slightly longer period), thus improving the thermal efficiency. In addition, some chemical and physical properties of F–T diesel fuel, such as the absence of aromatics and composition based on paraffins and the lower boiling point, seem to be beneficial in improving the combustion process.

Modal-averaged emission results were computed for F–T diesel by weighing each result (in g/kWh) based upon the power of its associated mode (FT 99.1) and then the results were normalised with respect to diesel. As seen in Figure 5.127, F–T diesel yielded reductions in all regulated emissions. From the normalised results, it may be observed that the reductions in THC and CO seem to be most impressive. However, it should be noted that their

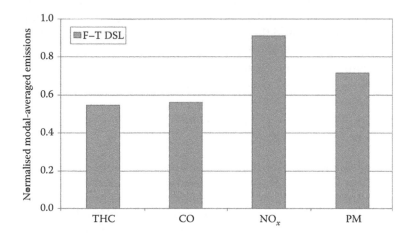

**FIGURE 5.127**
Normalised modal-averaged emissions for F–T diesel. (Adapted from A. S. Cheng and R. W. Dibble, *Society of Automotive Engineers, Inc.*, SAE No.1999-01-3606.)

absolute levels are small (e.g., the reduction in modal-averaged emissions for THC is from 0.146 to 0.081 g/kWh). More important from the standpoint of the diesel engine are the $NO_x$ and PM emissions. A 9% reduction in $NO_x$ emissions was observed while PM decreased by 28%. The reduction in PM with F–T diesel may be attributed to the negligible aromatics and sulphur content of this synthetic fuel.

### 5.2.5 Dimethyl Ether

With the concerns of diminishing petroleum reserves, dimethyl ether (DME) is gathering more attention as a viable alternative to diesel. The advantages of DME over conventional diesel include decreased emissions of $NO_x$, hydrocarbons and carbon monoxide. DME combustion does not produce soot. The operation of a DME engine requires a new storage system and a new fuel delivery system. A DME fuel storage tank must be twice the size of a conventional diesel fuel tank due to its lower energy density as compared to diesel fuel. The most challenging aspects of a DME engine are related to its physical properties and not to its combustion characteristics. The viscosity of DME is lower than that of diesel by a factor of about 20 causing an increased amount of leakage in pumps and fuel injectors. There are also lubrication issues with DME, resulting in the premature wear and eventual failure of pumps and fuel injectors. Additives were used to increase the lubricity of DME, and the commonly used additives were those developed for reformulated diesel. Fundamental research on improving DME wear and lubricity is ongoing.

#### 5.2.5.1 Advantages of DME as Alternative Fuel

- High oxygen content: Together with the absence of any C–C bonds, high oxygen content in the fuel leads to high oxidation rates of particulates, which would therefore reduce the smoke forming tendency under CI engine operation.
- Low boiling point: Leads to quick evaporation when a liquid-phase DME spray is injected into the engine cylinder.
- High cetane number: A high cetane number and low critical temperature, 400 K (127°C), allow the DME injected into the cylinder to evaporate immediately.

The fuel injection delay for the DME and diesel engines versus the BMEP is illustrated in Figure 5.128, in which the two engines have the same fuel delivery advance angle. It can be seen that the fuel injection delay, in crank angle, increases with increasing engine speed for both versions, this delay angle reflecting the pressure wave propagation in the fuel pipe. It is well known that the period of propagation is mainly dominated by the length of the fuel pipe and fuel compressibility, and therefore the fuel injection delay remains almost unchanged in time but increases in crank angle with

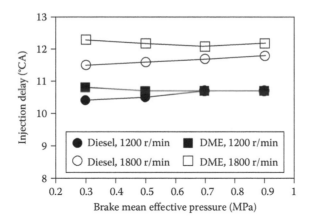

**FIGURE 5.128**
Comparison of injection delay. (Adapted from Z. H. Huang et al., *Journal of Automobile Engineering*, 213, 647–652, 1999.)

increasing engine speed. It can also be seen that the DME engine shows a longer fuel injection delay than the diesel engine at the same engine speed due to its higher specific heat.

From Figure 5.129, it can be seen that the ignition delays of the two fuels show similar trends with engine speed and load, a significant difference at 1200 rev/min and very similar values at 1800 rev/min. It is obvious that

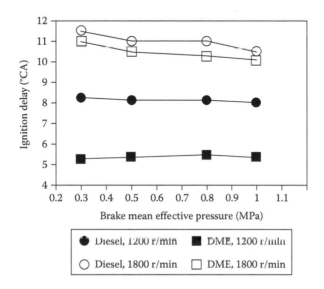

**FIGURE 5.129**
Comparison of ignition delay. (Adapted from Z. H. Huang et al., *Journal of Automobile Engineering*, 213, 647–652, 1999.)

DME has a greater sensitivity of ignition delay to engine speed. The DME engine demonstrates a shorter ignition delay as compared to the diesel-fuelled engine because DME has a high cetane number, a low auto-ignition temperature and good atomisation and ignition properties. All these factors favour a reduction in ignition delay. The short period of ignition delay of the DME engine reduces the amount of fuel injected into the cylinder during this period and results in a reduction in peak pressure and rate of pressure rise, thus reducing the engine mechanical load and combustion noise. Injection delay is defined as the delay period between static injection timing and dynamic injection timing, whereas ignition delay is the time difference between the start of combustion and the start of injection.

Figure 5.130 shows the rate of pressure rise for the DME and diesel engines. It is found that, at the same engine speed and load, the rate of pressure rise shows great differences with the diesel engine both in value and in crank angle. The rate of pressure rise of the DME engine is 0.28–0.47 MPa/°CA, which is at the level of a gasoline engine, and the crank angle at the maximum rate of pressure rise exceeds that after the top dead centre. This low rate of pressure rise results in a reduction in combustion and mechanical noise and realises the smoother combustion of the CI engine.

Figure 5.131 gives the rate of heat release for the DME and diesel engines under the same load. It can be seen that the heat release curves of the DME and diesel engines show a similar curve pattern, although the rate of heat release for the DME engine starts late owing to its long injection delay. However, its end point of combustion arrives earlier than that of the diesel engine. The periods of premixed combustion of the two engines show no

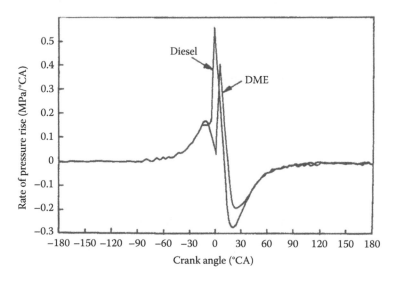

**FIGURE 5.130**
Comparison of rate of pressure rise.

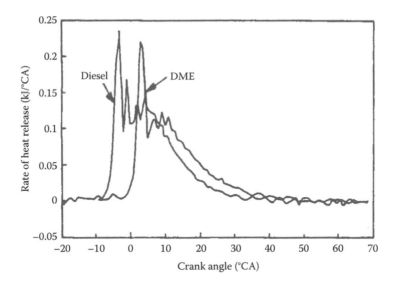

**FIGURE 5.131**
Comparison of heat release. (Adapted from Z. H. Huang et al., *Journal of Automobile Engineering*, 213, 647–652, 1999.)

difference, but the rate of diffusion combustion of the DME engine is much faster than that of the diesel engine.

The good atomisation and evaporation of DME promotes the rapid mixing of DME with the surrounding air, and the excessive oxygen in the combustion chamber during the diffusion combustion period also increases the rate of diffusive combustion. It can be concluded that a rapid rate of diffusive combustion of DME is realised by the fast rate of mixing processes as compared to the poor mixing processes of diesel fuel. Consequently, the DME engine has a controlled rate of premixed combustion and a faster rate of diffusive combustion, with a shorter overall combustion duration than the diesel engine.

Figure 5.132 illustrates the variation of power output against engine speed for different fuels at various speeds (speed characteristics at full load). It can be seen that the power output for DME operation is almost comparable with that of diesel operation at medium and high speeds.

DME has only C–H and C–O bonds but no CKC bond; therefore, the emissions produced from combustion, such as smoke, are expected to be lower than those from diesel operation. An engine with two kinds of vegetable oil can realise smokeless combustion under all operating conditions because they oxygenate DME and the fast diffusion combustion suppresses smoke formation. In addition, the amount of vegetable oil is very small, and so it can be regarded as if there is almost no smoke emission for DME–castor oil and DME–rapeseed oil, as shown in Figure 5.133.

Figure 5.134 shows $NO_x$ emission curves for different fuels at 1800 rev/min. It can be seen that $NO_x$ emission for a DME–vegetable oil operation is

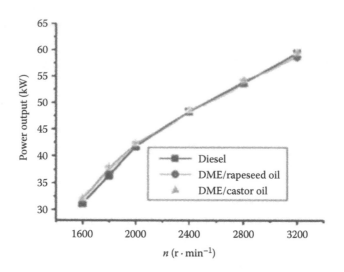

**FIGURE 5.132**
Comparison of the power outputs of the engine fuelled with different fuels. (Adapted from Y. Wang and L. B. Zhou, *Journal of Automobile Engineering*, 221, 1467–1473, 2007.)

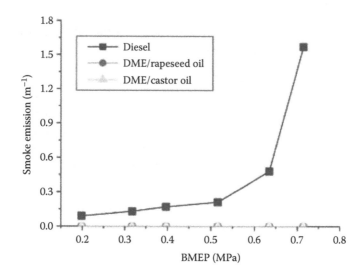

**FIGURE 5.133**
Smoke emission from different fuels at 1800 rev/min.

lower than that for diesel operation. This can be attributed to two facts. On the one hand, DME has a higher cetane number and excellent auto-ignition characteristics, and so the ignition delay is shorter and the premixed combustion is faster than that of diesel fuel, which makes the maximum in-cylinder pressure and temperature lower and prohibits the formation of $NO_x$.

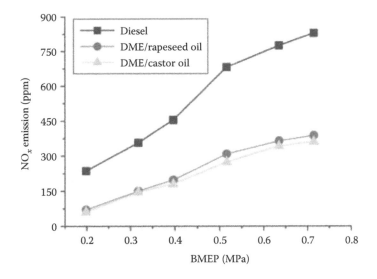

**FIGURE 5.134**
NO$_x$ emission from different fuels at 1800 rev/min. (Adapted from Y. Wang and L. B. Zhou, *Journal of Automobile Engineering*, 221, 1467–1473, 2007.)

On the other hand, the real injection timing of DME fuel is delayed because the propagation time of the pressure wave in DME fuel is longer than that in diesel fuel, which can reduce the NO$_x$ emissions effectively.

Figures 5.135 and 5.136 illustrate CO and HC emission curves for different fuels. It is clear that the CO emission from the DME engine is higher than

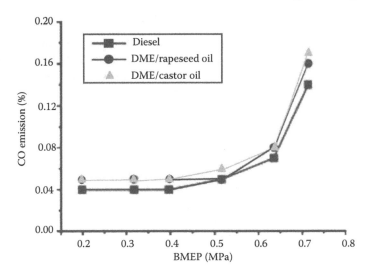

**FIGURE 5.135**
CO emissions from different fuels at 1800 rev/min.

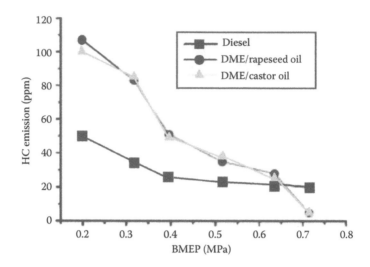

**FIGURE 5.136**
HC emissions from different fuels. (Adapted from Y. Wang and L. B. Zhou, *Journal of Automobile Engineering*, 221, 1467–1473, 2007.)

that from a diesel engine. This can be explained by the secondary injection expected from the higher residual pressure and oscillations encountered in the fuel system when using DME. Similar to the behaviour of CO emission, the exhaust HC concentration from a DME engine is higher than that from a diesel-fuelled engine as shown in Figure 5.136.

### 5.2.6 Dual-Fuel Engine Fuelled with Biodiesel and Hydrogen

Biodiesel could be an alternate and sustainable fuel for diesel engines. It has a superior inherent ignition quality as the cetane number of biodiesel (>58) is higher than that of petrodiesel (50). The biodiesel-blended diesel at any percentage would lead to an enhanced fuel quality of diesel. The use of biodiesel in diesel engines could significantly reduce emissions such as CO, HC and particulate emissions [30–34]. A noticeable decrease in particulate emissions of about 20–40% is reported. However, $NO_x$ emission increased marginally by about 2–4% for 20% biodiesel-blended fuel (B20) [32]. Even though the use of biodiesel in diesel engines could reduce carbon-based emissions (CO, HC and PM) significantly, it has to be reduced further to very lower levels to meet very stringent emission norms in the future.

Carbon-based emission could be reduced drastically to a very low or zero level by hydrogen use in IC engines. Hydrogen is one of the best fuels for SI engines as it has a higher octane number, flame speed and diffusivity. But it can also be used in diesel engines under the dual fuel mode. It means that the main fuel such as diesel and biodiesel will be injected using the main injection system, and the supplementary fuel usually in a gaseous state such as hydrogen, CNG and LPG will be inducted through an intake manifold. The supplementary fuel will be ignited by the main fuel. The main advantages of dual fuel operation with gaseous fuel in diesel engines are the higher brake thermal efficiency due to the benefits of the diesel engine's higher compression ratio and flexible fuel use, and also it needs only a slight engine modification. As the control of particulate emission from diesel engines is one of the major challenges, the gaseous fuel use in IC engines will generally be less in smoke/particulate emissions. The combination of biodiesel and hydrogen use could be expected to give drastic carbon-based emission reductions with significant performance improvements to a diesel engine. It may be highlighted that the control of $CO_2$ emission from IC engines to tackle global warming consequences is a big challenge in the future. So hydrogen emerges as an environmental-friendly auto fuel in all aspects.

The karanja-derived biodiesel (B20) was used for blending with commercially available petrodiesel. The blended fuel was injected using the main injection system, whereas hydrogen was inducted with different percentages (5–56% by energy basis) using a discrete induction-based diaphragm-operated solenoid valve system. The complete set-up of the test rig is shown in Figure 5.137. The injection timing of 230 BTDC was kept constant for all experimental tests. The compressed hydrogen (200 bar) was first fed into a flame trap that was filled with water to prevent backfire hazards and then it was inducted into the intake manifold using a gas induction kit. The kit

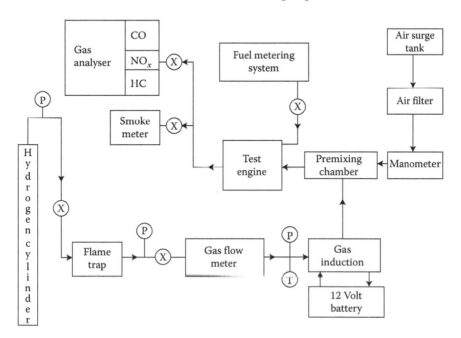

**FIGURE 5.137**
Typical experimental setup.

consists of three stages. When gas enters the first stage, its pressure is reduced to 3.8–4.8 bar. Similarly, in the second stage, its pressure is reduced to 0.8–1.5 bar. The third stage is a solenoid valve part system that is operated with a 12V DC power supply. A solenoid switch is integrated with a diaphragm. Whenever a suction stroke is present, the diaphragm pulls inwards due to vacuum pressure, resulting in the turning of the solenoid switch. Thus, hydrogen is induced in a discrete manner, which ensures the gas induction only during the suction stroke, and in the remaining strokes, the system will be off. The flow rate of hydrogen was measured using a rotameter, whereas the biodiesel blends were measured using a gravimetric method. The air flow rate was measured using a manometer. The CO, $CO_2$, $NO_x$ and HC emissions were measured using an AVL 4000 di-gas analyser. Smoke opacity was measured using an AVL smoke meter.

The brake thermal efficiency increases with an increase in the hydrogen energy share up to a power output of 3.24 kW (74% load). However, it decreases at 94% load. The brake thermal efficiency was the highest at 38% with a 22% hydrogen energy share at 74% load (3.24 kW). It increased from 23% with base diesel to 28% with base B20. Further, the increase up to 38% was observed with a B20+ 22% hydrogen energy share. The diesel engine with a wide range of hydrogen (0–56% energy share) can only be operated up to a 47% load (2.05 kW). In case of higher loads (74% and 94%), hydrogen can be used only up to 22% and 15%, respectively (Figure 5.138). Backfiring was observed beyond the hydrogen percentage at higher loads.

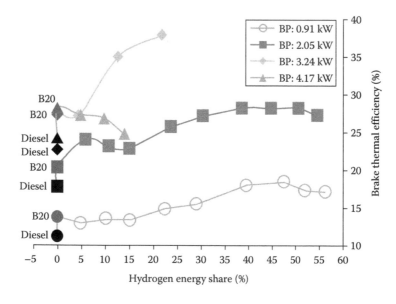

**FIGURE 5.138**
Variation of brake thermal efficiency with hydrogen energy share for different brake power output.

As shown in Figure 5.139, CO emission decreased drastically with the entire hydrogen energy share at higher loads. There was no noticeable change in CO reduction at lower loads as the level was already very low for the base diesel and base B20 cases. HC emission decreases significantly at all loads (Figure 5.140). $NO_x$ emission decreases slightly at lower loads.

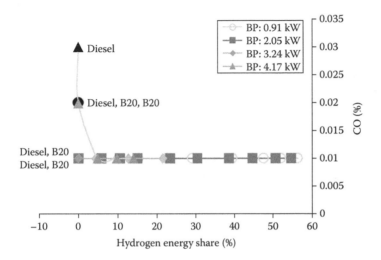

**FIGURE 5.139**
Variation of CO emission with different percentage of hydrogen energy share.

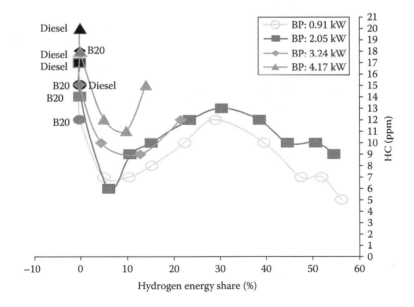

**FIGURE 5.140**
Variation of HC emission with different hydrogen energy share for different brake powers.

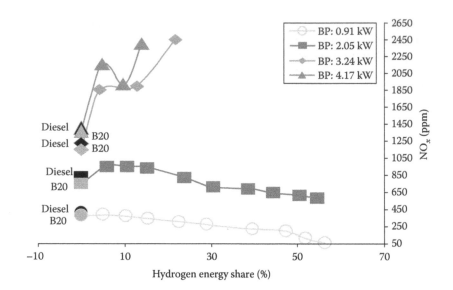

**FIGURE 5.141**
Variation of $NO_x$ with different percentage of hydrogen energy share for different brake powers.

However, it increases significantly at higher loads. It increased at 3.24 kW from 1225 ppm with diesel to 2447 ppm with 22% of the hydrogen energy share (Figure 5.141). Smoke decreases drastically at higher loads. However, it increases beyond about 8% hydrogen energy share at lower loads (Figure 5.142). $CO_2$ emission decreases drastically at all loads (Figure 5.143). A diesel engine with a dual fuel mode emits a higher level of smoke emission due to poor combustion as the in-cylinder temperature is relatively low under part-load operation, resulting in a poor oxidation process. This results in high smoke at lower loads in the dual fuel mode.

The $CO_2$ decreases drastically with an increase in the CNG energy share as CNG has a lower carbon content than diesel/biodiesel. On the whole, it can be concluded that even though it is difficult to specify the optimum hydrogen percentage as it changes with different loads, in general, the use of about 10% hydrogen (energy basis) could give beneficial results in terms of higher thermal efficiency and lower level of emissions at all loads. The high $NO_x$ emission is a major problem with the use of hydrogen. Further studies, such as injection timing optimisation and hydrogen injections, will be needed to assess the full potential of the use of hydrogen in diesel engines.

### 5.2.7 Homogeneous Charge Compression Ignition Engine

HCCI is a concept based on combustion initiation by compression heating. The HCCI engine combines features of both the SI and CI engines, promising the high efficiency of a diesel engine with virtually a very low $NO_x$ and

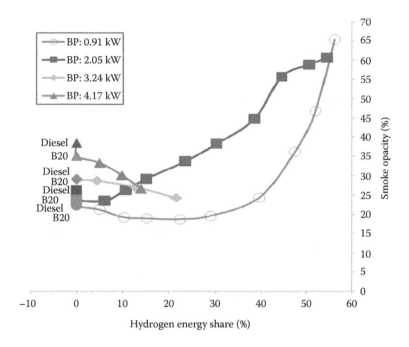

**FIGURE 5.142**
Variation of smoke opacity (%) with different percentage of hydrogen energy share for different brake powers.

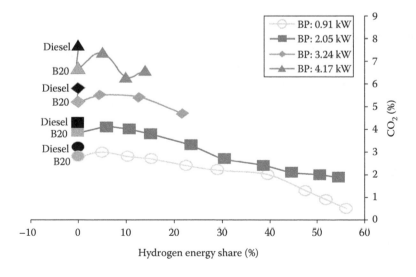

**FIGURE 5.143**
Variation of $CO_2$ emission with different percentage of hydrogen energy share for different brake powers.

particulate emissions [35]. The defining characteristic of HCCI is that the ignition occurs at several places at a time, which makes the fuel/air mixture burn nearly simultaneously. There is no direct initiator of combustion. This makes the process inherently challenging to control. However, with advances in microprocessors and a physical understanding of the ignition process, HCCI can be controlled to achieve gasoline engine emissions along with the efficiency of a diesel engine.

In the HCCI process, the entire fuel/air mixture ignites and burns nearly simultaneously, resulting in high peak pressures and energy release rates. To withstand higher pressures, the engine has to be structurally stronger, which means heavier. Several strategies were proposed to lower the rate of combustion. Two different blends of fuel can be used. In that way, the two fuels will ignite at different points of time, resulting in lesser combustion speed. The problem with this idea is the requirement to set up an infrastructure to supply the blended fuel. Dilution, for example, with exhaust, reduces the pressure and combustion rate at the cost of work production.

Increasing the power output of the HCCI engine is an inherently challenging issue. In HCCI, the entire mixture burns nearly simultaneously. Increasing the fuel/air ratio will result in even higher peak pressures and heat release rates. Also, increasing the fuel/air ratio (also called the equivalence ratio) increases the tendency to knock. In addition, many of the viable control strategies for HCCI require thermal preheating of the charge, which reduces its density. This leads to a reduction in the mass of the air/fuel charge in the combustion chamber, thereby lowering the power.

Diesel HCCI should therefore be considered mainly as a possibility to reduce emissions, rather than to improve fuel consumption. Since the mixture in the HCCI mode is homogeneously distributed in the combustion chamber, there is no diffusive combustion of a rich and high-temperature combustion zone. This results in low levels of both soot and $NO_x$. The low peak temperatures prevent the formation of higher $NO_x$. However, they also lead to the incomplete burning of fuel, especially near the walls of the combustion chamber. This leads to higher carbon monoxide and hydrocarbon emissions. An oxidising catalyst would be effective in removing the regulated species since the exhaust is still rich in oxygen.

### 5.2.8 Premixed Charged Compression Ignition

The premixed charge compression ignition (PCCI) mode of operation involves the preparation of a premixed charge outside the cylinder. A partial amount of the total fuel supply is injected in the intake manifold (may be modified to a chamber) where it is mixed with the intake air and the mixture enters the combustion chamber. The rest of the fuel is injected as usual. The premixing of the fuel with the intake air raises the equivalence ratio of the charge entering, and hence the overall non-homogeneity is reduced in the combustion chamber. The PCCI mode operation is shown in Figure 5.144.

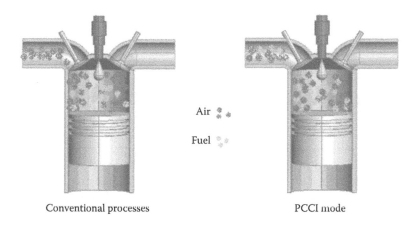

Conventional processes                                      PCCI mode

**FIGURE 5.144**
Conventional CI engine operating under normal an PCCI mode.

Port injection of diesel fuel is very difficult since the environment is too cold for the fuel to vaporise. In the diesel engine, the combustion and emission characteristics are greatly influenced by the quality of atomisation and, in particular, by the fuel–air mixture present in the combustion chamber [36].

Various methods were tried to achieve proper vaporisation of the fuel in the intake manifold. Processes such as hot and cold EGR, preheating the air and large premixing chamber were utilised. Each process has its own set of advantages as well as disadvantages. For example, preheating the air will not only increase the fuel atomisation rate but also decrease the air density, thereby drastically affecting the volumetric efficiency. Hot EGR, if employed, will increase the fuel vaporisation, but it would also raise the net chamber temperature, thereby increasing the chance of $NO_x$ production, and hence the EGR quantity would necessarily require an automatic control mechanism if it is to be used under different load conditions.

In the present case, simple premixing is employed very close to the intake port. Also, the injection of the pre-mixed fuel is timed just after the intake valve opening so that the major quantity of fuel directly enters the cylinder during the suction stroke. This will help to reduce the wall wetting issues in the intake manifold. The partial premixing is achieved by using two injectors, namely, the main injector and an auxiliary injector. The performance and emission characteristics of the engine with a partial PCCI mode are studied and its results compared with the conventional diesel mode operation.

The results of the experimental investigations were analysed and it was found that the partial PCCI mode operation did result in a better performance than the conventional engine. The reductions in emissions were the primary area of investigation and the area of interest. It is involved in testing the feasibility of a partial PCCI concept in achieving the simultaneous reduction of $NO_x$ and smoke.

The results were analysed under two main headings:

- Performance analysis
- Emission analysis

### 5.2.8.1 Performance Analysis

The major advantage of the premixing concept is the higher degree of homogeneity achieved. But since the fuel is injected in the intake manifold, it essentially leads to a drop in the amount of air entering the cylinder. As a result, the volumetric efficiency decreases with an increasing premix ratio with diesel as fuel are shown in Figure 5.145. This disadvantage is offset by the increased fuel–air homogeneity achieved by premixing.

Volumetric efficiency decreases with an increase in load due to a decrease in the density of air by higher residual gas temperature. The volumetric efficiency decreases more in case of partial homogeneous charge compression ignition (PHCCI) as some of the air is replaced by fuel addition into the intake manifold during the suction stroke.

The variation in the BMEP with reference to the equivalence ratio for diesel is shown in Figure 5.146. The better utilisation of the fuel in the engine leads to a very small variation in the BMEP, and with an increasing premix

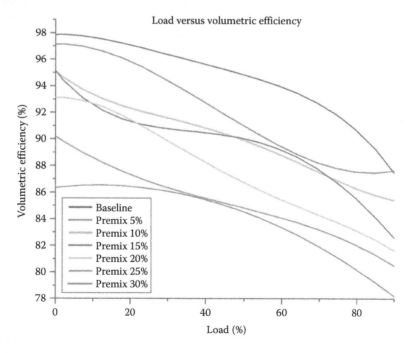

**FIGURE 5.145**
Variation in volumetric efficiency.

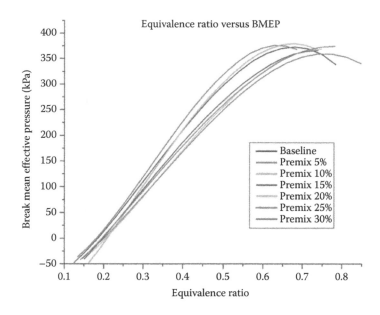

**FIGURE 5.146**
Variation in break mean effective pressure.

fraction it remains more or less constant. This means that the engine performs equally well and, in fact, better in certain instances than the conventional engine.

A small dip in the BMEP is observed at a higher premix fraction; this can be attributed to the early occurrence of the peak pressure as a larger portion of the mixture reaches a combustible range even before the piston approaches the TDC. This early peak pressure tends to oppose the upward movement of the piston leading to a loss of effective power.

The drop in the BMEP is also evident in the thermal efficiency plot as shown in Figure 5.147; with an increasing premix content, the thermal efficiency first increases and then drops for a higher premix fraction.

The variation has been in the range of about 4–7%, with a positive response at a lower premixing range than at higher ends. The efficiency has decreased at the higher ends owing to a drop in the effective amount of fuel vapourising during the auxiliary injection process. Figure 5.148 shows the BSFC with reference to the brake power produced for diesel fuel. As can be seen from the figure, the fuel consumption first reduces and then rises to produce the same power as compared to the baseline data.

The similarity in the trends followed by the BSFC, power and BMEP shows that the fuel atomisation and the level of premixing achieved decreases with increased premix fuel quantity. A drop in the volumetric efficiency owing to the intake manifold injection is also a factor to be considered. The space restriction encountered in the intake manifold limits the amount of space

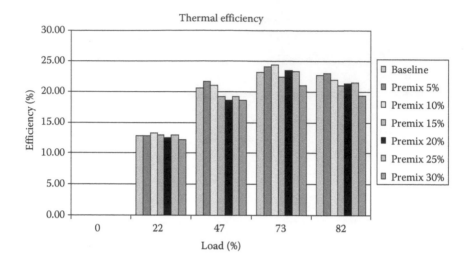

**FIGURE 5.147**
Thermal efficiency under various premix fractions.

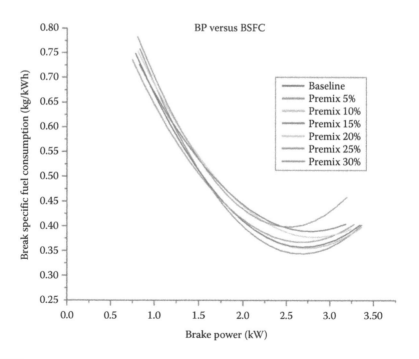

**FIGURE 5.148**
Fuel consumption with reference to the power produced.

available for the fuel jet to travel, at higher fuel content, before it is entrapped by the air. As such, with increasing fuel quantity being injected into the intake manifold, issues like wall impingement come into the picture and this has a pronounced impact on the performance of the engine.

### 5.2.8.2 Emissions Characteristics

The premixing of fuel and air in the intake manifold brings about a pre-cooling effect as the mixture temperature is slightly reduced due to fuel vaporisation. This leads to a more homogeneous combustion and hence lower temperatures. Improved homogeneity also ensures that more air is now utilised by the fuel rather than the nitrogen in the mixture. Both these factors tend to bring down the $NO_x$ level. It may be noted that the $NO_x$ reduction range is about 200% at no load to about 40% at higher loads. As seen from Figure 5.150, the occurrence of peak values also shifts towards higher power values as compared to the baseline (diesel) data. This can be attributed to the increase in temperature due to a higher quantity of the fuel getting burnt. Nevertheless, the peak values are still decreasing with higher premix ratios.

It can be extrapolated from the experimental results as shown in Figure 5.149 that the same trend could continue to the almost zero $NO_x$ level if the spray characteristics of the auxiliary injection are improved. Otherwise, it may lead to deteriorating spray characteristics resulting in higher smoke

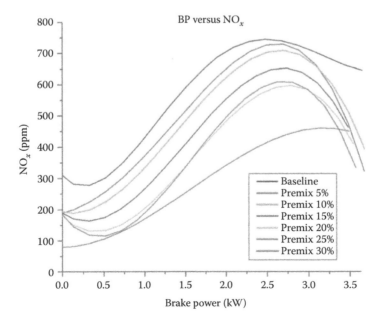

**FIGURE 5.149**
Variation of $NO_x$ emissions with brake power.

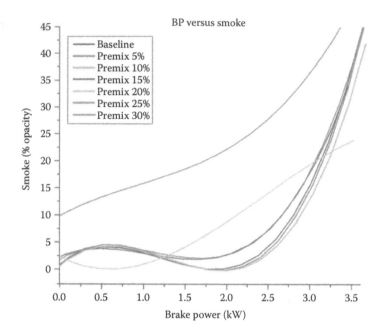

**FIGURE 5.150**
Brake power versus smoke emissions.

emission as shown in Figure 5.150. Smoke emission decreases significantly with the PHCCI mode up to a 10% premixed charge as compared to base diesel. The values at 15% and 20% premixed charges are closer to base diesel. However, it increased to 30% premixed charge due to poor BSFC (as already seen) or brake thermal efficiency. The reason for the trend is the poor mixing of air with relatively more quantity of fuel and also a change in the start of combustion as the premixed charge could take combustion relatively earlier. Two main reasons can be attributed to this increased fuel flow, which has resulted in a decreased air mass owing to a lower volumetric efficiency and reduction in fuel atomisation. Reduced fuel atomisation has more impact than the reduced volumetric efficiency because even at the lowest volumetric efficiency of about 79%, the air–fuel ratio was about 22:1 higher than the stoichiometric value of 16.5:1, proving that there is still excess air for the complete combustion of the fuel. The reduced fuel atomisation, however, will lead to liquid fuel droplets, which result in a higher mass and hence momentum. This tends to cut across the air flow and impinge on the manifold or cylinder walls. Smaller atomised fuel particles are carried with the air flow into the cylinder, leading to better combustion. Hence, the premixed quantity can be increased to still higher values with reduced smoke emissions if means are provided to achieve better atomisation.

The reduction in excess air has also resulted in an increase in HC and CO emissions with increasing premix fractions as shown in Figures 5.151

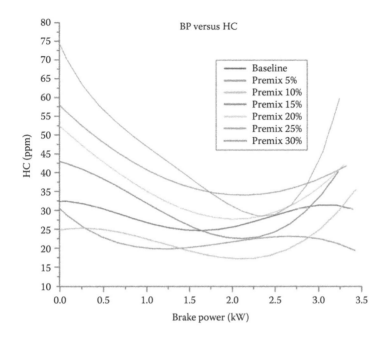

**FIGURE 5.151**
Brake power versus HC emission.

and 5.152. The HC and CO emissions increase with an increase in the pre-mixed charge due to the over-leaning of the mixture at a nearby cylinder wall, which could not sustain the combustion as it was beyond the flamma-bility limit of diesel. The levels are higher at the 30% premixed charge due to poorer combustion, which is confirmed by the higher BSFC and smoke emission. This is due to the presence of higher temperatures at these loads due to the higher fuel consumption. The CO gets converted to $CO_2$ in the after-burning process. At these higher loads, the cylinder conditions ensure that more fuel is atomised in the cylinder leading to a higher amount of HC being consumed than exhausted.

The plot in Figure 5.153 shows the emissions of smoke and $NO_x$ with respect to the equivalence ratio for diesel fuel. As is evident, the combined plot effectively shows the objective of simultaneous reduction of $NO_x$ and smoke. The $NO_x$ emissions decreased considerably and the peak emissions shifted towards higher equivalence ratios. This implies that within the nor-mal operating equivalence ratios as obtained from the baseline data, the $NO_x$ emission reductions were achieved.

With a conventional engine, this should lead to a corresponding increase in the smoke content, which rightly so happens in the baseline case. But with premixed fractions, the smoke content has also reduced. This can be attrib-uted to an increased homogeneity achieved by the premixing, leading to an improved thermal efficiency.

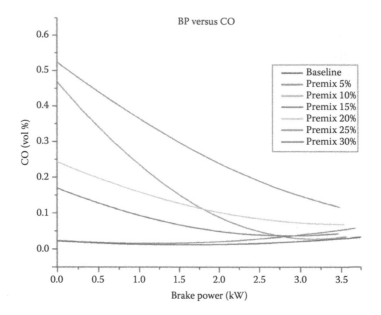

**FIGURE 5.152**
Brake power versus carbon monoxide emission.

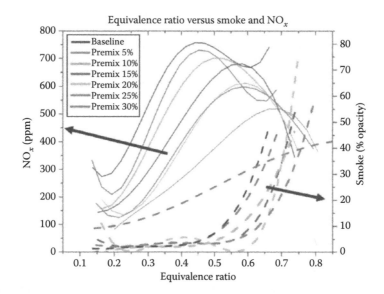

**FIGURE 5.153**
Variation in $NO_x$ and smoke with equivalence ratio.

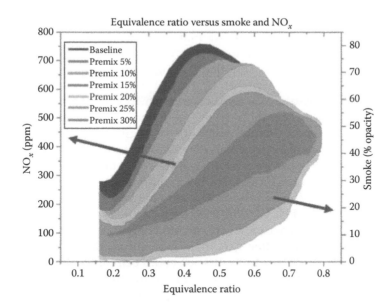

**FIGURE 5.154**
Emission area plot depicting trend of simultaneous $NO_x$ and smoke reduction with increasing premix fractions.

The combined area under the $NO_x$ and smoke plot (Figure 5.154), if taken as the emissions range, has shifted diagonally towards lower $NO_x$ and smoke values, albeit with an increased equivalence ratio.

This increase in the equivalence ratio is not because of increased fuel consumption but due to reduced volumetric efficiency. The utility of this reduction of $NO_x$ and smoke has been substantiated with an improvement in thermal efficiency with nearly similar fuel consumption as the baseline engine operation.

## 5.2.9 Partial HCCI

Technologies such as common rail direct injection (CRDI) and turbocharger are suitable for higher-rated power engines but not suitable economically for small-capacity engines. Certain technologies such as oxygen enrichment and an $H_2$ additive would reduce smoke/PM emissions drastically with negative effects of a dramatic increase in $NO_x$ emission. Among the methods, PHCCI is a viable technique for smoke/PM emission reduction for small-capacity engines.

The formation of $NO_x$ in diesel engines is a function of higher temperature, oxygen concentration and residence time. During combustion, the air containing nitrogen and oxygen reacts at high temperature, leading to the formation of $NO_x$. The localised temperature in heterogeneous combustion is relatively higher than the homogeneous charge combustion resulting in a higher $NO_x$ emission. There are many techniques available to reduce $NO_x$ emissions from diesel engines. Techniques such as EGR, water injection, water–diesel emulsion, injection timing optimisation, split injection, enhancement of premixed charge by PHCCI and after-treatment devices could be used to reduce $NO_x$ emissions. Even though these technologies reduce $NO_x$ emission significantly by reduction in combustion temperature, there is a drop in thermal efficiency of engines. The after-treatment technologies such as selective catalyst reduction (SCR), urea injection and De-$NO_x$ catalyst have the potential to reduce $NO_x$ emission, but the issues such as durability, reaction temperature requirement, initial cost, maintenance, ultra-sulphur fuel requirement, ammonia slip and handling have to be properly designed. PHCCI is an effective technology to reduce $NO_x$ emission significantly by increasing the premixed charge.

The $CO_2$ emission reduction gets more momentum due to global warming and climate change. The $CO_2$ emission could be reduced significantly by increasing fuel economy. Hence, there is a need to conduct studies for reducing these emissions as well as for improving the fuel economy. Most of the technologies reduce $NO_x$ and smoke emission with fuel penalty. In case of PHCCI, the degree of charge homogeneity increases and it is like a constant volume combustion resulting in higher fuel economy.

The new combustion concept, namely PHCCI, has taken the advantage of working principles of both the SI and CI engines. Here, the mixture preparation is like an SI engine and the combustion is like an CI engine. The PHCCI engine operates at nearly a constant volume combustion, resulting in high thermal efficiency and improved fuel economy. Lower $NO_x$ could be achieved due to the localised mixture being relatively lean by homogeneous nature and less residence time by shortening the combustion duration. Particulate emission can be reduced significantly due to homogeneous charge combustion. Even though HCCI has the advantage of a high emission reduction potential and improved fuel economy, it has many challenges such as obtaining the homogeneous mixture and controlled auto-ignition.

A stationary single-cylinder, four-stroke diesel engine was used for the experimental study. The engine had a bore of 102 mm, a stroke of 116 mm and a compression ratio of 19.5 developing 7.4 kW (10 bhp) at 1500 rpm. An 80 kW eddy current-type dynamometer was used for loading and measuring the engine power. The main injector was operated at a nozzle opening pressure of 250–260 bar and the auxiliary injector was at 3 bar. A digital oscilloscope was used for monitoring the auxiliary injection duration and the start of the injection. AVL's di-gas analyser was used to measure CO, HC and $NO_x$. A piezoelectric-type transducer was used for measuring the in-cylinder pressure. The inlet and exhaust gas temperatures were measured using a K-type thermocouple and displayed with the digital indicator. About 0–30% of fuel was injected into the engine intake manifold during the suction stroke, whereas the remaining fuel was injected into the engine cylinder using the main conventional injection system during the compression stroke. The typical valve timing diagram for the conventional and PHCCI modes is shown in Figures 5.155 and 5.156.

The pressure and crank angle signal was acquired using the data acquisition system (AVL) with the help of a crank angle encoder and a pressure sensor, respectively. The heat release rate, percentage of mass fraction burning rate and peak pressure were calculated using the measured pressure–crank angle data. The heat release rate was calculated using the first law of thermodynamics with a measured input parameter of in-cylinder pressure (Equations 5.20 through 5.22).

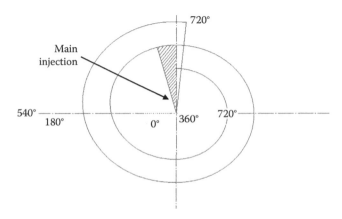

**FIGURE 5.155**
Valve timing diagram with main injection.

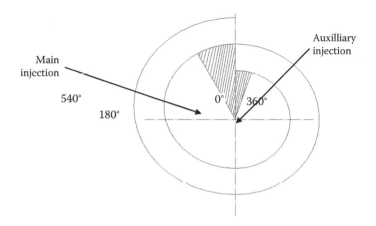

**FIGURE 5.156**
Valve timing diagram for main injection and auxiliary injection for PHCCI mode of operation (hatched).

$$\frac{dQ}{d\theta} = \left(\frac{1}{\gamma - 1}\right) \times \left\{\left[\gamma \times p \times \frac{dV}{d\theta}\right] + \left[V \times \frac{dp}{d\theta}\right]\right\} + \frac{dW}{d\theta} \qquad (5.20)$$

where

$$\frac{dW}{d\theta} = h_c \times A \times (T - T_w) \qquad (5.21)$$

where
$h_c$ = heat transfer coefficient obtained

$$h_c = \frac{130 \times (p_c)^{0.8} \times (v_p + 1.4)^{0.8}}{V^{0.06} \times T_g^{0.4}} \qquad (5.22)$$

### 5.2.9.1 Auxiliary Injector Assembly

Injector selection was a major challenge as the right combinations of the flow rate and spray characteristics were essential to meet the objectives. At a too low flow rate, the fuel would not atomise enough to be used in premixing; on the other hand, a higher flow rate would result in a higher momentum for the fuel particle as compared to the air, which will fail to entrap the fuel. This fuel would then hit the walls of the manifold, resulting in wall impingement. Hence, a proper combination of the flow rate and pump pressure was required. For this purpose, an injector was selected and various pump pressures were tried to achieve a suitable flow rate setting. Injector position is shown in Figure 5.157. The angle of the injector was about 70° with reference to the axis of the intake manifold. The angle was arrived at based on a consideration of wall impingement by a simple geometrical drawing analysis. The combination of the auxiliary fuel injection pump pressure and the injector position ensured the absence of wall impingement.

*Mounting of TDC Position Sensor on Rocker Arm*

The engine rocker arm cover was modified for fixing the proximity sensor to sense the TDC position, which was the input signal to the electronic control unit (ECU) as shown in Figures 5.158 and 5.159. The pick-up has a capability of sensing any metal components within a maximum of an 8 mm gap between the sensor and the inlet rocker arm. The intake rocker arm obtains a drive from the engine cam through the pushrod. The proximity pick-up generates a signal while the intake rocker arm would move up during the suction stroke.

*Injector control unit*

The injector operation was controlled by using an ECU developed specially for this purpose. Using the baseline data for fuel consumption and the

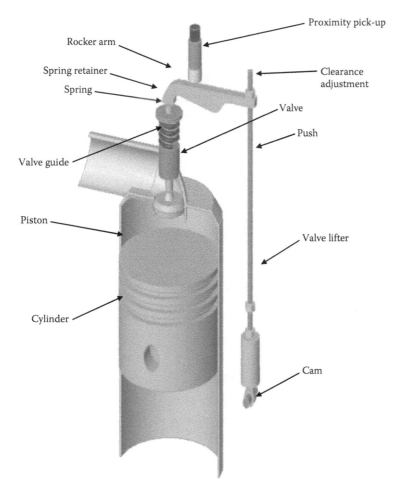

**FIGURE 5.157**
TDC position sensor (proximity pick-up) for providing input signal to the ECU.

injector flow rate, the duration of injector operation was determined. The timing was set as the IVO and different premix fuel quantities were obtained by varying the time for which the injector was energised. The ECU's circuit is shown in Figure 5.160. The cost of the system is about Rs. 5000 (~$100). As the cost of the development of ECU is not significant, it therefore has more feasibility to implement the PHCCI technology of the diesel engine.

### 5.2.9.2 Optimisation of Auxiliary Injection Duration for Achieving Partial HCCI

The engine torque variation is constant up to an injection duration of 4 ms as shown in Figure 5.161. It decreases beyond the 4 ms injection duration.

**FIGURE 5.158**
Gap between proximity pick-up and rocker arm for finding out BDC position.

**FIGURE 5.159**
Gap between proximity pick-up and rocker arm for finding out TDC position.

Torque is the main function of the BMEP and swept volume of the engine. BMEP depends on the volumetric efficiency, thermal efficiency, calorific value of fuel, density of air and fuel–air ratio. The main reason for the decrease in torque is that too early combustion leads to the loss of effective power as the peak pressure, which is developed before TDC opposes the piston motion during the compression stroke.

The peak in-cylinder pressure increases with an increase in injection duration (Figure 5.162). The reason for the increase in pressure is that more fuel accumulation takes place up to ignition delay. The in-cylinder pressure variation is shown in Figure 5.163. Air is inducted during the suction stroke and it

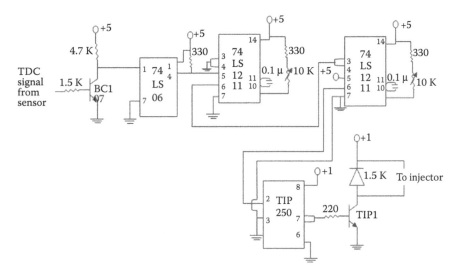

**FIGURE 5.160**
A typical circuit diagram of electronic circuit unit.

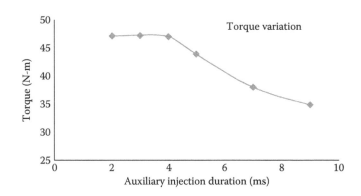

**FIGURE 5.161**
Torque reduction with respect to the increase in auxiliary injection of diesel at 100% load.

is compressed during the compression stroke. At the end of the compression stroke, the fuel is injected and then combustion occurs. But in case of PHCCI, the auxiliary fuel is injected during the suction stroke and enough time is available to mix the auxiliary fuel with air before the main injection starts at the end of the compression stroke. A premixed charge is prepared during the ignition delay period in conventional diesel engines. The premixed charge enhances further with the PHCCI mode by auxiliary fuels.

The heat release rate for base diesel, diesel (main)–diesel (auxiliary) with 10 ms and diesel (main)–diesel (auxiliary) with 14 ms for the premixed combustion is clearly seen in Figure 5.164. In case of PHCCI, the start of combustion

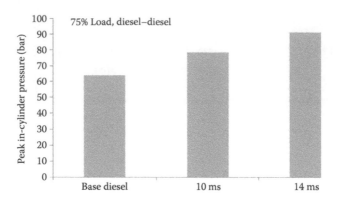

**FIGURE 5.162**
Variation of peak in-cylinder pressure for different auxiliary injection durations.

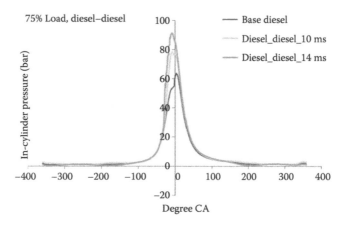

**FIGURE 5.163**
Increase in in-cylinder pressure variation with different auxiliary injection duration at 75% load.

will be advanced during compression stroke as compared to base diesel. The reason for the earlier combustion is due to the auxiliary fuel injected during the suction stroke. The early start of combustion is the main cause for the drop in thermal efficiency. The maximum auxiliary injection duration is up to 4 ms (7.26E − 05 kg/s), which is selected based on torque reduction and knocking. So it was decided to continue further studies within 4 ms.

## 5.3 Conclusion

Alternative fuels have a general problem of lesser energy content, which decreases power and torque relatively as compared to conventional fuels. On

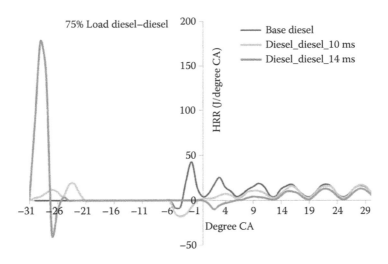

**FIGURE 5.164**
Heat release rate variation for different auxiliary injection duration at 75% load.

the other hand, alternative fuel has desired properties such as lower sulphur, olefins and aromatics as compared to conventional fuel.

**PROBLEMS**

**Example 5.1**

The values obtained from a test bed for a four-cylinder, four-stroke engine is given below:

Shaft speed N = 2500 rev/min
Fuel consumption $m_f$ = 2 g/s
Calorific value = 42 MJ/kg
Stroke L = 100 mm
Bore D = 100 mm
Torque = 80 N m

Calculate the brake power, friction power, indicated power, mean effective pressure, brake thermal efficiency ($\eta_{BTh}$), indicated thermal efficiency ($\eta_{ITh}$) and mechanical efficiency ($\eta_{mech}$).

**Solution to Example 5.1**

BP
$$= 2\pi NT$$
$$= 2\pi \times (2500/60) \times (80)$$
$$= 20.94 \text{ kW}$$

FP
$$= \text{mass/s} \times \text{C.V.}$$

$$= 0.002 \text{ kg/s} \times 42\,000 \text{ kJ/kg}$$
$$= 84 \text{ kW}$$

| | |
|---|---|
| IP | = pLAN |
| Assuming p (MEP) | = 400 kPa |
| IP | = 400 × 0.1 × (π × 0.12/4) × (2500/60)/2 per cylinder |
| | = 6.54 kW per cylinder |
| For four cylinders IP | = 6.54 × 4 = 26.18 kW |
| $\eta_{BTh}$ | = 20.94/84 |
| | = 24.9% |
| $\eta_{ITh}$ | = 26.18/84 |
| | = 31.1% |
| $\eta_{mech}$ | = 20.94/26.18 |
| | = 80% |

### Example 5.2

An Otto cycle works with air at 100 kPa and 20°C, which is compressed isoentropically and a typical otto cycle is shown in Figure 5.P1. The air is then heated at constant volume to 1500°C. The air comes back to the original state. The compression ratio of the engine is 8:1. Calculate the following:

1. The thermal efficiency
2. The heat input per kg of air
3. The network output per kg of air
4. The maximum cycle pressure

$$C_v = 718 \text{ kJ/kg} \quad \gamma = 1.4 \quad R = 287 \text{ J/kg K}$$

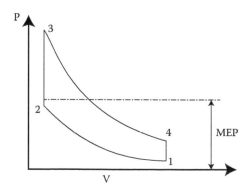

**FIGURE 5.P1**   Otto cycle.

**Solution to Example 5.2**

The mean effective pressure (MEP) is the average pressure such that

| | |
|---|---|
| W(net) | = enclosed area |
| | = MEP $\times$ A $\times$ L |
| W(net) | = MEP $\times$ swept volume |
| | |
| Efficiency ($\eta$) | = $W_{net}/Q_{in}$ |
| | = $1 - ((T_4 - T_1)/(T_3 - T_2))$ |
| | |
| $T_1$ | = 20 + 273 |
| | = 293 K |
| | |
| $T_3$ | = 1500 + 273 |
| | = 1773 K |
| | |
| Compression ratio ($r_c$) | = 8 |
| Efficiency ($\eta$) | = $1-(r_c)^{\gamma-1}$ |
| | = $1-8^{0.4}$ |
| | = 56.5% |

For the adiabatic compression 1–2:

$$T_1 V_1^{\gamma-1} = T_2 V_2^{\gamma-1}$$

| | |
|---|---|
| $T_2$ | = 673.1 K |
| $Q_{in}$ | = $m \times c_v \times (T_3 - T_2)$ |
| | = $1 \times 718 \times (1773 - 673.1)$ |
| | = 789.7 kJ/kg |
| Wnet | = $Q_{in} \times \eta$ |
| | = 446.2 kJ/kg |
| | |
| From the gas law, $P_3$ | = $(P_1 \times V_1 \times T_3)/(T_1 \times V_3)$ |
| | = 4.84 MPa |

**Example 5.3**

A four-stroke SI engine runs at 2500 rpm. The engine swept volume is 0.003 m³. The air is supplied at 0.52 bar and 15°C with a volumetric efficiency of 0.8. The air–fuel ratio is 12. The calorific value is 46 MJ/kg. Calculate the heat release by combustion.

**Solution to Example 5.3**

| | |
|---|---|
| Capacity | = 0.003 m³ |
| Volume induced | = 0.003 $\times$ (2500/60)/2 |
| | = 0.0625 m³/s |

| | |
|---|---|
| Using the gas law pV | $= mRT$ |
| Ideal air m | $= pV/RT$ |
| | $= 0.52 \times 105 \times 0.0625/(287 \times 288)$ |
| | $= 0.03932$ kg/s |
| Actual air m | $= 0.03932 \times 0.8$ |
| | $= 0.03146$ kg/s |
| Mass of fuel $m_f$ | $= 0.03146/12$ |
| | $= 0.002621$ kg/s |
| Heat released | $=$ calorific value $\times m_f$ |
| | $= 46\,000$ kJ/kg $\times 0.002621$ kg/s |
| | $= 120.5813$ kW |

**Example 5.4**

Determine BMEP at rated power, bore-to-stroke ratio, piston speed, power-to-volume ratio for an automotive engine. The engine details are given below:

| | |
|---|---|
| Bore (B), mm | 80.5 |
| Stroke (L), mm | 88.3 |
| Compression ratio (rc) | 10:01 |
| Rated speed (N), rpm | 4400 |
| Rated torque (T) N m | 174 |

If the fuel consumption rate of the engine is 0.0135 kg/s and the engine draws air of 0.0011 kg/s, calculate the thermal efficiency at the rated power, specific fuel consumption and volumetric efficiency.

**Solution to Example 5.4**

| | |
|---|---|
| Power | $= 2 \times \pi \times N \times T$ |
| | $= (2 \times \pi \times 4400 \times 174)/60$ |
| | $= 98$ kW |
| Displacement volume ($V_d$) | $= \pi \times B^2/4$ |
| | $= \pi \times 0.00805^2/4$ |
| | $= 1797 \times 10^{-6}$ m$^3$ |
| BMEP @ rated power | $= (P \times n \times 10^3)/(V_d \times N)$ |
| | $= 98 \times 2 \times 10^3/(1797 \times 10^{-6} \times 4400)$ |
| | $= 10.90$ bar |
| B/L | $= 80.5/88.3$ |
| | $= 0.91$ |
| Piston speed (Sp) | $= 2 \times L \times N$ |

$$= 2 \times 0.0883 \times 4400$$
$$= 17.66 \text{ m/s}$$

Power-to-volume ratio
$$= 98/1797 \times 10^{-6}$$
$$= 0.0545 \text{ kW/cc}$$

Brake thermal efficiency of the engine at rated power (BTE) $= P/(m_f \times CV)$
$$= 98/(0.0135 \times 40 \times 1000)$$
$$= 18.15\%$$

(assuming CV of the fuel is 40 MJ/kg)

Specific fuel consumption (SFC)
$$= m_f/P$$
$$= (0.0135 \times 1000 \times 3600)/98$$
$$= 495 \text{ g/kWh}$$

Volumetric efficiency
$$= (2 \times ma)/(\rho \times Vd \times N)$$
$$= (2 \times 0.011)/(1.205 \times 1797 \times 10^{-6} \times 4400)$$
$$= 87.42\%$$

## Example 5.5

Calculate the instantaneous volume, rate of change of volume, in-cylinder temperature, heat release rate (using the first law of thermodynamics) and rate of pressure rise for the given measured in-cylinder pressure with respect to the crank angle data (column 1 and 2). Engine details are given below:

| | |
|---|---|
| Connecting rod length (l) | = 232 mm |
| Bore (b) | = 110 mm |
| Compression ratio | = 19.5:1 |
| Adiabatic index (k) | = 1.33 for air–fuel mixture |

## Solution to Example 5.5

Instantaneous volume of cylinder (V)

$$V = \frac{V_d}{r-1} + \frac{V_d}{2}[1 + R - \cos\theta - (R^2 - \sin^2\theta)^{1/2}]$$

Rate of change of volume

$$\frac{dV}{d\theta} = \frac{V_d}{2}\sin\theta[1 + \cos\theta(R^2 - \sin^2\theta)^{-1/2}]$$

From the first law of thermodynamics, heat release rate is given by

$$\frac{\partial Q}{d\theta} = \frac{1}{k-1}V\frac{dP}{d\theta} + \frac{k}{k-1}P\frac{dV}{d\theta}$$

where
  V = instantaneous volume of cylinder
  R = l/a
  l = connecting rod length = 232 mm
  a = crank radius
  r = compression ratio = 19.5:1
  s = stroke
  b = bore
  $V_d$ = swept volume
  Adiabatic index (k) = 1.3–1.36 for air–fuel mixture
  η = crank angle

| CA (degree) (1) | P (bar) (2) | V (θ) (m³) (3) | dp/dθ (4) | dV/dθ (5) | dQ/dθ (6) |
|---|---|---|---|---|---|
| −90 | 2.437 | 0.000587207 | 6.40E−02 | −8.25E−06 | −7.59E+00 |
| −89 | 2.48 | 0.000578923 | 4.30E−02 | −8.28E−06 | −7.76E+00 |
| −88 | 2.523 | 0.000570603 | 4.30E−02 | −8.32E−06 | −7.93E+00 |
| −87 | 2.587 | 0.000562249 | 6.40E−02 | −8.35E−06 | −8.16E+00 |
| −86 | 2.651 | 0.000553864 | 6.40E−02 | −8.38E−06 | −8.40E+00 |
| −85 | 2.683 | 0.000545451 | 3.20E−02 | −8.41E−06 | −8.53E+00 |
| −84 | 2.747 | 0.000537013 | 6.40E−02 | −8.44E−06 | −8.76E+00 |
| −83 | 2.822 | 0.000528552 | 7.50E−02 | −8.46E−06 | −9.02E+00 |
| −82 | 2.865 | 0.000520071 | 4.30E−02 | −8.48E−06 | −9.18E+00 |
| −81 | 2.929 | 0.000511572 | 6.40E−02 | −8.50E−06 | −9.40E+00 |
| −80 | 3.025 | 0.00050306 | 9.60E−02 | −8.51E−06 | −9.73E+00 |
| −79 | 3.057 | 0.000494536 | 3.20E−02 | −8.52E−06 | −9.84E+00 |
| −78 | 3.153 | 0.000486005 | 9.60E−02 | −8.53E−06 | −1.02E+01 |
| −77 | 3.228 | 0.000477468 | 7.50E−02 | −8.54E−06 | −1.04E+01 |
| −76 | 3.303 | 0.000468928 | 7.50E−02 | −8.54E−06 | −1.07E+01 |
| −75 | 3.399 | 0.00046039 | 9.60E−02 | −8.54E−06 | −1.10E+01 |
| −74 | 3.485 | 0.000451856 | 8.60E−02 | −8.53E−06 | −1.12E+01 |
| −73 | 3.56 | 0.000443329 | 7.50E−02 | −8.53E−06 | −1.15E+01 |
| −72 | 3.656 | 0.000434812 | 9.60E−02 | −8.52E−06 | −1.18E+01 |
| −71 | 3.763 | 0.00042631 | 1.07E−01 | −8.50E−06 | −1.21E+01 |
| −70 | 3.87 | 0.000417823 | 1.07E−01 | −8.49E−06 | −1.24E+01 |
| −69 | 3.987 | 0.000409357 | 1.17E−01 | −8.47E−06 | −1.28E+01 |
| −68 | 4.094 | 0.000400915 | 1.07E−01 | −8.44E−06 | −1.31E+01 |
| −67 | 4.233 | 0.000392499 | 1.39E−01 | −8.42E−06 | −1.35E+01 |
| −66 | 4.34 | 0.000384113 | 1.07E−01 | −8.39E−06 | −1.37E+01 |
| −65 | 4.458 | 0.00037576 | 1.18E−01 | −8.35E−06 | −1.41E+01 |
| −64 | 4.607 | 0.000367444 | 1.49E−01 | −8.32E−06 | −1.45E+01 |
| −63 | 4.757 | 0.000359168 | 1.50E−01 | −8.28E−06 | −1.49E+01 |
| −62 | 4.917 | 0.000350935 | 1.60E−01 | −8.23E−06 | −1.53E+01 |
| −61 | 5.046 | 0.000342749 | 1.29E−01 | −8.19E−06 | −1.56E+01 |

| CA (degree) (1) | P (bar) (2) | V ($\theta$) (m³) (3) | dp/d$\theta$ (4) | dV/d$\theta$ (5) | dQ/d$\theta$ (6) |
|---|---|---|---|---|---|
| −60 | 5.238 | 0.000334613 | 1.92E−01 | −8.14E−06 | −1.61E+01 |
| −59 | 5.409 | 0.00032653 | 1.71E−01 | −8.08E−06 | −1.65E+01 |
| −58 | 5.591 | 0.000318505 | 1.82E−01 | −8.03E−06 | −1.70E+01 |
| −57 | 5.783 | 0.000310539 | 1.92E−01 | −7.97E−06 | −1.74E+01 |
| −56 | 5.997 | 0.000302637 | 2.14E−01 | −7.90E−06 | −1.79E+01 |
| −55 | 6.211 | 0.000294802 | 2.14E−01 | −7.84E−06 | −1.84E+01 |
| −54 | 6.435 | 0.000287037 | 2.24E−01 | −7.76E−06 | −1.89E+01 |
| −53 | 6.671 | 0.000279346 | 2.36E−01 | −7.69E−06 | −1.94E+01 |
| −52 | 6.927 | 0.000271732 | 2.56E−01 | −7.61E−06 | −1.99E+01 |
| −51 | 7.195 | 0.000264198 | 2.68E−01 | −7.53E−06 | −2.05E+01 |
| −50 | 7.473 | 0.000256747 | 2.78E−01 | −7.45E−06 | −2.10E+01 |
| −49 | 7.75 | 0.000249383 | 2.77E−01 | −7.36E−06 | −2.16E+01 |
| −48 | 8.082 | 0.000242109 | 3.32E−01 | −7.27E−06 | −2.22E+01 |
| −47 | 8.403 | 0.000234928 | 3.21E−01 | −7.18E−06 | −2.28E+01 |
| −46 | 8.755 | 0.000227844 | 3.52E−01 | −7.08E−06 | −2.34E+01 |
| −45 | 9.119 | 0.000220859 | 3.64E−01 | −6.98E−06 | −2.41E+01 |
| −44 | 9.504 | 0.000213977 | 3.85E−01 | −6.88E−06 | −2.47E+01 |
| −43 | 9.921 | 0.000207201 | 4.17E−01 | −6.78E−06 | −2.54E+01 |
| −42 | 10.37 | 0.000200533 | 4.49E−01 | −6.67E−06 | −2.61E+01 |
| −41 | 10.819 | 0.000193978 | 4.49E−01 | −6.56E−06 | −2.68E+01 |
| −40 | 11.321 | 0.000187537 | 5.02E−01 | −6.44E−06 | −2.75E+01 |
| −39 | 11.834 | 0.000181213 | 5.13E−01 | −6.32E−06 | −2.83E+01 |
| −38 | 12.38 | 0.000175011 | 5.46E−01 | −6.20E−06 | −2.90E+01 |
| −37 | 12.968 | 0.000168932 | 5.88E−01 | −6.08E−06 | −2.98E+01 |
| −36 | 13.588 | 0.000162979 | 6.20E−01 | −5.95E−06 | −3.06E+01 |
| −35 | 14.251 | 0.000157155 | 6.63E−01 | −5.82E−06 | −3.14F+01 |
| −34 | 14.935 | 0.000151463 | 6.84E−01 | −5.69E−06 | −3.21E+01 |
| −33 | 15.705 | 0.000145905 | 7.70E−01 | −5.56E−06 | −3.30E+01 |
| −32 | 16.506 | 0.000140484 | 8.01E−01 | −5.42E−06 | −3.38E+01 |
| −31 | 17.319 | 0.000135202 | 8.13E−01 | −5.28E−06 | −3.46E+01 |
| −30 | 18.196 | 0.000130062 | 8.77E−01 | −5.14E−06 | −3.53E+01 |
| −29 | 19.158 | 0.000125067 | 9.62E−01 | −5.00E−06 | −3.62E+01 |
| −28 | 20.12 | 0.000120218 | 9.62E−01 | −4.85E−06 | −3.69E+01 |
| −27 | 21.2 | 0.000115517 | 1.08E+00 | −4.70E−06 | −3.76E+01 |
| −26 | 22.29 | 0.000110968 | 1.09E+00 | −4.55E−06 | −3.83E+01 |
| −25 | 23.466 | 0.000106573 | 1.18E+00 | −4.40E−06 | −3.90E+01 |
| −24 | 24.717 | 0.000102332 | 1.25E+00 | −4.24E−06 | −3.96E+01 |
| −23 | 26.011 | 9.8249E−05 | 1.29E+00 | −4.08E−06 | −4.01E+01 |
| −22 | 27.379 | 9.4325E−05 | 1.37E+00 | −3.92E−06 | −4.06E+01 |
| −21 | 28.823 | 9.05621E−05 | 1.44E+00 | −3.76E−06 | −4.10E+01 |
| −20 | 30.309 | 8.6962E−05 | 1.49E+00 | −3.60E−06 | −4.12E+01 |
| −19 | 31.859 | 8.35266E−05 | 1.55E+00 | −3.44E−06 | −4.13E+01 |
| −18 | 33.463 | 8.02573E−05 | 1.60E+00 | −3.27E−06 | −4.13E+01 |

| CA (degree) (1) | P (bar) (2) | V ($\theta$) (m³) (3) | dp/d$\theta$ (4) | dV/d$\theta$ (5) | dQ/d$\theta$ (6) |
|---|---|---|---|---|---|
| −17 | 35.109 | 7.71557E−05 | 1.65E+00 | −3.10E−06 | −4.11E+01 |
| −16 | 36.83 | 7.42234E−05 | 1.72E+00 | −2.93E−06 | −4.08E+01 |
| −15 | 38.552 | 7.14617E−05 | 1.72E+00 | −2.76E−06 | −4.02E+01 |
| −14 | 40.326 | 6.8872E−05 | 1.77E+00 | −2.59E−06 | −3.95E+01 |
| −13 | 42.048 | 6.64555E−05 | 1.72E+00 | −2.42E−06 | −3.84E+01 |
| −12 | 43.833 | 6.42132E−05 | 1.79E+00 | −2.24E−06 | −3.71E+01 |
| −11 | 45.522 | 6.21464E−05 | 1.69E+00 | −2.07E−06 | −3.55E+01 |
| −10 | 47.222 | 6.0256E−05 | 1.70E+00 | −1.89E−06 | −3.37E+01 |
| −9 | 48.836 | 5.85428E−05 | 1.61E+00 | −1.71E−06 | −3.16E+01 |
| −8 | 50.354 | 5.70078E−05 | 1.52E+00 | −1.54E−06 | −2.92E+01 |
| −7 | 51.744 | 5.56517E−05 | 1.39E+00 | −1.36E−06 | −2.65E+01 |
| −6 | 53.017 | 5.4475E−05 | 1.27E+00 | −1.18E−06 | −2.36E+01 |
| −5 | 54.118 | 5.34784E−05 | 1.10E+00 | −9.97E−07 | −2.04E+01 |
| −4 | 55.069 | 5.26624E−05 | 9.51E−01 | −8.16E−07 | −1.70E+01 |
| −3 | 55.775 | 5.20273E−05 | 7.06E−01 | −6.35E−07 | −1.34E+01 |
| −2 | 56.299 | 5.15734E−05 | 5.24E−01 | −4.54E−07 | −9.65E+00 |
| −1 | 56.545 | 5.1301E−05 | 2.46E−01 | −2.72E−07 | −5.82E+00 |
| 0 | 56.502 | 5.12102E−05 | −4.30E−02 | −9.08E−08 | −1.94E+00 |
| 1 | 56.384 | 5.1301E−05 | −1.18E−01 | 9.08E−08 | 1.93E+00 |
| 2 | 55.433 | 5.15734E−05 | −9.51E−01 | 2.72E−07 | 5.70E+00 |
| 3 | 55.401 | 5.20273E−05 | −3.20E−02 | 4.54E−07 | 9.50E+00 |
| 4 | 56.213 | 5.26624E−05 | 8.12E−01 | 6.35E−07 | 1.35E+01 |
| 5 | 60.126 | 5.34784E−05 | 3.91E+00 | 8.16E−07 | 1.85E+01 |
| 6 | 65.055 | 5.4475E−05 | 4.93E+00 | 9.97E−07 | 2.45E+01 |
| 7 | 65.568 | 5.56517E−05 | 5.13E−01 | 1.18E−06 | 2.91E+01 |
| 8 | 68.722 | 5.70078E−05 | 3.15E+00 | 1.36E−06 | 3.52E+01 |
| 9 | 69.043 | 5.85428E−05 | 3.21E−01 | 1.54E−06 | 4.00E+01 |
| 10 | 68.529 | 6.0256E−05 | −5.14E−01 | 1.71E−06 | 4.44E+01 |
| 11 | 69.01 | 6.21464E−05 | 4.81E−01 | 1.89E−06 | 4.93E+01 |
| 12 | 67.182 | 6.42132E−05 | −1.83E+00 | 2.07E−06 | 5.25E+01 |
| 13 | 67.278 | 6.64555E−05 | 9.60E−02 | 2.24E−06 | 5.70E+01 |
| 14 | 65.696 | 6.8872E−05 | −1.58E+00 | 2.42E−06 | 6.00E+01 |
| 15 | 65.14 | 7.14617E−05 | −5.56E−01 | 2.59E−06 | 6.37E+01 |
| 16 | 63.12 | 7.42234E−05 | −2.02E+00 | 2.76E−06 | 6.59E+01 |
| 17 | 62.072 | 7.71557E−05 | −1.05E+00 | 2.93E−06 | 6.88E+01 |
| 18 | 60.222 | 8.02573E−05 | −1.85E+00 | 3.10E−06 | 7.06E+01 |
| 19 | 58.18 | 8.35266E−05 | −2.04E+00 | 3.27E−06 | 7.19E+01 |
| 20 | 56.213 | 8.6962E−05 | −1.97E+00 | 3.44E−06 | 7.30E+01 |
| 21 | 53.744 | 9.05621E−05 | −2.47E+00 | 3.60E−06 | 7.31E+01 |
| 22 | 52.439 | 9.4325E−05 | −1.31E+00 | 3.76E−06 | 7.45E+01 |
| 23 | 49.788 | 9.8249E−05 | −2.65E+00 | 3.92E−06 | 7.38E+01 |
| 24 | 47.81 | 0.000102332 | −1.98E+00 | 4.08E−06 | 7.37E+01 |
| 25 | 45.715 | 0.000106573 | −2.10E+00 | 4.24E−06 | 7.32E+01 |

| CA (degree) (1) | P (bar) (2) | V ($\theta$) (m³) (3) | dp/d$\theta$ (4) | dV/d$\theta$ (5) | dQ/d$\theta$ (6) |
|---|---|---|---|---|---|
| 26 | 43.662 | 0.000110968 | −2.05E+00 | 4.40E−06 | 7.25E+01 |
| 27 | 42.197 | 0.000115517 | −1.47E+00 | 4.55E−06 | 7.25E+01 |
| 28 | 40.198 | 0.000120218 | −2.00E+00 | 4.70E−06 | 7.14E+01 |
| 29 | 38.648 | 0.000125067 | −1.55E+00 | 4.85E−06 | 7.08E+01 |
| 30 | 36.798 | 0.000130062 | −1.85E+00 | 5.00E−06 | 6.94E+01 |
| 31 | 35.323 | 0.000135202 | −1.48E+00 | 5.14E−06 | 6.86E+01 |
| 32 | 33.602 | 0.000140484 | −1.72E+00 | 5.28E−06 | 6.70E+01 |
| 33 | 32.212 | 0.000145905 | −1.39E+00 | 5.42E−06 | 6.60E+01 |
| 34 | 30.672 | 0.000151463 | −1.54E+00 | 5.56E−06 | 6.44E+01 |
| 35 | 29.603 | 0.000157155 | −1.07E+00 | 5.69E−06 | 6.37E+01 |
| 36 | 28.106 | 0.000162979 | −1.50E+00 | 5.82E−06 | 6.18E+01 |
| 37 | 27.069 | 0.000168932 | −1.04E+00 | 5.95E−06 | 6.09E+01 |
| 38 | 25.947 | 0.000175011 | −1.12E+00 | 6.08E−06 | 5.96E+01 |
| 39 | 24.749 | 0.000181213 | −1.20E+00 | 6.20E−06 | 5.80E+01 |
| 40 | 23.702 | 0.000187537 | −1.05E+00 | 6.32E−06 | 5.66E+01 |
| 41 | 22.793 | 0.000193978 | −9.09E−01 | 6.44E−06 | 5.55E+01 |
| 42 | 21.82 | 0.000200533 | −9.73E−01 | 6.56E−06 | 5.40E+01 |
| 43 | 20.943 | 0.000207201 | −8.77E−01 | 6.67E−06 | 5.28E+01 |
| 44 | 20.152 | 0.000213977 | −7.91E−01 | 6.78E−06 | 5.16E+01 |
| 45 | 19.372 | 0.000220859 | −7.80E−01 | 6.88E−06 | 5.04E+01 |
| 46 | 18.613 | 0.000227844 | −7.59E−01 | 6.98E−06 | 4.91E+01 |
| 47 | 17.95 | 0.000234928 | −6.63E−01 | 7.08E−06 | 4.80E+01 |
| 48 | 17.287 | 0.000242109 | −6.63E−01 | 7.18E−06 | 4.69E+01 |
| 49 | 16.688 | 0.000249383 | −5.99E−01 | 7.27E−06 | 4.59E+01 |
| 50 | 16.047 | 0.000256747 | −6.41E−01 | 7.36E−06 | 4.46E+01 |
| 51 | 15.544 | 0.000264198 | −5.03E−01 | 7.45E−06 | 4.38E+01 |
| 52 | 14.956 | 0.000271732 | −5.88E−01 | 7.53E−06 | 4.26E+01 |
| 53 | 14.497 | 0.000279346 | −4.59E−01 | 7.61E−06 | 4.17E+01 |
| 54 | 13.791 | 0.000287037 | −7.06E−01 | 7.69E−06 | 4.01E+01 |
| 55 | 13.524 | 0.000294802 | −2.67E−01 | 7.76E−06 | 3.97E+01 |
| 56 | 12.914 | 0.000302637 | −6.10E−01 | 7.84E−06 | 3.82E+01 |
| 57 | 12.604 | 0.000310539 | −3.10E−01 | 7.90E−06 | 3.76E+01 |
| 58 | 12.251 | 0.000318505 | −3.53E−01 | 7.97E−06 | 3.69E+01 |
| 59 | 11.845 | 0.00032653 | −4.06E−01 | 8.03E−06 | 3.59E+01 |
| 60 | 11.418 | 0.000334613 | −4.27E−01 | 8.08E−06 | 3.49E+01 |
| 61 | 11.075 | 0.000342749 | −3.43E−01 | 8.14E−06 | 3.40E+01 |
| 62 | 10.797 | 0.000350935 | −2.78E−01 | 8.19E−06 | 3.34E+01 |
| 63 | 10.487 | 0.000359168 | −3.10E−01 | 8.23E−06 | 3.26E+01 |
| 64 | 10.145 | 0.000367444 | −3.42E−01 | 8.28E−06 | 3.17E+01 |
| 65 | 9.846 | 0.00037576 | −2.99E−01 | 8.32E−06 | 3.09E+01 |
| 66 | 9.611 | 0.000384113 | −2.35E−01 | 8.35E−06 | 3.03E+01 |
| 67 | 9.29 | 0.000392499 | −3.21E−01 | 8.39E−06 | 2.94E+01 |
| 68 | 9.098 | 0.000400915 | −1.92E−01 | 8.42E−06 | 2.89E+01 |

| CA (degree) (1) | P (bar) (2) | V (θ) (m³) (3) | dp/dθ (4) | dV/dθ (5) | dQ/dθ (6) |
|---|---|---|---|---|---|
| 69 | 8.83 | 0.000409357 | −2.68E−01 | 8.44E−06 | 2.82E+01 |
| 70 | 8.606 | 0.000417823 | −2.24E−01 | 8.47E−06 | 2.75E+01 |
| 71 | 8.371 | 0.00042631 | −2.35E−01 | 8.49E−06 | 2.68E+01 |
| 72 | 8.157 | 0.000434812 | −2.14E−01 | 8.50E−06 | 2.62E+01 |
| 73 | 7.975 | 0.000443329 | −1.82E−01 | 8.52E−06 | 2.57E+01 |
| 74 | 7.783 | 0.000451856 | −1.92E−01 | 8.53E−06 | 2.51E+01 |
| 75 | 7.59 | 0.00046039 | −1.93E−01 | 8.53E−06 | 2.45E+01 |
| 76 | 7.419 | 0.000468928 | −1.71E−01 | 8.54E−06 | 2.39E+01 |
| 77 | 7.227 | 0.000477468 | −1.92E−01 | 8.54E−06 | 2.33E+01 |
| 78 | 7.077 | 0.000486005 | −1.50E−01 | 8.54E−06 | 2.28E+01 |
| 79 | 6.906 | 0.000494536 | −1.71E−01 | 8.53E−06 | 2.23E+01 |
| 80 | 6.746 | 0.00050306 | −1.60E−01 | 8.52E−06 | 2.17E+01 |
| 81 | 6.628 | 0.000511572 | −1.18E−01 | 8.51E−06 | 2.13E+01 |
| 82 | 6.478 | 0.000520071 | −1.50E−01 | 8.50E−06 | 2.08E+01 |
| 83 | 6.307 | 0.000528552 | −1.71E−01 | 8.48E−06 | 2.02E+01 |
| 84 | 6.19 | 0.000537013 | −1.17E−01 | 8.46E−06 | 1.98E+01 |
| 85 | 6.083 | 0.000545451 | −1.07E−01 | 8.44E−06 | 1.94E+01 |
| 86 | 5.944 | 0.000553864 | −1.39E−01 | 8.41E−06 | 1.89E+01 |
| 87 | 5.869 | 0.000562249 | −7.50E−02 | 8.38E−06 | 1.86E+01 |
| 88 | 5.719 | 0.000570603 | −1.50E−01 | 8.35E−06 | 1.80E+01 |
| 89 | 5.634 | 0.000578923 | −8.50E−02 | 8.32E−06 | 1.77E+01 |
| 90 | 5.505 | 0.000587207 | −1.29E−01 | 8.28E−06 | 1.72E+01 |

The instantaneous volume of cylinder and rate of pressure rise with respect to crank angle is shown in Figures 5.P2 and 5.P3. In-cylinder pressure rise and heat release rate with respect to crank angle are given in Figures 5.P4 and 5.P5.

### Example 5.6

Calculate the BSEC, BMEP and brake thermal efficiency of a diesel engine at 3.6kW power output for the use of diesel and biodiesel (B20). The engine has a 100 mm bore and an 87.5 mm stroke and it operates at a constant speed of 1500 rpm. The mass flow rate of diesel and biodiesel are 1.29 and 1.24 kg/h, respectively. Assume that the calorific values of diesel and biodiesel are 42.5 and 38 MJ/kg, respectively.

### Solution to Example 5.6

$D = 100$ mm; $L = 87.5$ mm $= 0.875$ m

$N = 1500$ rpm $= 1500/(2 \times 60)$ cycle/s $= 12.5$ c/s

$A = 3.14 \times 0.\,100 \times 0.100/4 = 0.00196$ m³

$BP = 3.6$ kW

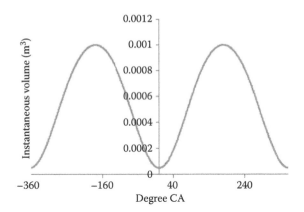

**FIGURE 5.P2**
Instantaneous volume of cylinder with respect to crank angle.

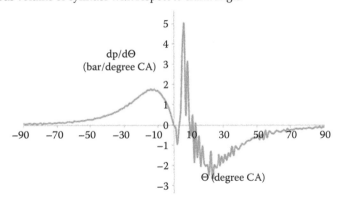

**FIGURE 5.P3**
Rate of pressure rise versus degree crank angle.

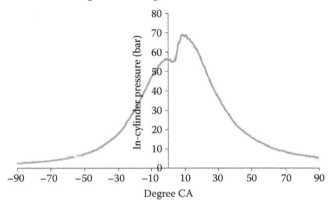

**FIGURE 5.P4**
In-cylinder pressure rise  versus degree crank angle.

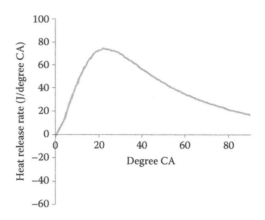

**FIGURE 5.P5**
Heat release rate versus degree CA.

*Diesel case*:

Brake thermal efficiency = BP/(mass of fuel flow rate × calorific value)
BP = 3.6 × 3600/(1.29 × 45000) × 100 = 24%
BP = 3.6 × 3600/(1.25)

*Biodiesel case*:

BP = 3.6 × 3600/(1.24 × 38000) × 100 = 27.5%
BMEP = BP/(L × A × N)
        = 3.6/(0.875 × 0.00196 × 12.5) = 4.19 bar
BSFC = mass flow rate of fuel/brake power
BSFC for diesel = 1.29 × 1000/3.6 = 358 g/kWh
BSEC for diesel = mass flow rate of fuel × calorific value/brake power
           = 1.29 × 42,500/(3600 × 3.6) = 4.2 kJ/kW
BSFC for B20 = 1.29/3.6 = 344 g/kWh
BSEC for B20 = 1.24 × 1000 × 38000/(3600 × 3.6) = 3.6 kJ/kW

**Example 5.7**

Hydrogen is produced as a by-product from an alkali industry. The availability of hydrogen gas is 1000 m³ per day and the industry is planning to generate supplementary electrical power for 6 h per day using an IC engine genset. The hydrogen calorific value, hydrogen density, engine's thermal efficiency, engine speed and BMEP are 120 mJ/kg, 0.07 kg/m³, 34%, 3000 rpm and 10 bar, respectively. Calculate the possible power output and number of cylinders of an SI engine.

**Solution to Example 5.7**

Available hydrogen = 1000 m³/day

Mass flow rate of hydrogen for 6 h = 1000/(6 × 3600) = 0.0463 m³/s

Thermal efficiency = brake power/(mass flow rate of fuel × calorific value)

Brake power = 0.34 × 0.0463 × 120000 = 1889 kW

BP = BMEP × swept volume × speed/2 × number of cylinder

Swept volume × number of cylinders = BP × 2/(BMEP × speed)

   = 1889 × 1000 × 2 × 60/(10 × 100000 × 3000)

   = 0.076 m³ = 76 L capacity of engine

Assuming data of 150 mm diameter and 150 mm stroke length per cylinder, the swept volume per cylinder will be 0.00283 m³/cylinder or 2.83 L/cylinder.

Number of cylinders = 0.076/0.00283 = 27 cylinders

The standard engines with a number of cylinders are 4, 6, 8, 12, 14, 24, 36, 48 and so on. The engine should be designed with number of cylinder of 24 for 1.8 MW power generation for hydrogen fuel. So the engine selection for the industry is 1889 kW or 1.8 MW.

**UNSOLVED PROBLEMS**

1. A four-stroke SI engine runs at 3000 rev/min. The engine capacity is 4 L. The air is supplied at 0.7 bar and 10°C with a volumetric efficiency of 0.5. The air–fuel ratio is 13. The calorific value is 45 MJ/kg. Calculate the heat release by combustion.

2. A four-stroke SI engine gave the following results during a test:

   Number of cylinders = 4

   Bore of cylinders = 80 mm

   Stroke = 90 mm

   Speed = 4000 rpm

   Fuel consumption rate = 0.245 kg/min

   Calorific value = 43 MJ/kg

   Torque = 60 N m

   Calculate the following:

   (a) Brake power

   (b) Mean effective pressure

   (c) Indicated power

   (d) Mechanical efficiency

3. The BMEP for a four-cylinder, four-stroke SI engine is 8.4 bar. The total capacity is 1.4 dm$^3$ (L). The engine is run at 4500 rpm. Calculate the brake power. There are 15 kW of mechanical losses in the engine. Calculate the indicated mean effective pressure. The volumetric efficiency is 85% and the brake thermal efficiency of the engine is 28%. The air drawn into the engine is at 15°C and 1.01 bar. The fuel has a calorific value of 43.5 MJ/kg. Calculate the air–fuel ratio.

4. In an Otto cycle, air is drawn in at 20°C. The maximum cycle temperature is 1800°C. The volume compression ratio is 10:1. Calculate the following:

   (a) The thermal efficiency

   (b) The heat input per kilogram of air

   (c) The net work output per kilogram of air

5. The hydrogen flow rate and diesel flow rate for a hydrogen-fuelled dual-fuel engine are 0.0323 g/s and 0.22 g/s, respectively, at 3.77 kW load and 1500 rpm. Determine the brake thermal efficiency and hydrogen energy share.

6. Calculate the instantaneous volume, rate of change of volume, in-cylinder temperature, heat release rate (using the first law of thermodynamics) and rate of pressure rise for the given (bolded) in-cylinder pressure with respect to crank angle data. The engine details are as follows:

   Connecting rod length (l) = 254 mm

   Bore (b) = 90 mm

   Compression ratio = 14

   Adiabatic index (k) = 1.35 for air–fuel mixture

| Degree CA | P (bar) | Degree CA | P (bar) | Degree CA | P (bar) | Degree CA | P (bar) | Degree CA | P (bar) | Degree CA | P (bar) |
|---|---|---|---|---|---|---|---|---|---|---|---|
| 231 | 1.151 | 276 | 2.035 | 321 | 7.027 | 366 | 24.757 | 411 | 19.298 | 456 | 7.183 |
| 232 | 1.151 | 277 | 2.087 | 322 | 7.287 | 367 | 25.693 | 412 | 18.674 | 457 | 7.079 |
| 233 | 1.151 | 278 | 2.139 | 323 | 7.547 | 368 | 26.785 | 413 | 18.154 | 458 | 6.975 |
| 234 | 1.203 | 279 | 2.139 | 324 | 7.859 | 369 | 27.981 | 414 | 17.634 | 459 | 6.923 |
| 235 | 1.203 | 280 | 2.191 | 325 | 8.067 | 370 | 29.281 | 415 | 17.114 | 460 | 6.819 |
| 236 | 1.203 | 281 | 2.243 | 326 | 8.431 | 371 | 30.633 | 416 | 16.646 | 461 | 6.715 |
| 237 | 1.203 | 282 | 2.295 | 327 | 8.743 | 372 | 32.037 | 417 | 16.178 | 462 | 6.611 |
| 238 | 1.203 | 283 | 2.347 | 328 | 9.107 | 373 | 33.493 | 418 | 15.762 | 463 | 6.507 |
| 239 | 1.203 | 284 | 2.399 | 329 | 9.419 | 374 | 34.896 | 419 | 15.294 | 464 | 6.455 |
| 240 | 1.203 | 285 | 2.451 | 330 | 9.783 | 375 | 36.3 | 420 | 14.878 | 465 | 6.351 |
| 241 | 1.255 | 286 | 2.555 | 331 | 10.147 | 376 | 37.6 | 421 | 14.514 | 466 | 6.299 |
| 242 | 1.255 | 287 | 2.555 | 332 | 10.563 | 377 | 38.848 | 422 | 14.15 | 467 | 6.247 |
| 243 | 1.255 | 288 | 2.659 | 333 | 10.979 | 378 | 39.94 | 423 | 13.786 | 468 | 6.143 |
| 244 | 1.255 | 289 | 2.711 | 334 | 11.343 | 379 | 40.824 | 424 | 13.422 | 469 | 6.091 |

| | | | | | | | | | | | |
|---|---|---|---|---|---|---|---|---|---|---|---|
| 245 | 1.307 | 290 | 2.763 | 335 | 11.759 | 380 | 41.552 | 425 | 13.058 | 470 | 6.039 |
| 246 | 1.307 | 291 | 2.763 | 336 | 12.226 | 381 | 42.02 | 426 | 12.798 | 471 | 5.987 |
| 247 | 1.307 | 292 | 2.815 | 337 | 12.694 | 382 | 42.332 | 427 | 12.486 | 472 | 5.935 |
| 248 | 1.307 | 293 | 2.919 | 338 | 13.11 | 383 | 42.332 | 428 | 12.174 | 473 | 5.831 |
| 249 | 1.359 | 294 | 2.971 | 339 | 13.578 | 384 | 42.176 | 429 | 11.915 | 474 | 5.779 |
| 250 | 1.359 | 295 | 3.075 | 340 | 14.098 | 385 | 41.656 | 430 | 11.603 | 475 | 5.727 |
| 251 | 1.359 | 296 | 3.179 | 341 | 14.566 | 386 | 41.136 | 431 | 11.343 | 476 | 5.675 |
| 252 | 1.411 | 297 | 3.231 | 342 | 15.034 | 387 | 40.356 | 432 | 11.135 | 477 | 5.623 |
| 253 | 1.411 | 298 | 3.335 | 343 | 15.554 | 388 | 39.576 | 433 | 10.875 | 478 | 5.571 |
| 254 | 1.411 | 299 | 3.439 | 344 | 16.022 | 389 | 38.588 | 434 | 10.667 | 479 | 5.519 |
| 255 | 1.463 | 300 | 3.543 | 345 | 16.49 | 390 | 37.652 | 435 | 10.407 | 480 | 5.467 |
| 256 | 1.463 | 301 | 3.647 | 346 | 17.01 | 391 | 36.56 | 436 | 10.199 | 481 | 5.415 |
| 257 | 1.463 | 302 | 3.751 | 347 | 17.478 | 392 | 35.572 | 437 | 9.991 | 482 | 5.363 |
| 258 | 1.515 | 303 | 3.855 | 348 | 17.894 | 393 | 34.48 | 438 | 9.783 | 483 | 5.311 |
| 259 | 1.515 | 304 | 4.011 | 349 | 18.31 | 394 | 33.441 | 439 | 9.627 | 484 | 5.259 |
| 260 | 1.567 | 305 | 4.115 | 350 | 18.726 | 395 | 32.349 | 440 | 9.419 | 485 | 5.207 |
| 261 | 1.567 | 306 | 4.271 | 351 | 19.142 | 396 | 31.309 | 441 | 9.263 | 486 | 5.155 |
| 262 | 1.619 | 307 | 4.375 | 352 | 19.506 | 397 | 30.269 | 442 | 9.055 | 487 | 5.155 |
| 263 | 1.619 | 308 | 4.479 | 353 | 19.818 | 398 | 29.333 | 443 | 8.899 | 488 | 5.103 |
| 264 | 1.671 | 309 | 4.635 | 354 | 20.13 | 399 | 28.345 | 444 | 8.743 | 489 | 5.051 |
| 265 | 1.671 | 310 | 4.791 | 355 | 20.39 | 400 | 27.409 | 445 | 8.587 | 490 | 4.999 |
| 266 | 1.723 | 311 | 4.947 | 356 | 20.65 | 401 | 26.525 | 446 | 8.431 | 491 | 4.999 |
| 267 | 1.723 | 312 | 5.103 | 357 | 20.91 | 402 | 25.641 | 447 | 8.275 | 492 | 4.947 |
| 268 | 1.775 | 313 | 5.311 | 358 | 21.118 | 403 | 24.809 | 448 | 8.119 | 493 | 4.895 |
| 269 | 1.775 | 314 | 5.467 | 359 | 21.326 | 404 | 24.029 | 449 | 8.015 | 494 | 4.843 |
| 270 | 1.827 | 315 | 5.675 | 360 | 21.586 | 405 | 23.301 | 450 | 7.911 | 495 | 4.843 |
| 271 | 1.879 | 316 | 5.883 | 361 | 21.898 | 406 | 22.522 | 451 | 7.755 | 496 | 4.791 |
| 272 | 1.879 | 317 | 6.091 | 362 | 22.262 | 407 | 21.846 | 452 | 7.651 | 497 | 4.739 |
| 273 | 1.931 | 318 | 6.247 | 363 | 22.73 | 408 | 21.17 | 453 | 7.547 | 498 | 4.687 |
| 274 | 1.983 | 319 | 6.507 | 364 | 23.249 | 409 | 20.494 | 454 | 7.443 | 499 | 4.687 |
| 275 | 1.983 | 320 | 6.767 | 365 | 23.977 | 410 | 19.87 | 455 | 7.339 | 500 | 4.635 |

# References

1. E. Conte and K. Boulouchos, A quasi-dimensional model for estimating the influence of hydrogen-rich gas addition on turbulent flame speed and flame front propagation in IC-SI engines, *Journal of Engines, SAE Transactions*, Paper no. 2005-01-022, 3, 438–448, 2005.
2. A. Kumar, M. K. Gajendra Babu and D. S. Khatri. An investigation of potential and challenges with higher ethanol-gasoline blend on a single cylinder spark ignition research engine, *Journal of SAE Special Publications*, SAE Paper number 2009-01-0137, SP-2241, p. 1, 2009.
3. J. B. Heywood, *Internal Combustion Engine Fundamentals*, International Edition, McGraw Hill Book Company, New York, 1989.
4. A. N. Lipatnikov and C. Jerzy, A simple model of unsteady turbulent flame propagation, *Journal of Engines, SAE Transactions*, Paper no. 972993, 3, 2411–2452, 1997.

5.  H. Willems and R. Sierens, Modeling the initial growth of the plasma and flame kernel in SI engines, *Journal of Engineering for Gas Turbines and Power, ASME Transactions,* 125(2), 479–484, 2003.

6.  W. Reckers, H. Schwab, C. Weiten, B. Befrui, Ing. R. Kneer, T. Zentrum and Luxemburg Delphi Corporation, Investigation of Flame Propagation and Cyclic Combustion Variations in a DISI Engine Using Synchronous High-Speed Visualization and Cylinder Pressure Analysis, *International Symposium on Internal Combustion Diagnostics,* AVL Deutscwand, 2002.

7.  H. Alois, R. Martin and E. Winklhofer, Flame and radiation measurement techniques for dynamic emission calibration in gasoline engines, *PTNSS Congress,* 2007.

8.  R. Budack, M. Kuhn, P. Wastl, S. Adam, R. Wurms and T. Schladt Audi AG, Systematic development of a turbocharged combustion process using the VisioTomo method, *International Symposium on Internal Combustion Diagnostics,* AVL Deutscwand, 2002.

9.  T. D. Fansler, M. C. Drake and Warren, Optical diagnostics applied to spark-ignited direct-injection engine development, *International Symposium on Internal Combustion Diagnostics,* AVL Deutscwand, 2002.

10. Ch. Beidl, H. Pholipp, W. Piock and E. Winklhofer, Experiences with Visio fiber technologies in developing modern combustion system, *International Symposium on Internal Combustion Diagnostics,* AVL Deutscwand, pp. 6–25, June 2002.

11. M. Bahattin Celik, B. Ozdalyan and F. Alkan, The use of pure methanol as fuel at high compression ratio in a single cylinder gasoline engine, *Fuel* 90, 1591–1598, 2011.

12. F. N. Alasfour, $NO_x$ emission from a spark ignition engine using 30% iso-butanol-gasoline blend: Part 1 - preheating inlet air, *Applied Thermal Engineering* 18, 245–256, 1998.

13. J. Dernotte, C. Mounaim-Rousselle, F. Halter and P. Seers, Evaluation of butanol–gasoline blends in a port fuel-injection, spark-ignition engine, *Oil and Gas Science and Technology—Rev. IFP,* 65, 345–351, 2010.

14. R. Hari Ganesh et al., Hydrogen fuelled spark ignition engine with electronically controlled manifold injection: An experimental study, *Renewable Energy* 33, 1324–1333, 2008.

15. K. Koppula, Study of spark ignition engines for use of gasoline and CNG using timed manifold injection, M.Tech Thesis, CES, IIT Delhi, May 2008.

16. M. I. Jahirul, H. H. Masjuki, R. Saidur, M. A. Kalam, M. H. Jayed and M. A. Wazed, Comparative engine performance and emission analysis of CNG and gasoline in a retrofitted car engine, *Applied Thermal Engineering* 30, 2219–2226, 2010.

17. M. Gumus, Effects of volumetric efficiency on the performance and emissions characteristics of a dual fuelled (gasoline and LPG) spark ignition engine, *Fuel Processing Technology* 92, 1862–1867, 2011.

18. M. A. Ceviz and F. Yuksel, Cyclic variations on LPG and gasoline-fuelled lean burn SI engine, *Renewable Energy* 31, 1950–1960, 2006.

18. R. Muthu Shanmugam et al., Performance and emission characterization of 1.2 L MPI engine with multiple Fuels (E10, LPG and CNG), *SAE International Journal of Fuels and Lubricants,* SAE no. 2010-01-0740, 3.

20. J. B. Heywood, *Internal Combustion Engine Fundamentals,* McGraw-Hill, New York, 1988.

21. A. K. Agarwal, Biofuels (alcohols and biodiesel) applications as fuels for internal combustion engines, *Progress in Energy and Combustion Science* 33, 233–271, 2007.
22. D. G. Kesse, Global warming—Facts, assessment, countermeasures. *Journal of Petroleum Science Engineering*, 26, 157–68, 2000.
23. M. D. Mann, M. E. Collings and P. E. Botros, Nitrous oxide emissions in fluidized-bed combustion: Fundamental chemistry and combustion testing, *Energy and Combustion Science*, 18, 447–461, 1992.
24. B. Nelter, *Algae Based Biodiesel*, Massachusetts Institute of Technology, April 2008.
25. J. Hira, Control of $NO_x$ emission in a biodiesel fuelled diesel engine using cooled exhaust gas recirculation, M.Tech Thesis, CES, IIT Delhi, May 2010.
26. H. Yongcheng, Z. Longbao, W. Shangxue and L. Shenghua, Study on the performance and emissions of a compression ignition engine fuelled with Fischer–Tropsch diesel fuel, *Journal of Automobile Engineering*, 220, 827–835, 2006.
27. A. S. Cheng and R. W. Dibble, Emissions performance of oxygenate-in-diesel blends and Fischer-Tropsch diesel in a compression ignition engine, *Society of Automotive Engineers, Inc.*, SAE No.1999-01-3606.
28. Z. H. Huang, H. W. Wang, H. Y. Chen, L. B. Zhou and D. M. Jiang, Study of combustion characteristics of a compression ignition engine fuelled with dimethyl ether, *Journal of Automobile Engineering*, 213, 647–652, 1999.
29. Y. Wang and L. B. Zhou, The effect of two different kinds of vegetable oil on combustion and emission characteristics of a vehicle dimethyl ether engine, *Journal of Automobile Engineering*, 221, 1467–1473, 2007.
30. K. A. Subramanian, S. K. Singal, M. Saxena and S. Singal, Utilization of liquid biofuels in automotive diesel engines: An Indian perspective, *International Journal of Biomass and Bioenergy*, 29(1), 65–72, 2005.
31. C. A. Sharp, S. A. Howell and J. Jobe, The effect of biodiesel fuels on transient emissions from modern diesel engines, Part-I regulated emissions and performance, Society of Automotive Engineers, SAE paper No. 2002-01-1967, 2002.
32. J. Krrahl, A. Munack, M. Bahadir, L. Schumacher and N. Elser, Review: Utiliation of rapeseed oil: Exhaust gas emissions and estimation of environmental effects, SAE Paper No. 962096, 1996.
33. J. P. Szybist, A. L. Boehman, R. L. McComick and J. D. Taylor, Evaluation of formulation of strategies to eliminate the biodiesel $NO_x$ effect, *Fuel Processing Technology*, 86(10), 1109–1126, 2005.
34. J. Sheehan, V. Camobreco, J. Duffield, M. Graboski and H. Shapouri, An overview of biodiesel and petroleum diesel life cycles, NREL/TP-580-24772, May 1998.
35. X.-C. Lu, W. Chen and Z. Huang, A fundamental study on the control of the HCCI combustion and emissions by fuel design concept combined with controllable EGR. Part 1. The basic characteristics of HCCI combustion, *Fuel* 84, 1074–1083, 2005.
36. E. Delacourt, B. Desmet and B. Besson, Characterisation of very high pressure diesel sprays using digital imaging techniques, *Fuel*, 84, 859–867, 2005.

# 6

# *Fuel Quality Characterisation for Suitability of Internal Combustion Engines*

## 6.1 General Introduction

The twin pillars of fuel quality and engine technology play a vital role in the improved performance and emission reduction of an internal combustion engine. The upgradation of fuel quality continues periodically throughout the world for meeting stringent emission norms as well as high fuel economy for the benefit of the consumer. The alternative fuel or formulated fuel would be characterised mainly by the following four methods:

- Fuel quality study
- Test bench experimental study on an internal combustion engine
- Chassis dynamometer study on an automotive vehicle
- Field trial study on an automotive vehicle

The development of fuel quality and engine technology primarily involves these four studies for meeting the stringent emission norms and assessing the potential of a new or alternative fuel. The importance of each study is explained in the upcoming sections.

## 6.2 Fuel Quality Study

*Spark ignition engine*: The main fuel quality requirements are octane number, distillation characteristics, density, viscosity, sulphur, aromatics and olefin. Alternative liquid fuels such as ethanol, methanol and butanol and gaseous fuels such as liquefied petroleum gas (LPG), compressed natural gas (CNG), biogas, producer gas and hydrogen should meet the specific desired standards for these fuels or in comparison with the base gasoline fuels.

*Compression ignition engine*: The main fuel quality requirements are cetane number, distillation characteristics, density, viscosity, sulphur and aromatics. The alternative fuels such as biodiesel, Fisher–Tropsch diesel, dimethyl ether and diethyl ether should meet the desired standards. If any one of the fuel qualities does not meet the desired level, the performance and emission characteristics of the internal combustion engine will deteriorate.

The fuel quality is upgraded in refineries with economical consideration. For example, removing aromatics in diesel would give the benefit of reducing polyaromatic hydrocarbon (PAH) emissions. On the other hand, the fuel cost would be high, which is not economically feasible. Hence, the refineries can reduce the aromatic level up to a certain point. Vehicular technologies such as common rail direct injection (CRDI) could reduce PAH by enhancing combustion efficiency. Hence, the combination of fuel quality upgradation and advancement of engine technology could lead to a sustainable development of the oil and transport sector. A wider consultation between the original engine manufacturer (OEM) and oil companies will be required to carry out fuel quality improvement. Some of the important fuel properties that should be considered for an internal combustion engine are explained in Section 6.3.

## 6.3 Measurement of Fuel Properties

Fuels have 13 properties:

1. Calorific value
2. Viscosity
3. Flash point
4. Cloud point
5. Pour point
6. Vapour index
7. Octane and cetane ratings
8. Distillation characteristics
9. Aromatic content
10. Ash content
11. Carbon residue
12. Corrosivity
13. Testing methods

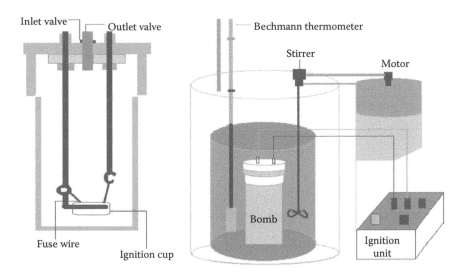

**FIGURE 6.1**
Bomb calorimeter.

## 6.3.1 Calorific Value

The calorific value of a fuel is also known as the heat of combustion and is expressed as megajoules/kilogram. The heat of combustion at a constant volume of a liquid or solid fuel containing only the elements carbon, hydrogen, oxygen, nitrogen and sulphur is the quantity of heat liberated when a unit mass of the fuel is burnt in oxygen in an enclosure of constant volume; the products of combustion are gaseous carbon dioxide, nitrogen, sulphur dioxide and liquid water, with the initial temperature of the fuel and the oxygen and the final temperature of the products at 25°C. The heat of combustion is a measure of the energy available from a fuel. Knowledge of this value is essential when considering the thermal efficiency of equipment for producing either power or heat.

The heat of combustion is determined by burning a weighed sample in an oxygen-bomb calorimeter (Figure 6.1) under controlled conditions. The temperature increase is measured by a temperature-reading instrument that allows the precision of the test method to be met. The heat of combustion is calculated from temperature observations before, during and after combustion, with proper allowance for thermochemical- and heat-transfer corrections. Either isoperibol or adiabatic calorimeters may be used.

### 6.3.1.1 Procedure

Weigh the sample cup to 0.01 mg on a semimicro analytical balance. Place a piece of pressure-sensitive tape across the top of the cup, trim around the edge with a razor blade and seal tightly. Place a 3 × 12-mm strip of tape creased in the middle and sealed by one edge in the centre of the tape disk to give a flap arrangement. Weigh the cup and tape. Remove from the balance with forceps. Fill a hypodermic syringe with the sample. The volume of sample necessary to produce a temperature rise is equivalent to approximately 30,000 J, which can be estimated as

$$V = (W \times 0.0032)/(Q \times D) \tag{6.1}$$

where
   $V$ = volume of sample to be used, cubic centimetre
   $W$ = energy equivalent of the calorimeter, joules per degree centigrade
   $Q$ = approximate heat of combustion of the sample, megajoules per kilogram
   $D$ = density, grams per cubic centimetre, of the sample.

A sample is added to the cup by inserting the tip of the needle through the tape disk at a point so that the flap of the tape will cover the puncture upon removal of the needle. The flap is sealed down by pressing lightly with a metal spatula. The cup with the tape and sample is weighed again. Care is taken throughout the weighing and filling operation to avoid contacting the tape or the cup with bare fingers. The cup is placed in the curved electrode and the fuse wire is arranged so that the central portion of the loop presses down on the centre of the tape disk. The bomb is assembled by tightening the cover securely. The bomb is connected to the oxygen cylinder and oxygen is slowly admitted until a pressure of 3.0 MPa is attained. The bomb is not purged to remove the entrapped air. The bomb is disconnected from the oxygen cylinder and the valve cover is replaced (warning: a violent explosion may occur). Care should be taken not to overcharge the bomb. If oxygen introduced into the bomb does exceed 4.0 MPa, combustion should not be initiated.

The stirrer motor is started on the calorimeter and the calorimeter controller is turned on. The manual control switch of the controller is used to bring the jacket temperature in close agreement with the bucket temperature. The controller is allowed to automatically control the temperature and after a time of 15 min, equilibrium is attained. At this point and also at the end point, the temperature of the jacket is controlled to the same temperature as the bucket, or slightly (0.005°C at most) below. Readings are taken at 1-min intervals until three consecutive readings show no change. The sample is fired by depressing the button of the ignition unit. The pilot light glows momentarily and the temperature starts increasing in about 15 s. (If the temperature does not start rising, the experiment must be discontinued.) The initial resistance is read and recorded. After 6 min from firing, temperature readings are taken at 1-min intervals. The three consecutive readings should not show any change, or the readings decrease. At this point, the final temperature is noted and recorded by estimating the value to the nearest 0.0005°C.

The controller and stirrer are turned off and the bomb is removed from the calorimeter. The needle valve is opened and the gas is allowed to escape at a uniform rate so as to reduce the pressure to atmospheric pressure in not less than 1 min. The bomb is opened and the interior is examined for unburnt carbon. If any trace of unburnt carbon is found, the experiment should be rejected. The interior of the bomb is washed, including the electrodes and the sample cup, with a fine jet of water and quantitatively the washings are collected in a 500 cm³ Erlenmeyer flask. A minimum amount of wash water is used, preferably less than 300 cm³. The washings are titrated with standard alkali solution using a methyl-red indicator. The sulphur content of the sample is determined to the nearest 0.02% sulphur as described in ASTM Test Method D129, D1266-IP 107, D2622, D3120, D4294 or D5453 depending upon the volatility of the sample.

### 6.3.1.2 Calculation

Using the data obtained, the temperature rise is calculated, $\Delta t$ (in degree centigrade) as follows:

$$\Delta t = t_f - t_i \tag{6.2}$$

where
$\Delta t$ = corrected temperature rise, degree centigrade
$t_f$ = final equilibrium temperature, degree centigrade
$t_i$ = temperature at the time of firing, degree centigrade

The gross heat of combustion is calculated by substituting in the following equations:

$$Q_g \ (\text{gross } t°C) = (\Delta t \times W - e1 - e2 - e3 - e4)/1000 \ M \tag{6.3}$$

$$Q_g \ (\text{gross, } 25°C) = Q_g \ (\text{gross, } t°C) + A \ (t - 25) \tag{6.4}$$

where
$Q_g$ (gross, °C) = gross heat of combustion at constant volume and final temperature of the experiment, expressed as megajoules per kilogram
$Q_g$ (gross, 25°C) = gross heat of combustion at constant volume, expressed as megajoules per kilogram
$\Delta t$ = corrected temperature rise, expressed as degree centigrade
$W$ = energy equivalent of the calorimeter, expressed as joules per degree centigrade
$M$ = mass of the sample, expressed as grams
$t$ = final temperature of combustion, expressed as degree centigrade
$e1$ = correction for the heat of formation of $HNO_3$, joules = cubic centimetre of standard (0.0866 $N$) NaOH solution used in titration ×5
$e2$ = correction for the heat of formation of a sulphuric acid ($H_2SO_4$), $J$ = 58.63% of sulphur in sample × mass of sample, expressed as grams

$e3$ = correction for the heat of combustion of a pressure-sensitive tape, $J$ = mass of tape, $g$ × heat of combustion of the tape, expressed as joules per gram

$e4$ = correction for heat of combustion of firing wire, $J$ = 1.133 mm of iron wire consumed = 0.963 mm of Chromel C wire consumed

$A$ = correction factor, expressed as megajoules per kilogram degree centigrade to correct from the final temperature of combustion to 25°C where values of factor $A$ are given in Table 6.1

$$Q_g p = Q_g + 0.006145\, H \tag{6.5}$$

where

$Q_g p$ = gross heat of combustion at constant pressure, megajoules per kilogram

$H$ = hydrogen content, mass percent

The net heat of combustion is calculated as follows:

$$Q_n \ (\text{net, } 25°C) = Q_g \ (Q_g \ \text{gross, } 25°C) - 0.2122 \times H \tag{6.6}$$

where

$Q_n$ (net, 25°C) = net heat of combustion at constant pressure, megajoules per kilogram

$Q_g$ (gross, 25°C) = gross heat of combustion at constant volume, megajoules per kilogram

$H$ = mass percent of hydrogen in the sample

When the percentage of hydrogen in the sample is not known, determine the hydrogen in accordance with ASTM Test Methods D1018 or D3701.

**TABLE 6.1**

Values of Factor $A$

| $Q_g$ (Gross, $t$°C) (MJ/kg) | $A$ (MJ/kg°C) | $Q_g$ (Gross, $t$°C) (MJ/kg) | $A$ (MJ/kg°C) |
|---|---|---|---|
| 43.00 | 0.00157 | 45.45 | 0.00271 |
| 43.25 | 0.00167 | 46.00 | 0.00282 |
| 43.50 | 0.00178 | 46.25 | 0.00292 |
| 44.00 | 0.00188 | 46.50 | 0.00302 |
| 44.25 | 0.00199 | 46.75 | 0.00313 |
| 44.50 | 0.00219 | 47.25 | 0.00333 |
| 44.75 | 0.00230 | 47.50 | 0.00344 |
| 45.00 | 0.00240 | 47.75 | 0.00354 |
| 45.25 | 0.00250 | 48.00 | 0.00365 |
| 45.50 | 0.00261 | | |

### 6.3.2 Viscosity

Viscosity is a measure of the resistance to flow of the fuel, and it will decrease as the fuel oil temperature increases. What this means is that a fluid with a high viscosity is heavier than a fluid with a low viscosity. A high-viscosity fuel may cause extreme pressures in the injection systems and will cause reduced atomisation and vapourisation of the fuel spray. The viscosity of diesel fuel must be low enough to flow freely at its lowest operational temperature, yet high enough to provide lubrication to the moving parts of the finely machined injectors. The fuel must also be sufficiently viscous so that leakage at the pump plungers and dribbling at the injectors will not occur. Viscosity also will determine the size of the fuel droplets, which, in turn, govern the atomisation and penetration qualities of the fuel injector spray.

There are two related measures of fluid viscosity—known as dynamic (or absolute) and kinematic viscosity. Absolute viscosity, or the coefficient of absolute viscosity, is a measure of the internal resistance. Dynamic (absolute) viscosity is the tangential force per unit area required to move one horizontal plane with respect to the other at unit velocity when maintained a unit distance apart by the fluid. It is denoted by $\mu$. The kinematic viscosity is the ratio of absolute or dynamic viscosity to density—a quantity in which no force is involved. Kinematic viscosity can be obtained by dividing the absolute viscosity of a fluid with its mass density. It is denoted by $v$:

$$v = \mu/\rho \tag{6.7}$$

where
  $v$ = kinematic viscosity
  $\mu$ = absolute or dynamic viscosity
  $\rho$ = density

The international system (SI) units of dynamic viscosity units are $Ns/m^2$, Pa s or kg/ms, where

$$1 \text{ Pa s} = 1 \text{ N s/m}^2 = 1 \text{ kg/m s}$$

The dynamic viscosity is also often expressed in the metric centimetre–gram–second (CGS) system as g/cm s, dyne s/cm² or poise (p), where 1 poise = dyne s/cm² = g/cm s = 1/10 Pa s.

The SI unit for kinematic viscosity is strokes (St):

$$1 \text{ St} = 10^{-4} \text{ m}^2/\text{s}$$

$$1 \text{ St} = 100 \text{ cSt}$$

$$1 \text{ cSt} = 10^{-6} \text{ m}^2/\text{s}$$

Generally two types of viscometers are used to measure the viscosity of a fuel:

1. Redwood viscometer
2. Saybolt viscometer

### 6.3.2.1 Redwood Viscometer

A redwood viscometer consists of a cylindrical oil cup furnished with a gauge point, a metallic orifice jet at the bottom having a concave depression from inside to facilitate a ball with stiff wire to act as a valve to start or stop oil flow as shown in Figure 6.2. The outer side of the orifice jet is convex, so that the oil under test does not creep over the lower face of the oil cup. The oil cup is surrounded by a water bath with a circular electrical immersion heater and a stirring device. Two thermometers are provided to measure the water bath

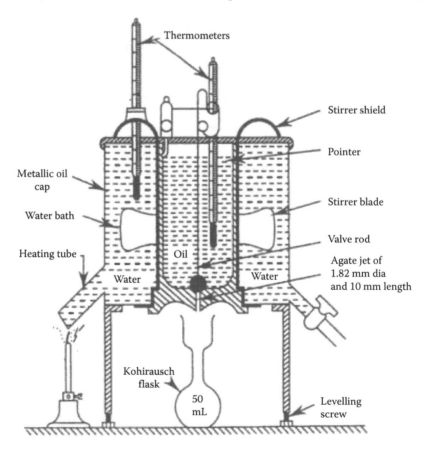

**FIGURE 6.2**
Redwood viscometer.

temperature and oil temperature under test. A round flat-bottomed flask of 50 mL marking is used to measure 50 mL of oil flow against time. The water bath with oil cup is supported on a tripod stand with levelling screws.

### 6.3.2.1.1 Procedure

The oil cup is cleaned with a solvent, preferably carbon tetrachloride (CTC), and it is wiped dry thoroughly with a paper napkin or a soft cloth (do not use cotton waste) and the orifice jet is wiped with a fine thread. The water bath is kept with the oil cup on the tripod stand and is levelled. Water is poured into the water bath up to 15–20 mm below the top portion. The ball (valve) is kept in position and clean filtered oil is poured as the sample (use a strainer not coarser than BS 100 mesh) to be tested into the oil cup up to the gauge point and it is covered with the lid. A 50-mL flask with clean dry is placed under the orifice jet of the oil cup. The ball (valve) is lifted and a stopwatch is simultaneously started and the oil is allowed into the receiving flask. The receiving flask (50 mL) is adjusted in such a way that the oil string coming out of the jet strikes the neck of the flask to avoid foaming (formation of air bubbles) on the oil surface. One has to wait till the oil level touches the 50-mL mark. The stopwatch is stopped and the time is recorded. The experiment is repeated at different temperatures above ambient.

### 6.3.2.2 Saybolt Viscometer

This apparatus mainly consists of a standard cylindrical oil cup surrounded with a water bath with an immersion heater and a stirring device (Figure 6.3). The apparatus is supplied with two stainless steel orifice jets, namely a universal jet and Furol jet, which can be fitted at the bottom of the oil cup as per our requirement. A rubber cork stopper arrangement is also provided at the bottom to facilitate the start and stop of the oil flow from the viscometer. Two thermometers are provided to measure the water bath temperature and oil temperature under test. A round flat-bottomed flask with a 60-mL marking on the neck is provided to measure 60 mL of oil flow against time. The oil cup with the water bath is supported on a stand with levelling screws.

### 6.3.2.2.1 Procedure

The oil cup is cleaned with a solvent, preferably CTC, and it is wiped dry thoroughly with a paper napkin or a soft cloth (cotton waste is not used) and the orifice jet is wiped with a fine thread. The water bath is kept with the oil cup on the tripod stand and is levelled. Water is poured into the water bath up to 15–20 mm below the top portion. The orifice opening is closed from the bottom with the rubber cork provided. The oil to be tested is poured into the strainer by keeping the strainer on the oil cup until the oil fills the cup as well as in the side well. The excess oil is withdrawn in the side well and the thermometers are positioned in the water bath and in the oil cup. A clean dry 60-mL flask is taken and placed under the orifice jet of the oil cup that centres

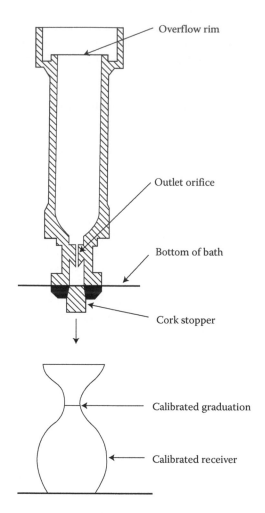

**FIGURE 6.3**
Saybolt viscometer.

it. The rubber cork is pulled open and a stop watch is simultaneously started and the oil is allowed into the receiving flask. The receiving flask (60 mL) is adjusted in such a way that the oil string coming out of the jet strikes the neck of the flask to avoid foaming (formation of air bubbles) on the oil surface. When the oil level reaches the 60-mL mark, the watch is stopped and the time is recorded in seconds. Repeat the experiment at different temperatures above ambient. Use a specific nozzle suitable for the lubricant or the oil.

*6.3.2.2.2 Calculation*

$$\text{Kinematic viscosity } (\nu) = (A \times t - (B/t)) \times 10^{-6} \text{ in m}^2/\text{s} \qquad (6.8)$$

where
   $A = 0.264$, $B = 190$, when $t = 40–85$ s
   $A = 0.247$, $B = 65$, when $t = 85–2000$ s

The *density* of the given oil is

$$\rho = (w_2 - w_1) \times (10^3/50) \text{ in kg/m}^3 \qquad (6.9)$$

where
   $w_1$ = weight of the measuring jar in grams
   $w_2$ = weight of 50 cc of oil in grams

$$\textit{Absolute viscosity } (\mu) = \rho \times v \text{ in Pa s or N s/m}^2 \qquad (6.10)$$

### 6.3.3 Flash Point

The flash point is the lowest temperature corrected to a barometric pressure of 101.3 kPa (760 mm Hg), at which the application of an ignition source causes the vapours of a specimen of the sample to ignite under specified conditions of the test. The flash point temperature is one measure of the tendency of the test specimen to form a flammable mixture with air under controlled laboratory conditions. It is only one of a number of properties that must be considered when assessing the overall flammability hazard of a material. It is used in shipping and safety regulations to define *flammable* and *combustible* materials. An Ables apparatus is used to determine the flash point that is shown in Figure 6.4.

#### 6.3.3.1 Description of Apparatus

In the Pensky–Martens closed cup flash point test, a brass test cup is filled with a test specimen and is fitted with a cover. The sample is heated and stirred at specified rates depending on what is being tested. An ignition source is directed into the cup at regular intervals with simultaneous interruption of stirring until a flash that spreads throughout the inside of the cup is seen. The corresponding temperature is its flash point. The Pensky–Martens closed cup is sealed with a lid through which the ignition source can be introduced periodically. The vapour above the liquid is assumed to be in reasonable equilibrium with the liquid. Closed cup testers give lower values for the flash point (typically 5–10 K) and are a better approximation of the temperature at which the vapour pressure reaches the lower flammable limit (LFL).

#### 6.3.3.2 Procedure

The apparatus is installed on a table near a 230-V, 50-Hz, 5-A single-phase power source. The electrical heater is kept on the table. The oil cup holder is

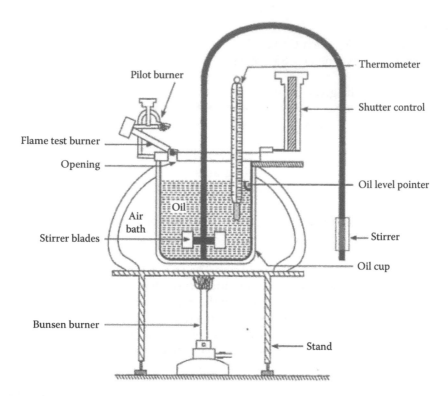

**FIGURE 6.4**
Ables apparatus.

positioned (air bath) on the heater. The oil cup is inserted into the bath and positioned in it. The oil to be tested is poured into the oil cup up to the mark. The lid is closed, the heater is connected to the electrical power source and the oil is heated at a slow and steady rate of 2°C/min with the help of the regulator. By stirring the oil with the stirring mechanism, a small flame is maintained on the wick. The flame is introduced to the oil surface by operating the circular handle, which makes the maintained flame that dips into the oil cup by opening the shutter. This is done at every 1/2 min, only after the sample oil reaches 15–17°C before the expected flash point. Record the temperature at which the first flash is recorded and that is reported as the flash point of the sample oil. To stop the experiment, switch off the heater and allow it to cool.

### 6.3.4 Cloud Point and Pour Point

The *cloud point* is the temperature of a liquid specimen when the smallest observable cluster of hydrocarbon (HC) crystals first occurs upon cooling under prescribed conditions. The purpose of the cloud point method is to detect the presence of wax crystals in the specimen; however, trace amounts of water and inorganic compounds may also be present. The intent of the cloud point method is to capture the temperature at which the liquids in the specimen begin to change from a single liquid phase to a two-phase system containing a solid and a liquid. The specimen is cooled at a specified rate and is examined periodically. The temperature at which a cloud is first observed at the bottom of the test jar is recorded as the cloud point. The cluster of wax crystals looks like a patch of whitish or milky cloud, hence the name of the test method. The cloud appears when the temperature of the specimen is low enough to cause wax crystals to form.

The *pour point* is the lowest temperature at which movement of the test specimen is observed under the prescribed conditions of the test. The pour point of a petroleum specimen is an index of the lowest temperature of its utility for certain applications.

### 6.3.4.1 Description of Apparatus

The apparatus for measuring cloud and pour points is shown in Figure 6.5, which consists of a cylindrical glass test jar with a 33.2–34.8 mm outside diameter and 115 and 125 mm height. The inside diameter of the jar may range

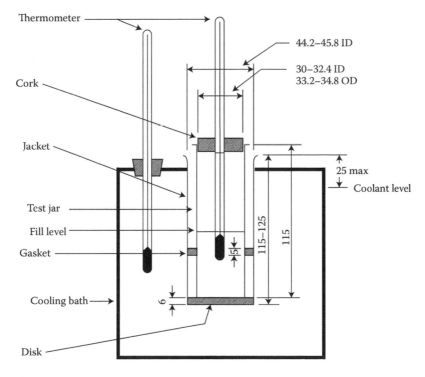

**FIGURE 6.5**
Apparatus for measuring cloud point and pour point.

from 30 to 32.4 mm within the constraint that the wall thickness must not be greater than 1.6 mm. The jar should be marked with a line to indicate the sample height 54 ± 3 mm above the inside bottom. This test jar is to be inserted in a metal or a glass jacket, watertight, cylindrical, flat bottom, about 115 mm in depth, with an inside diameter of 44.2–45.8 mm. It shall be supported free of excessive vibration and firmly in a vertical position in the cooling bath of 6.7 so that not more than 25 mm projects out of the cooling medium. Thermometers should be placed inside the apparatus as shown in Figure 6.5.

### 6.3.4.2 Procedure

#### 6.3.4.2.1 Cloud Point

The sample to be tested is brought to a temperature at least 14°C above the expected cloud point. Any moisture that may be present is removed by a method such as filtration through dry lintels filter paper until the oil is perfectly clear, but makes such filtration at a temperature of at least 14°C above the approximate cloud point. The sample is poured into the test jar to the level mark. The test jar is tightly closed by the cork carrying the test thermometer. High cloud and pour thermometers are used if the expected cloud point is above −36°C and a low cloud and pour thermometers are used if the expected cloud point is below −36°C. The position of the cork and the thermometer is so adjusted that the cork fits tightly, the thermometers and the jar are coaxial and the thermometer bulb is resting on the bottom of the jar. See that the disk, gasket and the inside of the jacket are kept clean and dry. The disk is placed in the bottom of the jacket. The disk and jacket shall have been placed in the cooling medium for a minimum of 10 min before the test jar is inserted. A jacket cover may be used while the empty jacket is cooling. The gasket is placed around the test jar, 25 mm from the bottom. The test jar is inserted in the jacket. The jar should not be placed directly into the cooling medium. The temperature of the cooling bath is maintained at 0 ± 1.5°C.

At each test thermometer reading that is a multiple of 1°C, the test jar is removed from the jacket quickly but without disturbing the specimen; after inspecting the cloud, it is replaced in the jacket. This complete operation shall require not more than 3 s. If the oil does not show a cloud when it has been cooled to 9°C, the test jar is to be transferred to a jacket in a second bath maintained at a temperature of −186 ± 1.5°C (see Table 6.2). The jacket should not be transferred. If the specimen does not show a cloud when it has been cooled to −6°C, the test jar is transferred to a jacket in a third bath maintained at a temperature of −336 ± 1.5°C. For the determination of very low cloud points, additional baths are required; each bath needs to be maintained in accordance with Table 6.2. In each case, the jar is transferred to the next bath, if the specimen does not exhibit the cloud point and the temperature of the specimen reaches the lowest specimen temperature in the range identified for the current bath in use, based on the ranges stated in Table 6.2. The cloud

**TABLE 6.2**

Bath and Sample Temperature Range

| Bath | Bath Temperature Setting (°C) | Sample Temperature Range (°C) |
|------|-------------------------------|-------------------------------|
| 1 | 0 ± 1.5 | Start to 9 |
| 2 | −18 ± 1.5 | 9 to −6 |
| 3 | −33 ± 1.5 | −6 to −24 |
| 4 | −51 ± 1.5 | −24 to −42 |
| 5 | −69 ± 1.5 | −42 to −60 |

point is reported, to the nearest 1°C, at which any cloud is observed at the bottom of the test jar, which is confirmed by continued cooling.

### 6.3.4.2.2 Pour Point

The specimen is poured into the test jar up to the level mark. If necessary, the specimen may be heated in a bath until it is just sufficiently fluid so that it can be poured into the test jar. Samples of residual fuels, black oils and cylinder stocks that have been heated to a temperature higher than 45°C during the preceding 24 h, or when the thermal history of these sample types is not known, shall be kept at room temperature for 24 h before testing. Samples that are known by the operator not to be sensitive to thermal history need not be kept at room temperature for 24 h before testing. The test jar is closed with the cork carrying the high-pour thermometer (Table 6.2). In the case of pour points above 36°C, use a higher-range thermometer such as IP 63C or ASTM 61C. The position of the cork and the thermometer are adjusted so that the cork fits slightly, the thermometer and the jar are coaxial and the thermometer bulb is immersed at the beginning of the capillary that is 3 mm below the surface of the specimen.

For the measurement of the pour point, the specimen is subjected in the test jar to the following preliminary treatment.

*6.3.4.2.2.1 Specimens Having Pour Points above −33°C*  The specimen is heated without stirring to 9°C above the expected pour point, but to at least 45°C, in a bath maintained at 12°C above the expected pour point, but below 48°C. The test jar is transferred to a bath maintained at 24.6 ± 1.5°C and observations are commenced for the pour point. When using a liquid bath, it is ensured that the liquid level is between the fill mark on the test jar and the top of the test jar.

*6.3.4.2.2.2 Specimens Having Pour Points of −33°C and Below*  The specimen is heated without stirring to at least 45°C in a bath maintained at 48.6 ± 1.5°C. The test jar is transferred to a bath maintained at 24.6 ± 1.5°C. When using a liquid bath, it should be ensured that the liquid level is between the fill mark on the test jar and the top of the test jar. When the specimen temperatures reach 27°C, the high cloud and pour thermometers are removed and they are placed in position. The test jar is transferred to the cooling bath.

The disk, gasket and the inside of the jacket are cleaned and dried. The disk is placed at the bottom of the jacket. The gasket is placed around the test jar, 25 mm from the bottom. The test jar is inserted into the jacket. A jar should never be placed directly in a cooling medium.

Pour points are expressed in integers that are positive or negative multiples of 3°C. Examine the appearance of the specimen when the temperature of the specimen is 9°C above the expected pour point (estimated as a multiple of 3°C).

At each test thermometer reading that is a multiple of 3°C below the starting temperature, the test jar is to be removed from the jacket. To remove condensed moisture that limits visibility, wipe the surface with a clean cloth moistened in alcohol (ethanol or methanol). The jar is tilted just enough to ascertain whether there is a movement of the specimen in the test jar. If movement of the specimen in the test jar is noted, then the test jar is to be replaced immediately in the jacket and a test is repeated for flow at the next temperature, 3°C lower. Typically, the complete operation of removal, wiping and replacement may require not more than 3 s. If the specimen has not ceased to flow when its temperature has reached 27°C, transfer the test jar to a jacket in a cooling bath maintained at 0.6 ± 1.5°C. As the specimen continues to get colder, transfer the test jar to a jacket in the next lower-temperature cooling bath in accordance with Table 6.2.

If the specimen in the jar does not show movement when tilted, hold the jar in a horizontal position for 5 s, as noted by an accurate timing device, and observe the specimen carefully. If the specimen shows any signs of movement before 5 s have passed, the test jar is to be replaced immediately in the jacket and the test is to be repeated for flow at the next temperature, 3°C lower.

## 6.4 Emission Characteristics of Internal Combustion Engines

*Carbon monoxide (CO)*: It indicates poorer combustion as the fuel–air mixture may be rich.

*Hydrocarbon*: It shows incomplete combustion.

*Oxides of nitrogen (NOx)*: Engine combustion with high-localised gas temperature inside the combustion chamber.

*Particulate matter (PM)*: It is due to poor combustion and high sulphur in fuel.

The engine parameters (as given below) need to be optimised using a bench study for effective use of an alternative fuel in an internal combustion engine.

*Engine design characteristics*: Compression ratio, bore and stroke of the engine cylinder.

*Injection characteristics*: Plunger stroke and diameter of the injection pump, plunger speed, pipeline length, in-line injection pressure, maximum injection pressure, nozzle-opening pressure, injection delay, injection rate shaping, static and dynamic injection timing and number of injection holes.

*Spray characteristics*: Break-up length, Sauter's mean diameter (SMD), spray penetration distance, spray cone angle, air entrainment fuel vapourisation and mixture preparation rate.

*Aerodynamic characteristics*: Swirl, tumble and squish velocity and volumetric efficiency.

*Geometric characteristics*: Intake and exhaust manifold and valves geometries, piston bowl shape and combustion chamber shapes.

*Operating characteristics*: Power, torque, brake mean effective pressure (BMEP), speed and load.

*Combustion characteristics*: Ignition delay, spark timing, premixed combustion phase, diffusion combustion phase, after-burning phase, heat release rate, flame speed, flame kernel growth rate, rapid burning angle and rate of pressure rise.

*Emission characteristics*: CO, HC, $NO_x$, PM.

## 6.5 Chassis Dynamometer Study

The performance and emission characteristics of an automotive vehicle are evaluated at actual operating simulated conditions using a chassis dynamometer that loads the vehicle as per the driving pattern. Driving cycles such as European, Japanese, Indian and American were designed to simulate the typical vehicle under actual operating conditions. The driving cycle is a velocity–time plot to describe the driving behaviour of a given city or a region. It consists of a sequence of several vehicle- operating conditions such as idle, acceleration, cruise and deceleration. The driving pattern varies from region to region and from city to city.

All the wheels of the vehicle are placed on the drums of a chassis dynamometer. The vehicle will be stationary but its wheel rotates as in normal driving. A driver drives the vehicle as per the driving cycle.

## 6.6 Control Volume Sampling

The emission measurements from raw exhaust gas do not represent the actual value as it dilutes with atmospheric air under real conditions. So there is a need to dilute the exhaust with ambient air to prevent condensation of water in the collection system. It is necessary to measure or control the total volume of exhaust plus dilution air and collect a continuously proportioned volume of the sample for analysis. The use of a full flow CVS system is mandatory in some legislation. The dilution of raw exhaust gas with some fresh ambient air is done in the ratio of about 4:1 (raw gas sample: fresh air). The portion of the diluted exhaust is stored in sample bags. Then the mass emission of CO, HC, $NO_x$, $CO_2$ and $CH_4$ is measured from the sample bag gas concentration.

A vehicle can be tested for a mileage build-up of 30,000 km or more using the dynamometer. The advantages are less time and greater economy as compared to field trials. The mileage build-up is used for the development of lubricants or alternative fuels for mostly research and development (R&D) purposes.

The unit of mass emission from a chassis dynamometer is in grams per kilometre where grams per kiloWatt–hour is from the test bench dynamometer.

## 6.7 Digas Analyser

CO, carbon dioxide ($CO_2$), HC, oxygen ($O_2$) and lambda ($\lambda$) were measured using an Anstalt für Verbrennungskraftmaschinen List (AVL) Digas analyser (type: Digas 4000). The Digas analyser and the smoke meter are shown in Figure 6.6.

## 6.8 CLD Analyser

$NO_x$ emission was measured using a chemiluminescence analyser (make: Emerson; type: NGA2000 CLD). The different views of a chemiluminescence detection (CLD) analyser are shown in Figure 6.7. The chemiluminescence reaction between ozone and nitric oxide was used to determine the presence of $NO_x$ in a sample gas. The chemiluminescence measurement involves the following reaction:

$$NO + O_3 = NO_2 + O_2 \tag{6.11}$$

$$NO_2^* = NO_2 + h\nu \text{ (red light)} \tag{6.12}$$

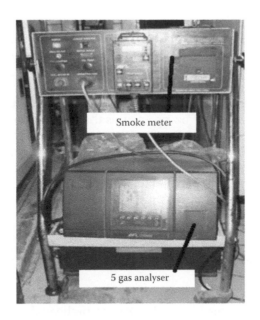

**FIGURE 6.6**
437C smoke meter and a Digas 4000 analyser.

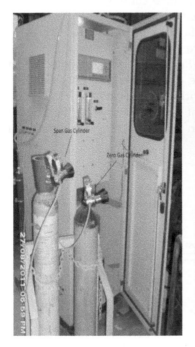 

**FIGURE 6.7**
Various views of a CLD NO$_x$ analyser.

In reaction 6.1, nitric oxide and ozone ($O_3$) readily react to form nitrogen dioxide in an electrically excited ($NO_2^*$) state. In reaction 6.2, the excited $NO_2^*$ immediately reverts to the ground state by emitting photons (red light). The light intensity is measured by a photodiode detector. An integral back pressure regulator and capillary tube are used to maintain the constant sample pressure and flow rate in the reaction chamber. Combining this controlled flow of the sample gas with an excess of ozone ensures that the intensity of the resultant chemiluminescence reaction 6.2 is directly proportional to the NO concentration in the sample. To measure the $NO_x$ ($NO/NO_2$) concentration, the sample gas is reacted with ozone, so that the $NO_2$ in the sample is converted into NO by a heated vitreous carbon bed and then NO is measured as described above. The $NO_x$ emission was measured throughout the experiment.

## 6.9 Biodiesel Reactor

The crude straight vegetable oil (SVO) was preheated in the temperature range of 90–100°C for removal of water content. The acid value of the oil was found to be about 34% by the titrimetric method. In the first stage, the high acid value crude oil was treated with an acid catalyst ($H_2SO_4$ concentration of 0.5% w/w of oil) and alcohol in the temperature range of 55–60°C to avoid soap formation during the transesterification process. After 3 h of reaction in the reactor (Figure 6.8), the mixture was allowed to settle for 2 h. Two layers such as the alcohol fraction (top layer) and transesterified oil (bottom layer) were formed due to their phase difference. Then transesterified oil was separated for further processing with the base catalyst. The reaction was repeated till the acid value was brought down to less than 1%. The theoretical mechanism of the transesterification process with the acid and base catalyst is given in Equation 6.13

$$
\begin{array}{lllll}
\text{Triglycerides} & + & \text{methanol} & = & \text{Mixture of fatty esters} & + & \text{glycerol}
\end{array}
\tag{6.13}
$$

Note: Catalyst*: Acid: $H_2SO_4$ and base: KOH

**FIGURE 6.8**
Biodiesel reactor.

In the second stage, the transesterification process was done with the base catalyst potassium hydroxide (KOH). Methanol and KOH were added to the acid-treated transesterified oil that was treated in the first stage. The reaction was carried out with methanol/oil at a molar ratio of 6:1 with the base catalyst (KOH concentration of 1% w/w of oil) at 65°C for 3 h. The processed oil was kept in a conical flask for separation of the monoester (biodiesel) and glycerol. In the third stage, the biodiesel was washed a number of times using hot distilled water till complete removal of the unreacted triglyceride, alcohol and salt. Finally, the oil was heated in the temperature range of 70–95°C to remove water moisture. After evaporation of the water moisture, the oil (B100) was blended with the commercially available diesel to obtain the B20 blend.

The existing fuel system was utilised to inject the diesel/biodiesel into the engine. An external fuel tank was used instead of the arrangement provided by the OEM. The fuel tank was placed on a rack beside the engine at a certain height above it. It was connected to the injection pump through a fuel filter with the help of a flexible pipe.

**FIGURE 6.9**
(a) Sensor of the airflow meter. (b) Transmitter of the airflow meter.

## 6.10 Airflow Meter

The airflow meter used for the study is called a vortex airflow meter (Endress Hauser make). It consists of a sensor (Figure 6.9a) and a transmitter (Figure 6.9b). This measuring principle is based on the fact that vortices are formed downstream of an obstacle in a fluid flow, for example, behind a bridge pillar. This phenomenon is commonly known as the Kármán vortex street. When the fluid flows past a bluff body in the measuring tube, vortices are alternately formed on each side of this body. The frequency of the vortex shedding down each side of the bluff body is directly proportional to the mean flow velocity and therefore to the volume flow. As they shed in the downstream flow, each of the alternating vortices creates a local low-pressure area in the measuring tube. This is detected by a capacitive sensor and is fed to the electronic processor as a primary, digitised, linear signal.

## 6.11 Gas Mass Flow Meter

The gas mass flow meter is a *Coriolis mass flow meter* (Endress Hauser make). It consists of a sensor (Figure 6.10a) and a transmitter (Figure 6.10b). It works on the Coriolis acceleration. The flow is guided into the *U*-shaped tube. When an oscillating excitation force is applied to the tube causing it to vibrate, the fluid flowing through the tube will induce a rotation or a twist to the tube because of the Coriolis acceleration acting in opposite directions on either side of the applied force. On the opposite side, the liquid flowing out of the meter resists having its vertical motion decreased by pushing up on the tube.

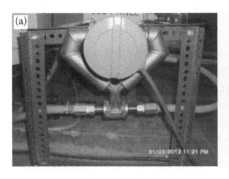 

**FIGURE 6.10**
(a) Sensor of the gas mass flow meter. (b) Transmitter of the mass flow meter.

This action causes the tube to twist. When the tube moves downwards during the second half of the vibration cycle, it twists in the opposite direction. This twist results in a phase difference (time lag) between the inlet side and the outlet side, and this phase difference is directly affected by the mass passing through the tube. An advantage of a Coriolis flowmeter is that it measures the mass flow rate directly, which eliminates the need to compensate for changing temperature, viscosity and pressure conditions.

## 6.12 Test Bench Experimental Study

The effect of the fuel on performance and emission characteristics of an internal combustion engine is studied under wide-open operating conditions. For example, the distillation property of a given fuel such as higher T10 would lead to cold start-ability. T10 refers to the temperature at which 10% of the fuel is distilled. A suggestion may be given to Oil Company to improve T10. If not possible, a cold-starting aid such as a glow plug may be used to overcome the problem. The development of any engine can be done using this study. The following parameters are studied using the test as given below.

### 6.12.1 Performance Characteristics of Internal Combustion Engines

Power and torque output: The maximum power output can be obtained at a wide-open throttle (WOT). For example, ethanol 85% (by volume) and 15% gasoline (E85) fuel would give lower power and torque output in a spark ignition engine as compared to the base gasoline. The reason is mainly due to the lower calorific value of E85 (29 MJ/kg) than gasoline (44 MJ/kg). The remedial measure will be optimisation of the injection system for getting the same power output or enhancing the swept volume.

*Fuel economy*: It is very important from the consumer point of view. The engine has to be optimised for maximum fuel economy.

*Lube oil temperature*: It indicates a suitability of the lubricant for the engine corresponding to a particular fuel.

*Exhaust gas temperature*: It is a qualitative integrator of engine performance periodically. If it exceeds or is below the normal value, the engine may have problems of lubricating oil performance deterioration, gas leakages through valves and piston rings and high deposits on combustion chamber surfaces.

## 6.13 Field Trial Study

A field trial of a vehicle needs to be conducted for a specific distance on an actual road. It is a real field study as compared to the other study mentioned above. Even though it is expensive and time consuming, the field trial test is very important for the evaluation of performance of the fuel and lubricants. The following methodology is generally adopted for a field trial study:

*Distance of mileage build-up*: The distance to be covered by the field study is generally taken on the basis of OEM warranty of the spares and accessories. The warranty for spares and accessories of a vehicle is generally given to the consumer for about 30,000 km of distance travel or 1 year, whichever is earlier. So the vehicle distance to be covered by the field trial test will be about 30,000 km.

*Road condition*: The road condition is chosen as steep hills, deserts, normal roads, muddy roads, stoned roads and paved roads. The combination of the road is dependent on the geographical configuration and vehicle fleet history of a country.

*Climatic condition*: It is very important to evaluate fuels and lubricants in different climatic conditions such as cold, severe cold, rain and heat.

*Payload*: The field trial of a vehicle should be with the maximum payload that is recommended by its manufacturer so that the effect of fuels and lubricants can be evaluated. The vehicle payload is often simulated with a high dense material such as filling of mud, sand, cement or heavy goods. It is an obvious reason that the effect of fuels and lubricants on vehicle performance will be relatively lesser with a lower payload.

*Lubricant analysis*: The vehicle's lubricant is periodically changed as per the manufacturer's instruction. Subsequently, the drained lubricant's performance, which includes physicochemical properties such as viscosity, viscosity index, total acid number (TAN) and total base number (TBN), is measured, in order to assess lubricant deterioration.

*Deposit rating*: The engine parts will be dismantled for the analysis of component performance. The deposits on engine components such as cylinder liner, piston top, piston ring, intake and exhaust valve, injector and spark plug will be rated as per the standard rating methods such as Coordinating Research Council (CRC) and Japanese Automobile Standard Organization (JASO). Deposit characteristics are very important as they influence the durability of vehicle performance. The deposit on the cylinder wall and piston reduces heat transfer due to the insulation of components. The aerodynamic characteristics such as squish will be affected as its non-uniform deposits may lead to poor mixing, resulting in reduced performance and emission characteristics. Similarly, coking in the injector hole results in a reduced flow rate of the fuel and alters the spray characteristics.

*Wear-and-tear analysis*: The wear in the metal analysis of drained lubricants indicates the wear and tear of the cylinder components. The dimensions of components are measured physically in a metrology laboratory.

Remedial measures will be suggested after examining the vehicle in all its aspects as mentioned above with the comparison of prefield trial conditions.

*Conclusions*: The performance of fuel quality needs to be evaluated using bench tests, chassis dynamometer tests and field trials. The analysis of fuel quality is the first step to evaluate the performance of any engine.

# 7

## Modelling of Alternative Fuelled Internal Combustion Engines

## 7.1 Introduction

Modelling and simulation are widely used to study an internal combustion (IC) engine under a wide operating range. Modelling is based on the laws of thermodynamics, heat transfer, fluid mechanics and chemical kinetics and makes use of comprehensive databases of engine parameters such as fuel properties, friction coefficients and piping parameters. Modelling of engines has been classified into a zero-dimensional, phenomenological model, a quasi-dimensional model and a multi-dimensional model. On the basis of requirements, any one-dimensional modelling could be chosen. Zero-dimensional models are used for the analysis of engine processes mostly using thermodynamic and basic conservation laws, whereas phenomenological and quasi-dimensional models provide better accuracy by analysis with additional details such as piston geometry and spark plug position. Among the models, multi-dimensional modelling is more accurate as it includes the effect of flow on the engine cylinder and combustion chamber geometry. The three-dimensional model takes a large amount of time to give the solution. However, in the computer era, it can be processed with greater speed using a high-speed computational machine. Computational fluid dynamic (CFD) software has mainly three parts, namely, pre-processor, solver and post-processor. Pre-processor defines the plenum of the analysis. Solver solves the equation and post-processor facilitates post-processing.

### 7.1.1 Objective of Conducting a Simulation

The primary objective of engine simulation is to narrow down the range of experimental investigations relating to the development of the engine and thereby minimise the cost and time.

Figure 7.1 shows the model methodology for IC engines. The models are basically derived from mass, momentum and the energy conservation equation. The models are in the nature of thermodynamic, fluid dynamics, heat transfer, chemical kinetics and combustion and they could be used for the

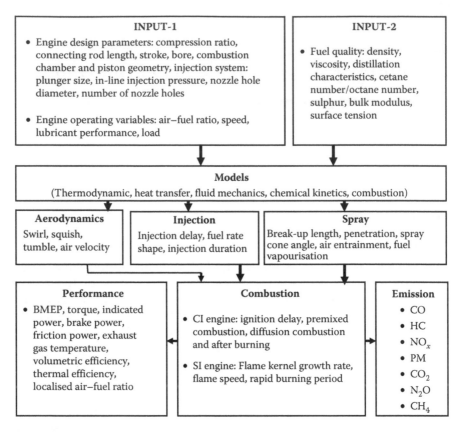

**FIGURE 7.1**
Model methodology for IC engines.

analysis of the desired subsystem or parameter. The engine design characteristics such as the compression ratio, connecting rod length, bore, stroke,
combustion chamber and piston geometry, injection pressure, number of
nozzle holes and size are given as input to the models. In addition to this,
engine operating variables such as the air–fuel ratio, speed, load and exhaust
gas temperature are also given to the models. Furthermore, the effects of fuel
quality parameters such as the density, viscosity, distillation characteristics,
cetane number or octane number, sulphur, bulk modulus, surface tension,
heat of formation and Reid vapour pressure play a critical role in the performance and emission characteristics of an IC engine. These engines are
normally optimised for either gasoline or diesel fuel. If the fuel is changed,
the performance of the engine will change. So these models could be used
for the analysis of the effect of fuel quality on the performance and emission characteristics of an IC engine, resulting in a reduced experimental system that enables fast product development. Further, it enables one to obtain
the data that may be difficult to measure. A typical example relates to the

time-varying gas composition and temperature inside the engine cylinder. A detailed analysis of simulation of alternative fuelled engines will not be given here as lack of space precludes us from doing so. However, a brief summary of the basic concepts is given here.

Engine simulation is usually started with the compression process that continues into the combustion, expansion, exhaust and intake processes. The basic steps for an alternative fuelled engine would be the same as for a conventional fuelled engine except for the combustion process. This is primarily due to the variation of fuel properties of an alternative fuel as compared to the conventional fuel. The need of model usage in real-time application is described through an actual example. Figures 7.2 and 7.3 show brake-specific

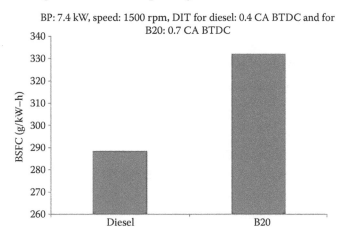

**FIGURE 7.2**
Comparison of BSFC for diesel and B20.

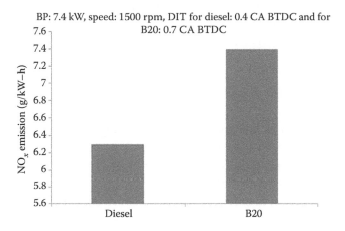

**FIGURE 7.3**
Comparison of $NO_x$ emission for diesel and B20.

fuel consumption (BSFC) and nitrogen oxide ($NO_x$) emission of a diesel engine in comparison with the biodiesel–diesel blend (B20). It is clearly seen from the figures that BSFC and $NO_x$ increased with B20. Many technologies such as exhaust gas recirculation, injection timing optimisation, injection system modification, combustion chamber modification and aerodynamic optimisation can reduce $NO_x$ emission. The screening of suitable technology to simultaneously reduce BSFC and $NO_x$ could be done rapidly using modelling and simulation. Otherwise, more experimental tests with many combinations will need to be conducted. It may not be possible to conduct some tests as they may have difficulties in measuring parameters. The time taken for the tests is high, resulting in a delay in the development of a product. So modelling and simulation play a vital role in developing suitable technology with minimum experimental tests.

### 7.1.2 Problem-Solving Process

The problem-solving process for a computational problem can be outlined as follows:

Define the problem

Create a mathematical model

Develop a computational method for solving the problem

Implement the computational method

Test and assess the solution

The boundaries between these steps can be blurred, and for specific problems, one or two of the steps may be more important than others. Nonetheless, having this approach and a strategy in mind will help us to focus our efforts as we solve problems.

### 7.1.3 Problem Definition

The first steps in problem solving include

- Recognise and define the problem precisely by exploring it thoroughly (may be the most difficult step).
- Determine what question is to be answered and what output or results are to be produced.
- Determine what theoretical and experimental knowledge can be applied.
- Determine what input information or data are available.

- If the problem is not well defined, considerable effort must be expended at the beginning in studying the problem, eliminating the things that are unimportant and focussing on the root problem.

### 7.1.4 After Defining the Problem

- Collect all data and information about the problem.
- Verify the accuracy of these data and information.
- Determine what information you must find: intermediate results or data may need to be found before the required answer or results can be found.

### 7.1.5 Mathematical Model

- To create a mathematical model of the problem to be solved.
- Determine what fundamental principles are applicable.
- Draw sketches or block diagrams to understand the problem better.
- Define the necessary variables and assign notation.
- Reduce the problem as originally stated into one expressed in purely mathematical terms.
- Apply mathematical expertise to extract the essentials from the underlying physical description of the problem.
- Simplify the problem only enough to allow the required information and results to be obtained.
- Identify and justify the assumptions and constraints inherent in this model.

### 7.1.6 Computational Method

A computational method for solving the problem is to be developed based on the mathematical model.

- Derive a set of equations that allow the calculation of the desired parameters and variables.
- Develop an algorithm, or step-by-step method, of evaluating the equations involved in the solution.
- Describe the algorithm in mathematical terms and then implement as a computer program.
- Carefully review the proposed solution, with thought given to alternative approaches.

## 7.2 Internal Combustion Engine Processes

The thermal efficiency of IC engines is given in Equation 7.1.

$$\text{Thermal efficiency} = \frac{W}{Q_s} = 1 - \frac{Q_r}{Q_s} \tag{7.1}$$

A spark ignition (SI) engine works with an Otto cycle, whereas a compression ignition (CI) engine works with a dual cycle. The brief details of the cycle are given below.

### 7.2.1 Otto Cycle

The air-standard cycle efficiency of an Otto cycle for SI IC engines is given in Equation 7.2. The pressure–volume (P–V) and temperature–entropy (T–S) diagrams are shown in Figures 7.4 and 7.5, respectively. The Otto cycle 1–2–3–4 consists of the following four processes: (i) process 1–2: reversible adiabatic compression of air, (ii) process 2–3: heat addition at constant volume, (iii) process 3–4: reversible adiabatic expansion of air, (iv) process 4–1: heat rejection at constant volume.

$$\text{Air standard efficiency}, \eta_{th} = \frac{\text{Net work done}}{\text{Net heat added}} \tag{7.2}$$

The P–V diagram shows how much heat is converted into power, whereas T–S shows how much heat is available for power conversion. The simple cycle analysis identifies the scope for further development of the SI engine.

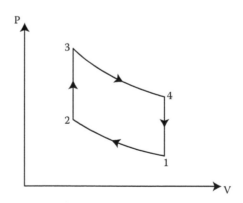

**FIGURE 7.4**
P–V diagram of the Otto cycle.

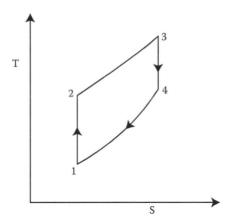

**FIGURE 7.5**
T–S diagram of the Otto cycle.

$$\text{Heat added} = mC_v(T_3 - T_2)$$

$$\text{Heat rejected} = mC_v(T_4 - T_1)$$

Then $\eta_{th} = 1 - (T_4 - T_1/T_3 - T_2)$ or the adiabatic compression 1–2: $T_1V_1^{\gamma-1} = T_2V_2^{\gamma-1}$ and for the adiabatic expansion 3–4: $T_3V_3^{\gamma-1} = T_4V_4^{\gamma-1}$.

The thermal efficiency of the Otto cycle is expressed as $\eta_{th} = 1 - (1/r^{\gamma-1})r =$ compression ratio $= (V_1/V_2) = (V_4/V_3)$. From the above equation, it can be observed that the efficiency of the Otto cycle is the function of the compression ratio for the given ratio of $C_p$ and $C_v$. It is clearly seen from the above equation that the efficiency increases with an increase in the compression ratio.

Mean effective pressure is defined as the ratio of the net work done to the displacement volume of the piston.

$$mep = \frac{W}{V_1 - V_2}$$

$$= \frac{\eta Q_{2-3}}{V_1 - V_2}$$

$$= P_1 r \left\{ \frac{(r^{\gamma-1} - 1)(r_p - 1)}{(r - 1)(\gamma - 1)} \right\}$$

Pressure ratio $r_p = P_3/P_2 = T_3/T_2$.

## 7.2.2 Diesel Cycle

The P–V and T–S diagrams of a diesel cycle are shown in Figures 7.6 and 7.7, respectively. This cycle consists of four processes: (i) reversible adiabatic

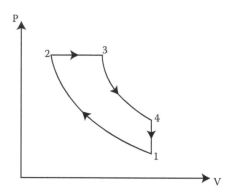

**FIGURE 7.6**
P–V diagram of a diesel cycle.

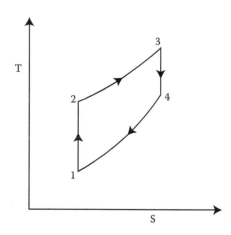

**FIGURE 7.7**
T–S diagram of a diesel cycle.

(isentropic) compression (step 1–2), (ii) isobaric heating (step 2–3), (iii) reversible adiabatic (isentropic) expansion (step 3–4) and (iv) ending with an isochoric cooling (step 4–1).

Consider 'm' kilogram of a working fluid

Heat supplied at constant pressure $Q_s = mC_p(T_3 - T_2)$

Heat rejected at constant volume $Q_r = mC_v(T_4 - T_1)$

Then $\eta_{th} = 1 - (C_v(T_4 - T_1)/C_p(T_3 - T_2))$ or the adiabatic compression 1–2: $T_1V_1^{\gamma-1} = T_2V_2^{\gamma-1}$ for the adiabatic expansion 3–4: $T_3V_3^{\gamma-1} = T_4V_4^{\gamma-1}$ or the isobaric heating 2–3: $V_2/T_2 = V_3/T_3$ expansion ratio during heating (also called the cut-off ratio), $V_3/V_2 = \beta$ thermal efficiency of the diesel cycle is expressed as $\eta = 1 - (1/r^{\gamma-1})[r_c^\gamma - 1/\gamma(r_c - 1)]$.

It can be observed from the above equation that the thermal efficiency of the diesel cycle is mainly the function of the compression ratio and cut-off ratio. If the thermodynamic coordinates such as pressure, volume, temperature and entropy at points 1–4 are known, heat availability and possible power output can be found out. However, this is a theoretical equation that is not sufficient for further development of the engines using the equations.

---

## 7.3 Governing Equations for IC Engines

As discussed, it is quite clear that air-standard efficiency is unable to give more insights into engine development; the following basic equations are required to analyse the engine processes. The basic governing equations applicable for both SI and CI engines are given in Table 7.1.

### 7.3.1 Chemical Kinetics

On the basis of the kinetic theory of gases, it is possible to calculate the rate of chemical reaction in both the forward (combustion) and reverse (dissociation) directions. Reaction rates are found to be directly proportional to the instantaneous concentration raised to some power of the reactant materials.

$$\text{Take the reaction of the type } A + B \Leftrightarrow C \tag{7.3}$$

The equation is an equilibrium balance. It is simply an atom balance between some of the species present at equilibrium. The instantaneous rate of the forward reaction after a given time period is found to be proportional to the product of the concentration of the reactants or

$$\text{Forward reaction rate} \propto [A]^a [B]^b = k_F [A]^a [B]^b \tag{7.4}$$

Here, [ ] indicates concentration, $k_F$ is the rate constant for the forward reaction, and $a$ and $b$ are indices, which in this case are unity. Similarly, for the reverse reaction

$$\text{Reverse reaction rate} = k_R[C]^c$$

and $k_R$ is the reverse reaction rate constant and $c$ is unity in this case.

At the dynamic equilibrium condition for any given temperature, the two rates are equal; hence

$$k_F [A]^a [B]^b = k_R[C]^c \tag{7.5}$$

**TABLE 7.1**

Basic Equations Applicable to IC Engines

| S. No. | Description | Equation |
|---|---|---|
| 1 | Equation of state for an ideal gas | $PV = mRT$ |
| 2 | Conservation of mass for an open system | $m = \sum_k m_k$ |
| 3 | First law of thermodynamics for any open system | $E = Q - W + \sum_k m_k$ |
| 4 | First law of thermodynamics employed in IC engines | $\dfrac{d(mu)}{d\theta} = \dfrac{dQ}{d\theta} - p\dfrac{dV}{d\theta} + \sum_k h_k \dfrac{dm_k}{d\theta}$ |
| 5 | Heat release rate | $\dfrac{\partial Q}{d\theta} = \dfrac{1}{k-1} V \dfrac{dP}{d\theta} + \dfrac{k}{k-1} p \dfrac{dV}{d\theta}$ |
| 6 | Adiabatic index of the air–fuel mixture | $k = c_p / c_v$ |
| 7 | Rate of change of cylinder volume | $\dfrac{dV}{d\theta} = \dfrac{V_d}{2} \sin\theta \left[1 + \cos\theta (R^2 - \sin^2\theta)^{-1/2}\right]$ |
| 8 | Instantaneous volume of the cylinder | $V = \dfrac{V_d}{r-1} + \dfrac{V_d}{2}\left[1 + R - \cos\theta - (R^2 - \sin^2\theta)^{1/2}\right]$ |
| 9 | Displacement volume ($V_d$) | $V_d = \dfrac{\pi B^2}{4} \times L$ |
| 10 | Heat loss to the cylinder walls | $Q = hA(T_g - T_w)$ |
| 11 | Combustion chamber surface area | $A = A_{ch} + A_p + \pi bx = \dfrac{\pi}{4}b^2 + \dfrac{\pi}{a}b^2 + \pi bx$ |
| 12 | Temperature of the gas | $T_g = (PV)/(mR)$ |
| 13 | Convective heat transfer coefficient [1] | $Nu = \left(\dfrac{h_c b}{k}\right) = a\left(\dfrac{\rho \bar{S}_p b}{\mu}\right)^b$ |
| 14 | Convective heat transfer coefficient [2] | $h = 3.26 * b^{-0.2} * P^{0.8} * T^{-0.55} * v^{0.8}$ |
| 15 | Average cylinder gas velocity without swirl | $v = c_1 \bar{S}_p + c_2 \dfrac{V_d T_r}{p_r V_r}(p - p_m)$ |
| 16 | Motoring pressure ($p_m$) | $P_{motor} = \dfrac{\left[\left(\dfrac{rV_d}{r-1}\right)^r P_a\right]}{V^r}$ |
| 17 | Compression ratio | $r_c = (V_d + V_c)/V_c$ |
| 18 | Equivalence ratio | $\phi = \dfrac{F/A}{(F/A)_s} = \dfrac{(A/F)_s}{A/F}$ |
| 19 | Mean piston speed | $S_p = 2LN$ |

**TABLE 7.1     (continued)**

Basic Equations Applicable to IC Engines

| S. No. | Description | Equation |
|---|---|---|
| 20 | Volumetric efficiency | $\eta_V = \dfrac{\dot{M}_i}{(N/2)V_d \rho_i}$ |
| 21 | Mass burnt fraction (Wiebe function) [3] | $f = 1 - \exp\left[-a\left(\dfrac{\theta - \theta_o}{\Delta\theta}\right)^n\right]$ |

so that

$$\frac{[C]}{[A][B]} = \frac{k_F}{k_R} = k' \tag{7.6}$$

Here, $k'$ is called the concentration equilibrium constant for the given reaction. Thus, the reaction products are located in the numerator and the reactants are located in the denominator of the equilibrium constant expressions.

## 7.3.2  General Combustion Equation

Generally, any hydrocarbon (HC) fuel will burn in the presence of air and dissociates from carbon monoxide (CO) and free hydrogen and oxygen [3]. The general combustion equation, incorporating six unknowns, will apply; thus

$$C_xH_y + m(O_2 + 3.76N_2) \rightarrow n_1CO_2 + n_2H_2O + n_3CO + n_4H_2 + n_5O_2 + n_6N_2 \tag{7.7}$$

$$\left.\begin{array}{ll} \text{Carbon balance:} & n_1 + n_3 = x \\ \text{Hydrogen balance:} & n_2 + n_4 = 0.5y \\ \text{Oxygen balance:} & n_1 + 0.5n_2 + 0.5n_3 + n_4 = m \\ \text{Nitrogen balance:} & n_5 = 3.76\,m \end{array}\right\} \tag{7.8}$$

Two more equations are required to solve the problem. These are the dissociation reactions involving carbon dioxide and water vapour and their corresponding equilibrium constants.

$$H_2 + \tfrac{1}{2}O_2 \Leftrightarrow H_2O \tag{7.9}$$

$$CO + \tfrac{1}{2}O_2 \Leftrightarrow CO_2 \tag{7.10}$$

$$k_{p,H_2O} = \frac{n_{H_2O}}{n_{H_2}(p(n_{O_2}/n_T))^{1/2}} = \frac{n_2}{n_4(p(n_5/n_t))^{1/2}} \tag{7.11}$$

**TABLE 7.2**

Equilibrium Constants

| Temperature (K) | $K_{CO_2}$ (atm)$^{1/2}$ | $K_{H_2O}$ (atm)$^{1/2}$ | $K_{WG}$ |
|---|---|---|---|
| 298.15 | 1.1641E45 | 11.169E39 | 9.3756E–06 |
| 300.00 | 575.44E42 | 6.1094E39 | 10.617E–06 |
| 500.00 | 10.593E24 | 76.913E21 | 7.2611E–03 |
| 1000.00 | 16.634E09 | 11.535E09 | 0.69343 |
| 1500.00 | 207.01E03 | 530.88E03 | 2.5644 |
| 2000.00 | 765.6 | 3.467E03 | 4.5290 |

*Source:* Adapted from J. B. Heywood, *Internal Combustion Engine Fundamentals*, McGraw-Hill, 1989.

$$k_{p,CO_2} = \frac{n_{CO2}}{n_{CO}(p(n_{O_2}/n_T))^{1/2}} = \frac{n_1}{n_3(p(n_5/n_T))^{1/2}} \tag{7.12}$$

where $p$ is the total pressure of the product mixture and $n_T$ is the total moles of the products present. The solution involves an iterative procedure so that there will not be a closed-form solution. Equilibrium constants for different temperatures are given in Table 7.2.

## 7.4 CI Engine Modelling for Alternative Fuels

### 7.4.1 Injection Characteristics

As the energy comes from fuel, the appropriate quantity of the fuel with respect to load has to be injected at the end of the compression stroke. The injection parameters such as fuel rate shaping, injection delay and injection duration need to be analysed. Some models of fuel injection characteristics for a CI engine are given in Table 7.3.

### 7.4.2 Spray Characteristics

The injected fuel has to be mixed with air. The fuel spray is characterised using the following processes:

1. Break-up length
2. Penetration distance
3. Spray cone angle
4. Air entrainment
5. Vapourisation

The models available for the calculation of spray characteristics of diesel engines for diesel fuel are given in Table 7.4.

**TABLE 7.3**

Injection Characteristics of Fuel for a CI Engine

| S. No. | Description | Equation |
|---|---|---|
| 1 | Mean value of fuel injection rate (kg/CA) [4,5] | $\bar{m}_{finj} = m_{ftot}/(z \cdot \Delta\varphi_{inj})$ |
| 2 | Total duration of fuel injection (CA) [4] | $\Delta\varphi_{inj} = (m_{ftot}/z)/\bar{m}'_{finj}$ |
| 3 | Mean spray velocity from each nozzle hole [5,6] | $\bar{u}_{inj} = \bar{m}_{finj}6N/(\rho_l \cdot F)$ or $\bar{u}_{inj} = c_d\sqrt{2\overline{\Delta p}_{inj}/\rho_l}$ |
| 4 | Mean pressure drop in the nozzle [4,5] | $\overline{\Delta p} = 0.5\rho_l(\bar{u}_{inj}/c_D)^2$ |
| 5 | Bulk modulus | $B = k * (dP/(dV/V))$ |
| 6 | Injection delay [4–8] | $\Delta\varphi_{injdel} = (L_p/a_s)6N$ |
| 7 | Pressure wave in the injector nozzle holes [9] | $\Delta p_w = a_s\rho_l c_{pump}(F_{pump}/F_{nozzle})$ |

**TABLE 7.4**

Models Available for Spray Characteristics of Diesel Fuel

*Break-Up Length*
1  $L_b = 15.8 * (\rho_l/\rho_a)^{0.5} * D_n$  Hiroyasu and Arai [10]
2  $L_b = 8.14 * D_n (We)^{0.5}$  [11]

*Spray Cone Angle*
1  $\theta = 0.03(L_n/D_n)^{-0.3}(\rho_a/\rho_l)^{0.1}Re_l^{0.7}$  Sitkei [12]
2  $Tan(\theta/2) = (1/A)4\pi(\rho_a/\rho_l)^{0.5} * 3^{0.5}/6$, where  Reitz [13]
   $A = 3 + (L_n/D_n)/3.6$
3  $\theta = 83.5(L_n/D_n)^{-0.22}(\rho_a/\rho_l)^{0.26}(D_n/d_n)^{0.15}$  Shimizu [12]
4  $\theta = 0.135(L_n/D_n)^{-\beta} * (\rho_l/\rho_a) * Re^{0.46}$, where  Yokota and Matsuka [12]
   $\beta = 0.0284(\rho_l/\rho_a)^{0.39}$
5  $Tan(\theta/2) = A * (\rho_a/\rho_l)^{0.5}$, where $A = 0.49$  Coghe [13]
6  $Tan(\theta/2) = (1.38 * 10^{-7} * V_i + 3.63 * 10^{-3}) * pa +$  Ishikawa [14]
   $2.86 * 10^{-5} * V_i + 1.88 * 10^{-3}$
7  $2\theta = 0.05(\rho_a * \Delta P * D_n^2/(\mu_a)^2)^{0.25}$  [15]

*Sauter Mean Diameter*
1  $X32 = A * (\Delta P)^{-0.135} * (\rho_a)^{0.121} * (Q)^{0.131}$, where  [16]
   $A = 25.1$ for a pintle nozzle, $A = 23.9$ for a hole nozzle
   and $A = 22.4$ for a throttling pintle nozzle
2  $X32 = 47 * 10^{-3} * (D_n/V_i) * (\sigma/\rho_n)^{0.25} * g^{0.2} * (1 + (33.1$  Tanasawa [17]
   $* 101.5)) * \mu_l/(\sigma * \rho_l * D_n)^{0.5}$
   Complete spray: $X32 = 0.14 * D_n * Re^{0.25} * We^{-0.32}$
   $* (\mu_l/\mu_a)^{0.37} * (\rho_l/\rho_a)^{0.17}$
   Ambient pressure: $X32 = 0.38 * D_n * Re^{0.25} * We^{-0.32}$
   $* (\mu_l/\mu_a)^{0.37} * (\rho_l/\rho_a)^{-0.47}$

*continued*

**TABLE 7.4    (continued)**

Models Available for Spray Characteristics of Diesel Fuel

| | | |
|---|---|---|
| 3 | $X32 = \max(X32^{LS}, X32^{HS})$ | [17] |
| | $X32^{LS} = 4.12 * D_n * Re^{0.12} * We^{-0.75} * (\mu_l/\mu_a)^{0.54} * (\rho_l/\rho_a)^{0.18}$ | |
| | $X32^{HS} = 0.38 * D_n * Re^{0.25} * We^{-0.32} * (\mu_l/\mu_a)^{0.37} * (\rho_l/\rho_a\rho_l/\rho_a)^{-0.47}$ | |
| 4 | $X32 = ((585 * \sigma^{0.5})/(VI * \rho_l^{0.5}\rho_l^{0.5})) + 597(\mu_l/(\sigma * \rho_l)^{0.5})^{0.45}$ | [18] |
| 5 | $SMR = C * (4 * \prod * \sigma * 3)/(\rho_a * V_i^2 * 2)$, where $C = 0.75$ | [19] |
| 6 | $X32 = C * 83.2 * (D_n/V_i) * (\sigma * g/\rho_a)^{0.25}((1 + 10.4 * \mu_l)/$ $(\sigma * \rho_l * D_n)^{0.5})$, where $C = 0.721$ for $D_n = 0.2$ and $C = 0.455$ for $D_n = 0.3$ | [20] |
| 7 | $SMD = 7.3 * \sigma^{0.6} * v^{0.2} * m^{0.25} * \Delta p^{-0.4}$ | |

*Penetration*

| | | |
|---|---|---|
| 1 | $S = 2.95 (\Delta P/\rho_a)^{0.25} * (D_n * t)^{0.5}$ for $t > t_b$ and $S = 0.39(2\Delta P/\rho_a)^{0.5} * t$ for $t < t_b$ | Hiroyasu and Arai [12] |
| | $t_{br} = 28.61\rho_l \cdot D_N / \sqrt{\rho_a \cdot \overline{\Delta p}}$ | |
| 2 | $S = 3.07(\Delta P/\rho_a)^{0.25} * (D_n * t)^{0.5} * (294/T_g)$ | [21] |

*Air Entrainment*

| | | |
|---|---|---|
| 1 | $m_a = 0.26(\rho_a * \rho_l)^{0.5} * D_n * S * V_i * \tan(\theta/2)$ | [11] |
| 2 | $m_a = 40 * V^{0.5} * D_n^{-0.5} * t^{1.5} * m_f$ | Wakuri [22] |
| 3 | $m_a = -(m_f * V_i/(dS/dt)^2) * d^2S/dt^2$ | Dohoy [23] |
| 4 | $m_a = (\pi/3) * (Tan(\theta/2))^2 * S^3 * \rho$ | [4] |

*Vapourisation*

| | | |
|---|---|---|
| 1 | $dD/dt = -K/2 * D$ where $K = 8 * \lambda_g * \ln(BT + 1)/(\rho_l * C_{pg})$, where $BT = C_{pg} * (T - T_s)/L$ | $D^2$ law [4] |

Rakopoulos et al. (2004) [4] were given a model for a spray cone angle, penetration and air entrainment with swirl

| | |
|---|---|
| Spray cone angle | $\theta_S = \theta (X/X_S)$ |
| Penetration distance | $X_S = X (1 + \pi R_S N X/30 U_{inj})^{-1}$ |
| Air entrainment | $m_S = (\pi/3) (Tan(\theta_S/2)) \rho_a X_S^3$ |

*Source:* Adapted from K. A. Subramanian and S. Lahane, *International Journal of Energy Resource*, 2011, DOI: 10.1002/er.1947 [24].

### 7.4.3  Ignition Delay

Ignition delay is defined as the elapsed time between the start of fuel injection and the start of combustion. It can be divided into two parts: physical and chemical processes. If ignition delay is longer, the engine will experience problems such as cold starting, white smoke and a high rate of pressure rise. To some extent, knocking will occur, resulting in the damage of engine components. Too long an ignition delay is not desirable in view of knocking as well as durable issues.

The most used form for defining ignition delay is given in Equation 7.13. The values of $A$ and $n$ are given in Table 7.5. The ignition delay of diesel engines can be calculated using Equation 7.13 with the inputs of in-cylinder pressure and constants of $A$ and $n$.

**TABLE 7.5**

Constants of $A$ and $n$

| Fuel | $A$ | $n$ |
|------|-----|-----|
| Diesel | 120 | 1.53 |
| B20 | 120 | 1.58 |

$$\tau = A \times P^{-n} \times e^{B/T} \tag{7.13}$$

The physical and chemical ignition delay model developed by Lahane and Subramanian [25,26] is given below:

### 7.4.3.1 Physical Delay

The physical processes are fuel spray atomisation, evaporation and mixing of fuel vapour with cylinder air. Good atomisation requires high fuel pressure, a smaller injector hole diameter, optimum fuel viscosity and high cylinder pressure (large divergence angle). The rate of vapourisation of the fuel droplets depends on the droplet diameter, velocity, fuel volatility, pressure and temperature of the air.

$$\tau_{Ph} = K \times \left( \underbrace{\frac{(\rho_l \times D_n)}{(0.5 \times \rho_a \times \rho_l \times V_i^2)^{0.5}}}_{\text{Break-up time}} + \underbrace{\frac{(dD \times X_{32} \times \rho_l \times C_{pg})}{(\lambda_g \times (ln(C_{pg}/L) \times (T - T_S) + 1)}}_{\text{Vapourisation}} \right) \tag{7.14}$$

### 7.4.3.2 Chemical Delay

Chemical processes similar to that described for the auto-ignition phenomenon in premixed fuel–air are more complex since heterogeneous reactions (reactions occurring on the liquid–fuel drop surface) also occur.

$$\tau_{Ch} = A \times P^{-n} \times e^{(E_a/R \times T)} \tag{7.15}$$

### 7.4.3.3 Total Ignition Delay

$$\tau = \tau_{ph} + \tau_{Ch} \tag{7.16}$$

$$\tau = \left( \left( K \times \left( \frac{(\rho_l \times D_n)}{(0.5 \times \rho_a \times \rho_l \times V_i^2)^{0.5}} + \frac{(dD \times X_{32} \times \rho_l \times C_{pg})}{(\lambda_g \times (ln((C_{pg}/L) \times (T - T_S)) + 1))} \right) \right) \right.$$

$$\left. + A \times P^{-n} \times e^{(E_a/R \times T)} \right) \tag{7.17}$$

### 7.4.4 Combustion Model

Combustion of the fuel in a diesel engine can be characterised using four stages: ignition delay, pre-mixed combustion phase, diffusion combustion phase and late combustion phase. According to the first law of thermodynamics, the heat given into the system is equal to the change in internal energy and the work done by the system.

$$\frac{dQ}{d\theta} = \frac{du}{d\theta} + \frac{pdv}{d\theta} + \frac{dh_w}{d\theta} \tag{7.18}$$

### 7.4.5 Extended Zeldovich Mechanism (NO$_x$ Model)

The NO$_x$ forms through three routes: thermal NO, prompt NO and fuel-bound NO [3]. The thermal NO mechanism is based on the extended Zeldovich mechanism as given in Equations 7.19 and 7.20, which involves atmospheric nitrogen and occurs during combustion thereafter in the post-flame gas region

$$N_2 + O \Leftrightarrow NO + N \tag{7.19}$$

$$N + O_2 \Leftrightarrow NO + O \tag{7.20}$$

$$N + OH \Leftrightarrow NO + H \text{ (Extended Zeldovich mechanism)} \tag{7.21}$$

Thermal NO is a post-flame phenomenon. Since combustion occurs at high pressures, the reaction zone (flame) is extremely thin and the residence time within the zone is short. Cylinder pressure rises during combustion; the charge will burn first and then be compressed to higher temperatures than the values reached after combustion. Thus, the NO formed in the post-flame gases would be much higher than the NO formed in the flame.

The rate of formation of NO is given by [3]

$$\frac{d[NO]}{dt} = k_1^+[O][N_2] + k_2^+[N][O_2] + k_3^+[N][OH]$$

$$- k_1^-[NO][N] - k_2^-[NO][O] - k_3^+[NO][H] \tag{7.22}$$

Since N is much less than the concentrations of other species of interest $N + OH = NO + H$ ($10^{-8}$ mol fraction), the steady-state approximation is appropriate: $d[N]/dt$ is set equal to zero. The NO formation rate becomes

$$\frac{d(NO)}{dt} = \frac{6.10^6}{\sqrt{T}} \exp\left(\frac{-69090}{T}\right)[O_2]_e^{0.5}[N_2]_e \tag{7.23}$$

### 7.4.6 Smoke/Soot Formation Model

The net soot formation rate is calculated by using the model proposed by Hiroyasu et al. [27,28], and it is modified by Lipkea and DeJoode [29]. Soot formation and a soot oxidation rate can be calculated using Equations 7.24 and 7.25. The net soot formation can be found using Equation 7.26.

$$\frac{dm_{sf}}{dt} = A_{sf} \, dm_f^{0.8} p^{0.5} \exp(-E_{sf}/(R_{mol}T)) \tag{7.24}$$

$$\frac{dm_{sc}}{dt} = A_{sc} m_{sn} (po_2/p)p^n \exp(-E_{sc}/(R_{mol}T)) \tag{7.25}$$

$$\frac{dm_{sn}}{dt} = \frac{dm_{sf}}{dt} - \frac{dm_{sc}}{dt} \tag{7.26}$$

## 7.5 Combustion in SI Engine

### 7.5.1 Spark Ignition

The gas exchange process in an SI engine is similar to that in a CI engine except for the charge. In a CI engine, air alone is inducted into the cylinder during a suction stroke, whereas an air–fuel mixture is drawn into the cylinder for an SI engine.

The combustion process begins with the formation of a laminar-burning kernel. As such, the kernel grows and when the flame surface departs from the ignition source, it begins to interact with the turbulent motion field of the cylinder charge. The flame surface is progressively corrugated and wrinkled by turbulence and after transition, the flame becomes fully turbulent.

One of the most important intrinsic properties of any combustible mixture is its burning velocity. Burning velocity is defined as the relative velocity, normal to the flame front, with which unburnt gas moves into the flame front and is transformed. This property depends upon the mixture composition, temperature of unburnt gas, pressure and other parameters. The burning velocity is important because it is the property that influences the flame shape and important flame stability characteristics, such as blow-off and flashback. The laminar burning velocity is of fundamental importance in analysing and predicting the performance of the combustion engine. Therefore, knowledge of the burning velocity is important for an improved understanding of the fundamental combustion processes, and for a direct practical application aimed at increasing fuel efficiency and reducing pollutant emission.

### 7.5.1.1 Characterisation of Combustion Process

Combustion is characterised by parameters such as the flame development angle, rapid burning angle and overall burning angle.

*Flame development angle:* The crank angle ($\theta$) interval between the spark discharge and the time at which a small but significant fraction of the cylinder mass has burnt or fuel chemical energy has been released.

*Rapid burning angle:* The crank angle interval required to burn the bulk of the charge is defined as the interval between the end of the flame development stage and the end of the flame propagation process.

*Overall burning angle:* It is the duration of the overall burning process. It is the sum of the flame development and rapid burning angle.

In addition to the heat release rate, combustion duration and ignition delay, mass fraction burnt rate is closely associated with flame speed, area of flame and unburnt charge density as given in Equation 7.27.

$$\text{Mass burnt rate: } m_k + \ell_u\, A_k\, S_l \tag{7.27}$$

Mass burnt rate can be calculated by Equation 7.27. The mass fraction is characterised using the flame speed and area. Flame speed is a function of design parameters and operating variables.

### 7.5.2 Quasi-Dimensional Two-Zone Model for an SI Engine

Multi-zone models are distinguished from zero-dimensional models as they include certain geometrical parameters such as the radius of a thin interface (the flame) separating burnt gases from unburnt gases in the basic thermodynamic approach [30]. Two-zone models consist of an unburnt zone and a burnt zone which are used by many researchers as it is an easy-to-do simulation. A combustion chamber is divided into two zones as shown in Figure 7.8.

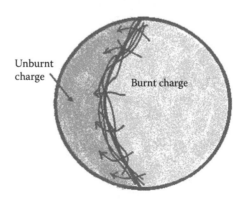

Unburnt charge

Burnt charge

**FIGURE 7.8**
Two-zone model.

The change in burnt temperature and the rate of pressure rise can be calculated using Equations 7.28 and 7.29 [16]:

$$\frac{dp}{d\theta} = \left( \frac{c_{v,u}}{c_{p,u}} - \frac{c_{v,b}}{R_b} \frac{R_u}{c_{p,u}} V_u + \frac{c_{v,b}}{R_b} V \right)^{-1}$$

$$\left\{ -\left( 1 + \frac{c_{v,b}}{R_b} \right) p \frac{dV}{d\theta} - c_{p,b} T_b \frac{dm_{l,b}}{d\theta} - \frac{R_u}{R_b} c_{p,b} T_u \frac{dm_{l,u}}{d\theta} - \frac{dQ}{d\theta} \right.$$

$$\left. - \left[ (u_b - u_u) - c_{v,b} \left( T_b - \frac{R_u}{R_b} T_u \right) \right] \frac{dm_x}{d\theta} + \left( \frac{c_{v,u}}{c_{p,u}} - \frac{c_{v,b}}{R_b} \frac{R_u}{c_{p,u}} \right) \frac{dQ_u}{d\theta} \right\} \quad (7.28)$$

$$\frac{dT_b}{d\theta} = \frac{p}{m_b R_b} \left[ \frac{dV}{d\theta} - \left( \frac{V_b}{m_b} - \frac{V_u}{m_u} \right) \frac{dm_x}{d\theta} + \frac{V_u}{m_u} \frac{dm_{l,u}}{d\theta} \right.$$

$$\left. + \frac{V_b}{m_b} \frac{dm_{l,b}}{d\theta} + \frac{V}{p} \frac{dp}{d\theta} - \frac{V_u}{T_u} \frac{dT_u}{d\theta} \right] \quad (7.29)$$

It has been seen that several models are available for predicting the performance of an IC engine. Further development of the model is needed for advanced engines for effective utilisation of alternative fuels in the engine. The selection of a model for particular analysis is a cumbersome process. The selection of a suitable model is key to obtaining the best results for particular problems. It may be noted that these models are mostly valid for diesel or gasoline or compressed natural gas (CNG)-fuelled IC engines. However, they can also be used for alternative fuels with slight modification as the model methodology is the same for all fuels. The accuracy of the model needs to be validated with measured experimental data of alternative fuels for confirming the accuracy of computational data.

## PROBLEMS

### Example 7.1

Estimate the average drop size (Sauter mean diameter) of the diesel fuel for a single-cylinder direct-injection diesel engine of 100 mm bore and 120 mm stroke. The cylinder pressure at the time of injection is 62.3 bar and the fuel is injected at 450.5 bar through a single-hole nozzle injector of 0.25 mm diameter. Assume the standard air properties at 1000 K: surface tension = 0.023 N/m, fuel density = 821.5 kg/m³, CR = 19.5 and R = 287 J/kg K.

### Solution to Example 7.1

$\Delta P = 450.5 - 62.3 = 38,820,000$ Pa

$V = (2 * \Delta P / \rho_l)^{0.5} = (2 * 38,820,000 / 821.5)^{0.5} = 307.5$ m/s

$Re = (\rho_l * V * D_n / \mu_l) = 821.5 * 307.57 * 0.19e - 3 / 0.00264 = 1.82e + 4$

$We = (\rho_1 * V^2 * D_n/\sigma) = 821.5 * (307.5)^2 * 0.19e - 3/0.023 = 6.41e + 5$
Assume $\mu_a = 8.21e - 5$ Ns/m$^2$ and $\rho_a = 21$ kg/m$^3$
Insert all known parameters in the following equation:

$$X_{32}^{LS} = 4.12 * D_n * Re^{0.12} * We^{-0.75} * (\mu_l/\mu_a)^{0.54} * (\rho_l/\rho_a)^{0.18}$$

$$X_{32}^{LS} = 1.4 \ \mu m$$

$$X_{32}^{HS} = 0.38 * D_n * Re^{0.25} * We^{-0.32} * (\mu_l/\mu_a)^{0.37} * (\rho_l/\rho_a)^{-0.47}$$

$$X_{32}^{HS} = 7.5 \ \mu m$$

$$\text{But } X_{32} = \text{Max } (X_{32}^{LS} 32LS, X_{32}^{HS})$$

Hence, the average diameter of the droplet is 7.5 μm
Spray penetration

$$S = 2.95 \times \left(\frac{\Delta P}{\rho_a}\right)^{0.25} \times (D_n \times t)^{0.5}$$

**Example 7.2**

Estimate the spray penetration for a single-cylinder direct-injection diesel engine. Cylinder bore: 102 mm, stroke: 116 mm, cylinder pressure: 65 bar, injection pressure: 600 bar and nozzle hole size: 0.2 mm for 4 ms injection duration, CR: 19.5, air density at the time of injection: 22.8 kg/m$^3$.

**Solution to Example 7.2**

$$S = 2.95 \times \left(\frac{\Delta P}{\rho_a}\right)^{0.25} \times (D_n \times t)^{0.5}$$

$$S = 2.95 \times \left(\frac{600 - 65}{22.8}\right)^{0.25} \times (0.2 \times 4e - 3)^{0.5}$$

Spray penetration $S = 103$ mm
Spray cone angle

$$\tan(\theta/2) = (1/A) \ 4\pi(\rho_a/\rho_l)^{0.5} * (3^{0.5}/6)$$

where $A = 3 + ((L_n/D_n)/3.6)$

**Example 7.3**

Find the spray cone angle for a diesel engine fuelled with diesel fuel (density: 821.5 kg/m$^3$). Engine data: $L/D = 4.3$, air density $= 26$ kg/m$^3$.

**Solution to Example 7.3**

$A = 3 + ((L_n/D_n)/3.6) = 3 + ((4.3)/3.6) = 4.19$
$Tan(\theta/2) = (1/A)\ 4\pi(\rho_a/\rho_l)^{0.5} * (3^{0.5}/6)$
$Tan(\theta/2) = (1/4.19)\ 4\pi(26/821.5)^{0.5} * (3^{0.5}/6)$
$Tan(\theta/2) = 0.1537$
$(\theta/2) = 0.1525$ rad
$\theta = 0.3051$ rad
$\theta = 17.49°$

Spray cone angle $\theta = 17.49°$

**Example 7.4**

Estimate the break-up length, spray penetration and spray cone angle for diesel, biodiesel and vegetable oil.

|  | Density (kg/m³) | Viscosity (Ns/m²) | $\Delta P$ (bar) | Injection Duration (ms) |
|---|---|---|---|---|
| Diesel | 820 | 0.00264 | 450 | 1 |
| Biodiesel | 870 | 0.00275 | 500 | 1.05 |
| Vegetable oil | 920 | 0.034 | 550 | 1.1 |

Air density: 26 kg/m³
Injector nozzle hole diameter: 0.16 mm

**Solution to Example 7.4**

Break-up length can be predicted using the following correlation:

$L_b = 15.8 * (\rho_l/\rho_a)^{0.5} * D_n$

Therefore, the break-up length for diesel is

$L_b = 15.8 * (\rho_l/\rho_a)^{0.5} * D_n = 15.8 * (820/26)^{0.5} * 0.00016$

$L_b = 14.19$ mm

Break-up length for biodiesel is

$L_b = 15.8 * (\rho_l/\rho_a)^{0.5} * D_n = 15.8 * (870/26)^{0.5} * 0.00016$
$L_b = 14.70$ mm

Break-up length for vegetable oil is

$L_b = 15.8 * (\rho_l/\rho_a)^{0.5} * D_n = 15.8 * (920/26)^{0.5} * 0.00016$
$L_b = 15.10$ mm

Spray penetration can be predicted using the Hiroyasu–Arai correlation:

$$S = 2.95 \times \left(\frac{\Delta P}{\rho_a}\right)^{0.25} \times (D_n \times t)^{0.5}$$

Spray penetration for diesel is

$$S = 2.95 \times \left(\frac{450}{26}\right)^{0.25} \times (0.00016 \times 0.001)^{0.5}$$

$$S = 42.3 \text{ mm}$$

Spray penetration for biodiesel is

$$S = 2.95 \times \left(\frac{500}{26}\right)^{0.25} \times (0.00016 \times 0.001)^{0.5}$$

$$S = 45 \text{ mm}$$

Spray penetration for vegetable oil is

$$S = 2.95 \times \left(\frac{550}{26}\right)^{0.25} \times (0.00016 \times 0.001)^{0.5}$$

$$S = 47.2 \text{ mm}$$

Spray cone angle can be predicted using Reitz correlation

$\text{Tan}(\theta/2) = (1/A) \, 4\pi \, (\rho_a/\rho_l)^{0.5} * (3^{0.5}/6)$
where $A = 3 + ((L_n/D_n)/3.6)$

So, the spray cone angle for diesel is

$A = 3 + ((L_n/D_n)/3.6) = 3 + ((4.3)/3.6) = 4.19$
$\text{Tan}(\theta/2) = (1/A) \, 4\pi \, (\rho_a/\rho_l)^{0.5} * (3^{0.5}/6)$
$\text{Tan}(\theta/2) = (1/4.19) \, 4\pi \, (26/820)^{0.5} * (3^{0.5}/6)$
$\text{Tan}(\theta/2) = 0.1539$
$(\theta/2) = 0.1527 \text{ rad}$
$\theta = 0.3054 \text{ rad}$
$\theta = 17.51°$
The spray cone angle for diesel $\theta = 17.49°$

So the spray cone angle for biodiesel is

$\text{Tan}(\theta/2) = (1/A) \, 4\pi \, (\rho_a/\rho_l)^{0.5} * (3^{0.5}/6)$
$\text{Tan}(\theta/2) = (1/4.19) \, 4\pi \, (26/870)^{0.5} * (3^{0.5}/6)$
$\text{Tan}(\theta/2) = 0.1494$
$(\theta/2) = 0.1483 \text{ rad}$
$\theta = 0.2966 \text{ rad}$
$\theta = 17°$
The spray cone angle for biodiesel is $\theta = 17°$

So the spray cone angle for vegetable oil is

$Tan(\theta/2) = (1/A) \, 4\pi \, (\rho_a/\rho_l)^{0.5}*(3^{0.5}/6)$
$Tan(\theta/2) = (1/4.19) \, 4\pi \, (26/920)^{0.5}*(3^{0.5}/6)$
$Tan(\theta/2) = 0.1453$
$(\theta/2) = 0.1443$ rad
$\theta = 0.2881$ rad
$\theta = 16.41°$
The spray cone angle for vegetable oil is $\theta = 16.41°$

**UNSOLVED PROBLEMS**

1. Plot the variation of Sauter mean diameter (SMD) with an injector nozzle hole diameter that varies between 0.16, 0.17, 0.18 and 0.2 mm. Use the following engine data:

   Bore: 110 mm
   Stroke: 122 mm
   Cylinder pressure: 75 bar
   CR: 19.5
   Fuel injection pressure: 500 bar
   Density of air: 25 kg/m³
   Fuel density: 820 kg/m³
   Surface tension: 0.023 N/m
   Assume standard air properties at 1000 K.

2. Find the SMD for three different fuels *A*, *B* and *C*. Use the same engine data as in Problem 2. Assume the standard air properties at 1000 K.

   |   | Density (kg/m³) | Viscosity (Ns/m²) | Surface Tension (N/m) |
   |---|---|---|---|
   | A | 820 | 0.00264 | 0.022 |
   | B | 840 | 0.00275 | 0.026 |
   | C | 860 | 0.00298 | 0.31 |

3. Using thermodynamic relations for pressure and temperature, find the SMD of diesel fuel for a single-cylinder diesel engine of 7.5 kW at 1500 rpm. The engine combustion pressure is 80 bar and fuel is injected at 550 bar through a single-hole-nozzle injector of 0.15 mm diameter. Assume suitable data. Cylinder bore: 102 mm, stroke: 116 mm and CR: 19.5.

   (*Hint: Find the volume at top dead centre [TDC]; for air density, use idle gas equation*)

4. Plot the effect of different injection pressures such as 400, 500 and 600 bar on SMD for a 0.17 mm nozzle hole injector. Use engine and fuels data from Problem 2.

5. Plot the effect of different injection pressures of 500, 700 and 800 bar on spray penetration. Use the same engine data as given in Problem 1.

6. Plot the effect of different nozzle hole diameters of 0.15, 0.18 and 0.2 mm on spray penetration. Use the same engine data as given in Problem 1.

7. Find the spray penetration for different air densities of 24, 30 and 34 kg/m³. Use the same engine data as given in Problem 1.

8. Estimate the spray cone angle for a diesel engine fuelled with biodiesel (density: 840 kg/m³). Engine cylinder pressure: 64 bar, cylinder bore: 102 mm, stroke: 116 mm, CR: 19.5, *L/D*: 4.2.

## Nomenclature

| | |
|---|---|
| $u_i$ | *i*th component of fluid velocity |
| $\rho$ | density of the fluid |
| $\dfrac{dp}{dx_i}$ | pressure gradient |
| $gi$ | gravitational acceleration |
| $C_{pg}$ | specific heat of gas (J/kg K) |
| $d_n$ | sac diameter (m) |
| $D_n$ | nozzle diameter (m) |
| $dP$ | in-line pressure difference |
| $dV$ | change in volume |
| $g$ | gravitational constant (m/s²) |
| $k$ | experimental bulk modulus constant |
| $L$ | latent heat of vapourisation (J/kg) |
| $L_b$ | break-up length (m) |
| $L_n$ | length of nozzle (m) |
| $m_a$ | mass of air entrainment (kg) |
| $m_f$ | mass of liquid (kg) |
| $m_S$ | mass of air entrainment with swirl (kg) |
| $Q$ | amount of fuel delivery (mm³/st) |
| $Re$ | Reynolds number |
| $R_S$ | swirl ratio |
| $S_X$ | penetration distance (m) |
| $t$ | injection time (s) |
| $t_b$ | break-up time (s) |
| $T_g$ | air temperature (K) |
| $T_s$ | droplet temperature (K) |
| $V_i$ | injection velocity (m/s) |
| $We$ | Webber number |
| $X_S$ | penetration distance with swirl (m) |
| $X_{32}$ | Sauter mean diameter (m) |
| $V$ | instantaneous volume of cylinder |
| $R$ | 1/*a* |

| | |
|---|---|
| $L$ | connecting rod length |
| $A$ | crank radius |
| $R$ | compression ratio |
| $s$ | stroke |
| $B$ | bore |
| $V_d$ | swept volume |
| $K$ | adiabatic index |
| $h_c$ | heat transfer coefficient |
| $B$ | bore |
| $k$ | thermal conductivity |
| $\mu$ | dynamic viscosity |
| $S_p$ | mean piston speed |
| $a$ | 0.35–0.8 for normal combustion |
| $b$ | 0.7 |
| $f$ | fraction of heat added |
| $\theta$ | crank angle |
| $\theta_0$ | angle of the start of the heat addition |
| $\Delta\theta$ | duration of the heat addition (length of burn) |
| $a$ | 5 |
| $n$ | 3 |
| $h$ | heat transfer coefficient |
| $A$ | exposed combustion chamber surface area |
| $T_g$ | temperature of the cylinder gas |
| $T_w$ | cylinder wall temperature |
| $M$ | molar mass of the fuel–air mixture |
| $m$ | mass of the fuel–air mixture |
| $\tilde{R}$ | universal gas constant |
| $P_{motor}$ | motor pressure |
| $T_a$ | ambient temperature, 298 K |
| $P_a$ | atmospheric pressure, 1 atm |
| $F_{pump}$ | cross-sectional area of the pump barrel |
| $F_{nozzle}$ | cross-sectional area total of the nozzle holes |
| $m_{ftot}$ | total duration of fuel injection in degrees crank angle |
| $m_{atot}$ | total air mass trapped in the cylinder |
| $n$ | exponent |
| $A_{sf}, A_{sc}$ | constants |
| $E_{sf}, E_{sc}$ | activation energies |
| $S_l$ | laminar burning velocity (m/s) |
| $\rho_u$ | mean density of unburned mixture (kg/m³) |
| $A_k$ | flame kernel surface (m²) |
| $W$ | work done (J) = $Q_s - Q_r$ |
| $Qs$ | heat supplied (J) |
| $Q_r$ | heat rejected (J) |
| $C_p$ | heat capacity at constant pressure |
| $C_v$ | heat capacity at constant volume. |

## Greek Letters

$\rho_l$     fuel density $(kg/m^3)$
$\rho_a$     air density $(kg/m^3)$
$\mu_l$     fuel viscosity $(Pa \cdot s)$
$\mu_a$     air viscosity $(Pa \cdot s)$
$\sigma$     surface tension of liquid $(N/m)$
$\theta$     spray cone angle (degree)
$\theta_S$     spray cone angle with swirl (degree)
$\lambda_g$     thermal conductivity of air $(W/m\ K)$
$\Delta P$     mean effective pressure drop across the nozzle (kPa)

## References

1. W. J. D. Annand, Heat transfer in the cylinders of reciprocating internal combustion engines, *Proceedings of the Institution of Mechanical Engineering*, 177, 973, 1963.
2. G. Woschni, A universally applicable equation for the instantaneous heat transfer coefficient in the internal combustion engine, *SAE Paper 67093*, 1967.
3. J. B. Heywood, *Internal Combustion Engine Fundamentals*, McGraw-Hill: New York, 1989.
4. C. D. Rakopoulos, D. C. Rakopoulos, E. G. Giakoumis and D. C. Kyritsis, Validation and sensitivity analysis of a two zone diesel engine model for combustion and emissions prediction, *Energy Conversion and Management*; 45, 1471–1495, 2004.
5. C. D. Rakopoulos, K. A. Antonopoulos and D. C. Rakopoulos, Multi-zone modeling of diesel engine fuel spray development with vegetable oil, bio-diesel or diesel fuels, *Energy Conversion and Management*, 47, 1550–1573, 2006.
6. C. D. Rakopoulos, K. A. Antonopoulos, D. C. Rakopoulos and D. T. Hountalas, Multi-zone modeling of combustion and emissions formation in DI diesel engine operating on ethanol–diesel fuel blends, *Energy Conversion and Management*, 49(4), 625–643, 2008.
7. C. D. Rakopoulos, Influence of ambient temperature and humidity on the performance and emissions of nitric oxide and smoke of high speed diesel engines in the Athens/Greece region, *Energy Conversion and Management*, 31, 447–458, 1991.
8. C. D. Rakopoulos, Olive oil as a fuel supplement in DI and IDI diesel engines, *International Energy Journal*, 17, 787–790, 1992.
9. N. Watson, A. D. Pilley and M. Marzouk, A combustion correlation for diesel engine simulation, SAE 800029, 1980.
10. T. Dan, T. Yamamota, J. Senda and H. Fujimoto, Effect of nozzle configurations for characteristics of non-reacting diesel fuel spray, *SAE Paper 970355*, 581–596, 1997. DOI: 10.4271/970355.
11. D. L. Siebers, Scaling liquid phase fuel penetration in diesel sprays based on mixing limited vaporization. *SAE Paper 1999-01-0528*, 703–728, 1999. DOI: 10.4271/1999-01-0528.

12. K. S. Varde and D. M. Popa, Diesel fuel spray penetration at high injection pressures. SAE Journal, SAE No. 830448 1983; Section 2:2.265–2.278. DOI: 10.4271/830448.
13. L. Araneo, A. Coghe, G. Brunello and G. E. Cossali, Experimental investigation of gas density effect on diesel spray penetration and entrainment, *SAE Journal*, SAE No. 1999-01-0525, Section 3:679–693, 1999. DOI: 10.4271/1999-01-0525.
14. N. Ishikawa and L. Zhang, Characteristics of air entrainment in a diesel spray, *SAE Journal of Engines*, SAE No. 1999-01-0522, Section 3:644–651, 1999. DOI: 10.4271/1999-01-0522.
15. M. Arai, M. Tabata and H. Hiroyasu, Disintegrating process and spray characterization of fuel jet injected by a diesel nozzle, *SAE Journal*, SAE No. 840275, Section 2:2.358–2.2.371, 1984. DOI: 10.4271/840275.
16. H. Hiroyasu and T. Kadota, Fuel droplet size distribution in diesel combustion chamber, *SAE Journal*, SAE No. 740715, Section 3:2615–2624, 1974. DOI: 10.4271/740715.
17. H. Hiroyasu and M. Arai, Empirical equations for the sauter mean diameter of a diesel spray, *SAE Journal*, SAE No. 890464, Section 3:868–877, 1989. DOI: 10.4271/890464.
18. L. C. Lichty, Internal Combustion Engines. Text book, 6th edition, McGraw-Hill: New York, 1951.
19. W. Kuo and R. C. Yu, Modeling of transient evaporating spray mixing processes effect of injection characteristics, *SAE Journal*, SAE No. 840226, Section 2:2.16–2.29, 1984. DOI:10.4271/840226.
20. T. Kamimoto and S. Matsuoka, Prediction of spray evaporation in reciprocating engines, *SAE Paper* 770413, 1792–1802, 1977. DOI: 10.4271/770413.
21. J. C. Dent and P. S. Mehta, Phenomenological combustion model for a quiescent chamber diesel engine, *SAE Paper* 811235, 3884–3902, 1981. DOI: 10.4271/811235.
22. S. Shundoh, T. Kakegawa, K. Tsujimura and S. Kobayashi, The effect of injection parameters and swirl on diesel combustion with high pressure fuel injection, *SAE Paper* 910489, 793–805, 1991. DOI: 10.4271/910489.
23. T. Chikahisa and T. Murayama, Theory and experiments on air entrainment in fuel sprays and their application to interpret diesel combustion processes, *SAE Paper* 950447, 787–797, 1995. DOI: 10.4271/950447.
24. K. A. Subramanian and S. Lahane, Comparative evaluations of injection and spray characteristics of a diesel engine using karanja biodiesel–diesel blends, *International Journal of Energy Resource*, 2011, DOI: 10.1002/er.1947.
25. S. Lahane and K. A. Subramanian, Analysis of physical and chemical ignition delay of a diesel engine for biodiesel–diesel blend (B20) using combustion characteristics, *8th International Symposium on Fuels and Lubricants (ISFL)*, March 5–7, New Delhi, 2012b.
26. S. Lahane and K. A. Subramanian, Modeling and CFD simulation of effects of spray penetration on piston bowl impingement in a DI diesel engine for biodiesel–diesel blend (B20), *ICES2012-81171, ASME 2012 ICED Spring Technical Conference, ICES2012*, held on May 6–9, Torino, Piemonte, Italy, 2012a.
27. H. Hiroyasu, T. Kadota and M. Arai, *Supplementary Comments: Fuel Spray Characterization in Diesel Engines*, Plenum Press: New York, pp. 369–408, 1980.
28. H. Hiroyasu, T. Kadota and M. Arai, Development and use of a spray combustion modeling to predict diesel engine efficiency and pollutant emissions, *Bulletin JSME*, 26(214), 569–576, 1983.

29. W. H. Lipkea and A. D. DeJoode, Direct injection diesel engine soot modeling: Formulation and results. *SAE Paper* 940670, 1994.
30. S. Verhelst and C. G. W. Sheppard, Multi-zone thermodynamic modeling of spark-ignition combustion—An overview, *Energy Conversion and Management*, 50, 1326–1335, 2009.

# 8

## Alternative Powered Vehicles

## 8.1 General Introduction

Even though a maximum share of automobiles is a combination of spark ignition (SI) and compression ignition engines, the other alternative of powered vehicles and multifuel capability vehicles are also used in automotive applications and these vehicles are explained briefly in this chapter.

1. Bi-fuel vehicle
2. Dual-fuel vehicle
3. Electric vehicle
4. Hybrid vehicle
5. Fuel cell vehicle

## 8.2 Bi-Fuel Vehicle

### 8.2.1 Introduction

The bi-fuel vehicle could operate on two fuels. The vehicles are designed for the main fuel such as gasoline. These vehicles are also designed for other fuel uses such as compressed natural gas (CNG), liquefied petroleum gas (LPG) or hydrogen. The vehicle could either run on gasoline or other alternative fuels as given above. For example, CNG infrastructure is available in New Delhi but CNG is not available in other neighbouring cities. The vehicle could operate with CNG in Delhi and the same vehicle could also operate with gasoline in other cities. The consumer has advantages for these vehicles where the cost of fuel is less than that of gasoline. The government is also encouraging the promotion of bi-fuel vehicles out of environmental concern.

The retrofitted device will be fitted in a bi-fuel vehicle for use of alternative gaseous fuels such as CNG, LPG, hydrogen and biogas. The layout of the bi-fuel vehicle is shown in Figure 8.1. A block diagram of the bi-fuel system is given in Figure 8.2. The most popular combination of fuel is gasoline and CNG or LPG. The initial starting of the engine is by gasoline and then the gaseous fuel will be inducted for running the vehicle. The solenoid valve shuts off the natural gas when the engine is not running. The bi-fuel vehicles are manufactured by the companies as given in Table 8.1.

The bi-fuel vehicle performance is lower than that of the conventional fuelled vehicle. For example, the compression ratio of an SI engine for gasoline is 9:1. But it should be increased from 9:1 to 10–11:1 for CNG fuel to improve the performance of the vehicle. However, the compression ratio cannot be increased for gasoline fuel that cannot be operated at a higher compression ratio due to knocking. So in a bi-fuelled vehicle, power and torque drop would occur for gaseous fuel as compared to gasoline fuel.

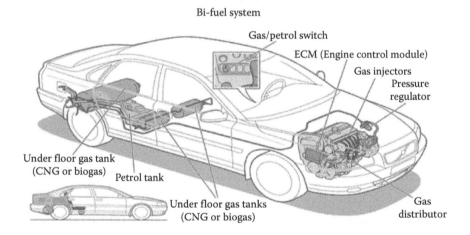

**FIGURE 8.1**
Layout of bi-fuel vehicle system. (Adapted from http://www.volvocars.com/in/Pages/default.aspx.)

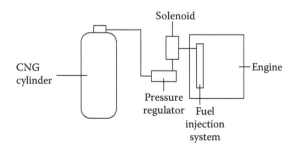

**FIGURE 8.2**
Block diagram of a CNG vehicle.

**TABLE 8.1**

Bi-Fuel Vehicles of Different Makes

| Vehicle Make | Model | Fuel Type | Availability |
|---|---|---|---|
| BYD | F3GMD | CNG/gasoline | Bolivia |
| Buick | Lucerne | CNG/gasoline | United States |
| Chevrolet | Optra | CNG/gasoline | Thailand |
| Chevrolet | Impala | CNG/gasoline | United States |
| Chevrolet | Aveo | CNG/gasoline | India |
| Chevrolet | Spark | CNG/gasoline | India |
| Citroen | Berlingo | CNG/gasoline | France, Germany |
| Citroen | Elysee | CNG/gasoline | France, Germany |
| DR | DR5 | CNG/gasoline | Italy |
| Fiat | Punto | CNG/gasoline | Germany, Italy |
| Fiat | Panda | CNG/gasoline | Germany, Italy, Spain |
| Fiat | Doblo | CNG/gasoline | Czech Republic, Germany, Italy, Spain |
| Fiat | Siena Tetrafuel | Methanol/CNG/ gasoline | Brazil, France, Germany |
| Ford | Focus 2.0 | CNG/gasoline | Germany |
| Ford | IKON Flair | CNG/gasoline | India |
| Honda | Civic GX | CNG/gasoline | United States |
| Hyundai | Santro Xing | CNG/gasoline | India |
| ICKO-Renault | Logan | CNG/gasoline | Iran |
| Lifan | 520 | CNG/gasoline | China, Peru |
| Lincoln | Town Car | CNG/gasoline | United States |
| Maruti | Alto Green Lxi | CNG/gasoline | India |
| Maruti | Eeco Green | CNG/gasoline | India |
| Maruti | SX4 Green | CNG/gasoline | India |
| Maruti | Wagon R Lxi Green | CNG/gasoline | India |
| Maruti | Zen Estilo Lxi Green | CNG/gasoline | India |
| Opel | Combo 1.6 | CNG/gasoline | Germany, Spain |
| Opel | Zafira 1.6 Turbo Flex | CNG/gasoline | Germany |
| Skoda | Octavia | CNG/gasoline | Germany, Spain |
| Volvo | V50 | CNG/gasoline/alcohol/ gasohol blends | Switzerland |
| Volvo | V70 | CNG/gasoline/alcohol/ gasohol blends | Switzerland |
| Volvo | V80 | CNG/gasoline/alcohol/ gasohol blends | Switzerland |

*Source:* Adapted from http://www.byd.com; http://www.citroen.com; http://www.drmo-tor.lt/modelll.php; http://www.gm.com; http://www.fiat.com/cgi-bin/pbrand.dll/FIAT_COM/home.jsp; http://www.ford.com; http://www.honda.com/; http://www.hyundai.com/in/en/main; http://www.renault.com/pages/index.aspx; http://www.lincoln.com; http://www.lifan.com/English; http://www.marutisuzuki.com; http://www.opel.com; http://www.skoda-auto.co.in/IND/Pages/homepage.aspx; http://www.volvocars.com/in/Pages/default.aspx.

### 8.2.1.1 Advantages of Bi-Fuelled Vehicles as Compared to Conventional Vehicles

- Lower fuel cost
- Lower emissions
- Engine life increases due to reduced carbon deposits on the combustion chamber surface
- Better fuel economy due to better mixing of gas with air and decreasing pumping work of the engine during suction stroke

### 8.2.1.2 Disadvantages as Compared to Conventional Vehicles

- Drop in power output and torque
- Additional infrastructure is needed for fuels during transportation, storage and distribution

## 8.3 Dual-Fuel Vehicles

### 8.3.1 Introduction

Dual-fuel engine means an engine system that is so designed that one fuel is inducted to the cylinder through intake manifold during suction stroke and the other fuel is injected into an engine cylinder using the main or conventional injector during a compression stroke. A schematic diagram of a dual-fuel engine is shown in Figure 8.3. The fuel inducted through the intake manifold can be a liquid or a gaseous fuel.

Dual-fuel technology enables a diesel engine to operate on a high proportion of natural gas. First, a dual-fuel engine is a diesel engine, unchanged in its basic thermodynamic operation. However, with dual fuel, diesel fuel is used only to initiate ignition of a metered charge of natural gas with air. Pilot fuel, a small amount of diesel, is always injected to initiate the reaction and ignition. It is injected using a conventional fuel injection system that is fitted on a diesel engine for dual-fuel operation and it acts as a main fuel (diesel) for 100% diesel mode if gas is not available. Once ignited, the gas and air charge burns rapidly and cleanly. By using diesel pilot ignition and retaining the diesel's high compression ratio, the gas combustion can be achieved at very lean air–fuel mixture ratios. Lean burn delivers high efficiency and low nitrogen oxides ($NO_x$) emissions. The high compression ratio of the diesel engine can be retained due to the high autoignition temperature of methane (the main constituent of natural gas). It is significantly more efficient than a spark-ignited engine due to

- Higher compression ratio.
- Very lean-burn combustion.

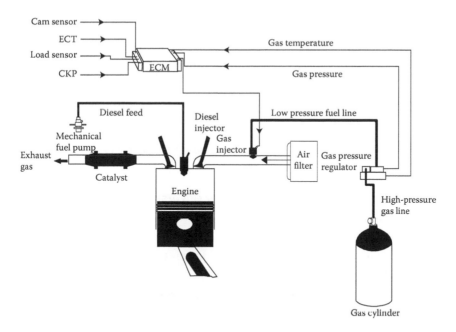

**FIGURE 8.3**
Schematic diagram of a dual-fuel engine.

- No throttle to cause additional pumping work on the engine.
- Higher torque due to high volumetric efficiency.
- Higher fuel economy and lower carbon dioxide ($CO_2$) emissions.
- Theoretically, dual-fuel configuration can be applied to any size of the diesel-cycle engine. Dual-fuel technology has the capability to utilise a widely available, low-carbon, low-emission alternative fuel—methane. The technology reduces the reliance on imported diesel fuel and increases the ability for efficient utilisation of liquid and gaseous biofuels.

### 8.3.2 Benefits of Dual-Fuel Technology

Dual-fuel technology offers the advantage of a higher compression ratio of the diesel engine and there is no derating of the power output due to the absence of a throttling system, resulting in a more improved performance and emission reduction than a dedicated spark-ignited natural gas engine. The benefits are summarised below:

- Lower operating cost than a diesel engine due to the cheaper cost of gaseous CNG fuel. It is valid in India as well as in other countries having CNG resources.

- Better fuel economy than a natural gas-fuelled dedicated diesel engine due to a higher compression ratio.
- No derating of power output.
- Higher thermal efficiency due to a higher compression ratio of the engine.
- Low emissions.

### 8.3.2.1 Compared to the Diesel Engine: Dual-Fuel Engine Delivers

- Lower emission of $NO_x$.
- Lower emission of particulate matter (PM)/smoke.
- Lower emissions of $CO_2$.
- Fuel flexibility.
- Dual-fuel engines can operate on 100% diesel fuel if gas is not available.
- Dual-fuel engines can operate with different gaseous fuels such as liquid natural gas (LNG) or CNG, LPG, producer gas, biogas and hydrogen.

## 8.4 Electric Vehicles

The first electric vehicle (EV) was built in the year 1832. In the 1990s, renewed interest was shown in the use of EVs because of increasing concerns about the environment and higher fuel costs. Further, its market has increased due to the increase in fuel economy under hybrid mode.

Battery-powered EVs use an electric motor for propulsion with batteries for electrical power storage. The energy in the batteries provides all motive and auxiliary power on board the vehicle. Batteries are recharged from grid electricity and brake-energy recuperation, and also potentially from non-grid sources, such as photovoltaic panels at recharging centres. EVs emit zero-localised pollution and very low noise. The main advantages of EVs are higher energy efficiency and lower system cost as compared to the conventional internal combustion engine (ICE) vehicles. The main drawback is their reliance on batteries that currently have very low energy and power densities as compared to liquid fuels [16].

The electric motor gets its power from a rechargeable battery. The components of an EV are shown in Figure 8.4. It consists of four main parts such as a potentiometer, batteries, a direct current (DC) controller and a motor (Figure 8.5).

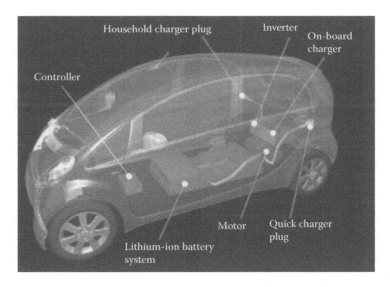

**FIGURE 8.4**
Components of an EV.

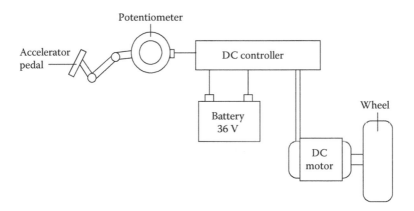

**FIGURE 8.5**
Block diagram of an EV.

*Potentiometer:* It is called a variable resistor and is hooked to the accelerator pedal.

*Batteries:* The batteries provide power for the controller. Three types of batteries are normally used. They are lead acid, lithium ion, and nickel–metal hydride batteries. The batteries range in voltage (power).

*DC controller*: The controller takes power from the batteries and delivers it to the motor. The controller can deliver zero power (when the car is stopped) and full power (when the driver floors the accelerator pedal), or any power level in between.

*Motor*: The motor receives power from the controller and turns a transmission. The transmission then turns the wheels, causing the vehicle to run.

A number of EVs have already been developed commercially and they are listed in Table 8.2.

## 8.4.1 Advantages of EVs

- Fuel can be harnessed from any source of electricity, which is available in most homes and businesses.
- The localised pollution is zero and it is ideal for urban transportation.

## 8.4.2 Disadvantages of EVs

- Limited energy storage in batteries.
- Battery efficiency is very low.
- The initial and maintenance costs of the battery are high.
- As the speed of the vehicle is limited, it is not suitable for highway transportation.

**TABLE 8.2**

List of Electric Vehicles in Production

| Vehicle | Country | Range (km) |
|---|---|---|
| BMW-Megacity | United States, Germany | 160 |
| BYD-E6 | China | 320–400 |
| Ford-Focus EV | United States | 160–240 |
| Mahindra-Reva | India | 80 |
| Mitsubishi-iMiev | Japan | 120 |
| Nissan-Leaf | United States | 160 |
| Renault-Twizy | United Kingdom | 160 |
| Volvo-C30 Electric | Sweden | 240 |

*Source:* Adapted from http://www.byd.com; http://www.citroen.com; http://www.drmotor.it/modelli.php; http://www.gm.com; http://www.fiat.com/cgi-bin/pbrand.dll/FIAT_COM/home.jsp; http://www.ford.com; http://www.honda.com/; http://www.hyundai.com/in/en/main; http://www.renault.com/pages/index.aspx; http://www.lincoln.com; http://www.lifan.com/English; http://www.marutisuzuki.com; http://www.opel.com; http://www.skoda-auto.co.in/IND/Pages/homepage.aspx; http://www.volvocars.com/in/Pages/default.aspx; C. C. Chan and K. T. Chau, *IEEE Transactions of Industrial Electronics*, 44(1), 3–13, 1997.

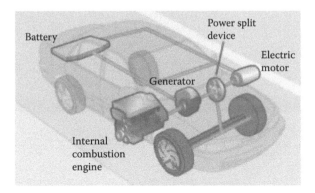

**FIGURE 8.6**
Components of hybrid vehicles.

## 8.5 Hybrid Vehicle

Hybrid vehicles are powered by the combination of an ICE and an electric motor. There are several types of hybrid vehicles and they are given below:

*Internal combustion engine:* Primary source of power; additional power: electric motor

*Electric motor:* Primary source of power; backup: internal combustion (IC) engine

These vehicles could also employ regenerative braking, in which energy captured from the brakes is used to recharge the battery. This allows the vehicle to get better gas mileage when driving in the city and in stop-and-go traffic. A number of hybrid vehicles have been developed. They include the Toyota Prius, Honda Civic Hybrid and the Ford Escape Hybrid [18]. Figure 8.6 depicts the layout of a hybrid vehicle.

### 8.5.1 Components

A hybrid vehicle has five major parts consisting of a battery, ICE, generator, power-split device and electric motor (Figures 8.6 and 8.7).

*Battery:* The batteries in a hybrid car are the energy storage device for the electric motor.

*Internal combustion engine:* A generator-coupled alternator supplies electricity to the motor while the vehicle is running. If the vehicle is idling, the electrical energy is saved in a battery that will be used again for the motor for running the vehicle.

*Generator:* The generator that is coupled with an IC engine generates electric power that is used for the battery and motor.

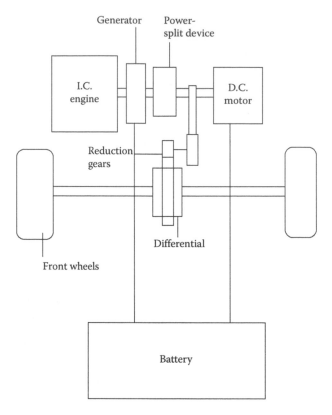

**FIGURE 8.7**
Block diagram of components of a hybrid vehicle.

*Power-split device*: The power-split device resides between the two motors and, together with the two motors, creates a type of continuously variable transmission.

*Electric motor*: The electric motor drives the wheel of a vehicle.

## 8.5.2 Working Principle

In a hybrid vehicle, the engine is initially started using a battery that is similar to a conventional ICE. When a driver presses the accelerator pedal, the generator converts an engine's shaft power into electricity that it uses to drive the motor-coupled wheel and stores it in the battery during idling. The battery provides power to the electric motor while the vehicle is running. The waste energy from wheel braking is also converted into electricity and then stored in the battery. The battery remains a power source for auxiliary systems, such as the air-conditioning and dashboard displays [20]. The comparison of EV with other vehicles is shown in Table 8.3. Some electrical vehicle models are shown in Table 8.4.

**TABLE 8.3**

Comparison between an IC Engine Vehicle, Hybrid Vehicle and Electric Vehicle

|  | IC Engine Vehicles | Hybrid Vehicles | Electric Vehicles |
|---|---|---|---|
| Efficiency | Converts 20% of the energy stored in gasoline to power the vehicle | Converts 20% of the energy stored in gasoline to power the vehicle [20] | Converts 75% of the chemical energy from the batteries to power the wheels [19] |
| Speed (kmph) | 200 kmph | 180 kmph | 50–150 kmph |
| Acceleration | 0–100 kmph in 8.4 s | 0–100 kmph in 6–7 s | 0–100 kmph in 4–6s [19] |
| Maintenance | 1. Wheels/tyres | Same as IC engine | Does not require as much maintenance because it does not use an IC engine |
|  | 2. Engine |  |  |
|  | 3. Fuel/gas |  |  |
|  | 4. Bodywork/paint |  |  |
|  | 5. Electrical |  |  |
|  | 6. Lights |  |  |
|  | 7. Dash-/instrument-warning lights |  |  |
| Mileage | Can go over 300 miles before refuelling. Typically gets 19.8 miles per gallon | Typically gets 48–60 mpg [20] | Can only go about 100–200 miles before recharging [19] |
| Cost | Cheaper | Costlier than ICE vehicles | Very extensive range of prices |

*Source:* Adapted from Electric vehicles (EVs), Retrieved January 31, 2010 from http://www.fueleconomy.gov/feg/evtech.shtml, 2009.

**TABLE 8.4**

List of Hybrid Vehicles in Production

| Vehicle | Country | Range (Combined mpg) |
|---|---|---|
| BMW X6 | United States, Germany | 18 |
| Chevrolet—Volt | United States | 60 |
| Honda CRZ | Japan | 37 |
| Honda—Insight | Japan | 41 |
| Honda—Civic Hybrid | Japan | 44 |
| Lincoln MKZ Hybrid | United States | 39 |
| Lexus—CT 200 h | Japan | 42 |
| Nissan Altima Hybrid | Japan | 34 |
| Toyota—Prius | China | 50 |
| Toyota—Camry Hybrid | China | 41 |

*Source:* Adapted from http://www.byd.com; http://www.citroen.com; http://www.drmotor.it/modelli.php; http://www.gm.com; http://www.fiat.com/cgi-bin/pbrand.dll/FIAT_COM/home.jsp; http://www.ford.com; http://www.honda.com/; http://www.hyundai.com/in/en/main; http://www.renault.com/pages/index.aspx; http://www.lincoln.com; http://www.lifan.com/English; http://www.marutisuzuki.com; http://www.opel.com; http://www.skoda-auto.co.in/IND/Pages/homepage.aspx; http://www.volvocars.com/in/Pages/default.aspx.

## 8.6 Fuel Cell Vehicles

### 8.6.1 Introduction

A fuel cell (FC) combines hydrogen and oxygen electrochemically to produce electricity and could be used to power an electric car. The attraction of FCs arises from zero-tailpipe emissions and from a high theoretical efficiency, unconstrained by the Carnot cycle unlike all heat engines. The theoretical efficiency is in the range of 70–90%. The hydrogen is not burnt in air and therefore does not produce any emissions. FCs are used to power National Aeronautical Space Association (NASA's) Space Shuttle, and large FCs for electricity generation (200 kW) are in the demonstration stage by power companies. FCs are also being developed for demonstration in electric buses. In steady-state operation at electrical utilities, the large FCs are currently achieving 40–50% efficiency, but are not yet cost competitive with other generation methods. Also, their durability and response to a rapidly varying power demand, as is the case in a vehicular use, have not been established.

Automotive FCs are therefore a very long range option. If a small FC with ~40–50 overall energy efficiency can be developed at an acceptable cost, it may provide an alternative to a battery-driven EV or may serve as an on-board recharger for a battery. The difficulty in developing an automotive FC must not be underestimated. It will probably be much more trying than developing an adequate battery, which already poses a formidable challenge for the automotive industry [21].

The use of FCs will drastically reduce the emission levels as compared to the conventional fuels. The advantage of this form of energy production over the ecological one is that the whole process is virtually silent and takes place with scarcely any mechanical wear particularly in the automotive sector. This shows that, in future, FCs will be one of the major energy suppliers for increasing the energy demands.

An FC is an electrochemical energy conversion device. It produces electricity from various external quantities of fuel (on the anode side) and an oxidant (on the cathode side). These react in the presence of an electrolyte. Generally, the reactants flow in and reaction products flow out while the electrolyte remains in the cell. FCs can operate virtually continuously as long as the necessary flows are maintained.

An FC runs like a battery, but it does not run down or require recharging. It will produce electricity and heat as long as fuel and an oxidiser are supplied.

As batteries, FCs are also called electrochemical devices as they have a positively charged anode, a negatively charged cathode and an ion-conducting material called an electrolyte. Generally, hydrogen is used as a fuel (anode) for FCs because of its abundant availability and non-polluting behaviour, which acts as a very strong contender in the desire for an alternative fuel. The best and most suitable solution is hydrogen drive.

### 8.6.2 Working Principle of Fuel Cells

An FC works according to a set of reactions called electrochemical redox reaction. The chemistry of FCs occurs in different steps as given below. Figure 8.8 shows the working of a hydrogen fuel cell.

*Step 1*: Hydrogen injection: Molecular hydrogen sets off the reaction when it is injected in the cell at relatively high pressures through compressed containers at the anode of the FC.

*Step 2*: Hydrogen releases electrons: Every hydrogen molecule (yellow) releases two electrons during this phase. The electrons between individual hydrogen atoms being no longer present, the

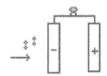

Step 1: Hydrogen injection

Step 2: Hydrogen releases electrons

Step 3: Electrons work

Step 4: Oxygen injection

Step 5: Oxygen absorbs electrons

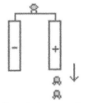

Step 6: Oxygen plus hydrogen equals water

**FIGURE 8.8**
Working of a hydrogen FC.

molecules break up leaving bare, separate hydrogen nuclei, known as protons.

*Step 3*: Electrons work: On release from the anode, the electrons flow to the cathode. As they are absorbed, they put to use their energy in productive ways, by lighting a bulb in this case. Hydrogen atoms/ protons travel towards the cathode in the process. At this stage, hydrogen is oxidised (surrenders its electrons) and hydrogen atoms get positively charged. But the cathode is also positively charged. Since, according to the laws of electrostatics, 'opposites attract and likes repel', the energy required to move positive electrons to the cathode causes a reduction in potential of the FC.

*Step 4:* Oxygen injection: Atmospheric air or oxygen in its molecular form is injected at the cathode.

*Step 5*: Oxygen absorbs electron: Molecular oxygen breaks apart as it absorbs electrons released by molecular hydrogen. Each oxygen atom absorbs two electrons since each oxygen atom is negatively charged twice.

*Step 6*: Oxygen + hydrogen = water: Positive protons and negative oxygen combine to form water; in this phase of the reaction, this is the most energetically favourable step in the reaction. The energy released by forming water drives this reaction. The water is then released from the reaction container as steam or liquid.

### 8.6.3  Types of Fuel Cells

FCs are classified primarily by the kind of electrolyte they employ. This determines the kind of chemical reactions that take place in the cell, the kind of catalysts required, the temperature range in which the cell operates, the fuel required and other factors. These characteristics, in turn, affect the applications for which these cells are most suitable. There are several types of FCs currently under development, each with its own advantages, limitations and potential applications. A few of the most promising types include

- Polymer electrolyte membrane
- Phosphoric acid
- Direct methanol
- Alkaline
- Molten carbonate
- Solid oxide
- Regenerative (reversible)

### 8.6.4 Parameters Affecting Power Generation in an FC

If the FC was perfect at transferring chemical energy into electrical energy, the ideal cell voltage (thermodynamic reversible cell potential) of the hydrogen FC would be at 25°C, 1 atm and 1.23 V. As the FC heats up to operating temperature, around 80°C, the ideal cell voltage drops to about 1.18 V. However, there are many limiting factors that reduce the FC voltage further. The voltage out of the cell is a good measure of electrical efficiency; the lower the voltage, the lower the electrical efficiency and the more the chemical energy that is released in the formation of water and transferred into heat.

The primary losses that contribute to a reduction in cell voltage are

1. *Activation losses:* Activation losses are a result of the energy required to initiate the reaction. This is a result of the catalyst. The better the catalyst, the lower the activation energy required. Platinum forms an excellent catalyst; however, there is much research underway for better materials. A limiting factor to the power density available from an FC is the speed at which the reactions can take place. The cathode reaction (the reduction of oxygen) is about 100 times slower than that of the reaction at the anode; thus, it is the cathode reaction that limits power density.

2. *Fuel crossover and internal currents:* Fuel crossover and internal currents are a result of fuel that crosses directly through the electrolyte, from the anode to the cathode without releasing electrons through the external circuit, thereby decreasing the efficiency of the FC.

3. *Ohmic losses:* Ohmic losses are a result of the combined resistances of the various components of the FC. This includes the resistance of the electrode materials, the resistance of the electrolyte membrane and the resistance of the various interconnections.

4. *Concentration losses (also referred to as 'mass transport'):* These losses result from the reduction of the concentration of hydrogen and oxygen gases at the electrode. For example, following the reaction, new gases must be made immediately available at the catalyst sites. With the buildup of water at the cathode, particularly at high currents, catalyst sites can become clogged, restricting oxygen access. It is therefore important to remove this excess water, and hence the term mass transport. A comparison of different FCs is given in Table 8.5.

### 8.6.5 Fuel Cell Vehicle

A layout of a fuel cell vehicle (FCV) is shown in Figure 8.9. In an FCV, the electrical power created by the chemical reaction between hydrogen and oxygen must be converted into mechanical energy. The typical FCV system consists of a fuel storage tank, FC stack, power controller unit and a traction

*Alternative Transportation Fuels*

**TABLE 8.5**

Comparison of Different Fuel Cells

| Fuel Cell Type | Common Electrolyte | Operating Temperature | Typical Stack Size | Efficiency | Application | Advantages | Disadvantages |
|---|---|---|---|---|---|---|---|
| Polymer electrolyte membrane | Perfluoro sulphonic acid | 50–100°C 122–212°F Typically 80°C | <1–100 kW | 60% transportation 35% stationary | Backup power Portable power Distributed generation Transportation Specialty vehicles | Solid electrolyte reduces corrosion and electrolyte management problems Low temperature Quick start-up | Expensive catalyst Sensitive to fuel impurities Low-temperature waste heat |
| Alkaline (AFC) | Aqueous solution of potassium hydroxide soaked in a matrix | 90–100°C 194–212°F | 10–100 kW | 60% | Military space | Cathode reduction, faster in an alkaline electrolyte, leads to high-performance low-cost components | Sensitive to $CO_2$ in fuel and air Electrolyte management |
| Phosphoric acid (PAFC) | Phosphoric acid soaked in a matrix | 155–200°C 302–392°F | 400–100 kW module | 40% | Distributed generation | High temperature enables combined heat and power (CHP) Increased tolerance to fuel impurities | Pt catalyst Long start-up time Low current and power |
| Molten carbonate (MCFC) | Solution of lithium, sodium, and/or potassium carbonate soaked in a matrix | 600–700°C 1112–1292°F | 300 kW–3 MW 300 kW module | 45–50% | Electrical utility Distributed generation | High efficiency Fuel flexibility Can use a variety of catalysts Suitable for CHP | High-temperature corrosion and breakdown of cell components Long start-up time Low-power density |
| Solid oxide (SOFC) | Yttria-stabilised zirconia | 700–1000°C 1202–1832°F | 1 kW–2 MW | 60% | Auxiliary power Electric utility Distributed generation | High efficiency Fuel flexibility Can use a variety of catalysts Solid electrolyte Suitable for CHP and CHHP Hybrid/gas turbine (GT) cycle | High-temperature corrosion and breakdown of cell components High-temperature operation requires a long start-up time and limits |

*Source:* Adapted from http://www1.eere.energy.gov/hydrogenandfuelcells/fuelcells/pdfs/fc_comparison_chart.pdf.

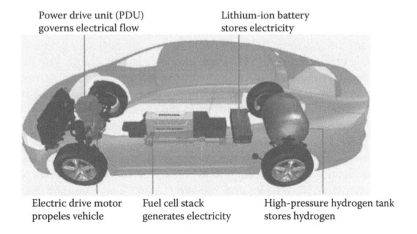

Power drive unit (PDU) governs electrical flow

Lithium-ion battery stores electricity

Electric drive motor propeles vehicle

Fuel cell stack generates electricity

High-pressure hydrogen tank stores hydrogen

**FIGURE 8.9**
Layout of an FCV power train. (Adapted from http://automobiles.honda.com/fcx-clarity/specifications.aspx?group=interior.)

motor. A layout of a fuel cell vehicle power train is shown in Figure 8.10. Some FCVs are designed to use a liquid fuel such as methanol or gasoline, which can be stored on board in a liquid form.

The vehicles using these fuels also need a reformer, a fuel processor that breaks the fuel into hydrogen, $CO_2$ and water. Although this process generates $CO_2$, it produces much less of it than the amount generated by conventional gasoline-powered vehicles. Similar to EVs, a traction motor is used as a drive train in FCVs instead of a combustion engine. The electricity produced by an FC is DC energy, and the voltage depends on the amount of FCs connected in series at the FC stack. A majority of the present-day FC systems are designed

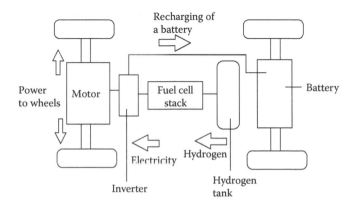

Recharging of a battery

Power to wheels

Motor

Fuel cell stack

Battery

Electricity

Hydrogen

Inverter

Hydrogen tank

**FIGURE 8.10**
Block diagram of an FCV.

for lower voltages, that is, from 12 to 100 V. Therefore, in most cases, a DC/DC boost converter is often needed to raise the voltage before further processing via a DC/alternate current (AC) inverter. The power controller unit is needed to manage the operation of the FC stack and the electric motor. The FC stack can be free breathing, which means it takes the oxygen from the surrounding air without any pressurising, or an additional air compressor can be used to control the oxygen or airflow according to the demand. Hydrogen supplied to the stack is always pressurised (up to 400–700 bars) to certain overpressure. Finally, the electrical energy is converted into motion by one or several electric motors. Lithium-ion batteries are preferred the conventional lead acid batteries for electric power storage in an FCV due to higher specific energy and energy density (105Wh/kg and 130 Wh/L) and lower weight (8.9 kg) as compared to that of lead acid batteries (33 Wh/kg, 99.25 Wh/L and 57.6 kg) [24]. Honda FCX Clarity (Figure 8.9) is one such type of FCV available in the market whose technical specifications are given in Table 8.7.

### 8.6.6 Performance

FCVs are more energy efficient than gasoline-powered vehicles. An FC uses about 40–60% of the available energy in hydrogen. IC engines use only about 20% of the energy available in gasoline, although this is expected to improve in the years to come.

There are two models of FCVs available currently but with limited distribution, and in these models, the fuel economy ratings illustrate the higher efficiency of FCVs. The Honda FCX Clarity from the model year 2011 has a fuel economy equivalent to 60 mpg of gasoline, while the 2011 Mercedes-Benz F-Cell has a fuel economy of 53 mpg. In comparison, the average fuel economy for passenger cars from the model year 2010 is 33.8 mpg for a gasoline vehicle and the most efficient hybrid electric vehicles (HEV) from the same model year has a fuel economy rating of 50 mpg. FC-powered vehicles that are commercially produced are listed in Table 8.6 and the technical specification of some selected FCVs is listed in Table 8.7.

### 8.6.7 Environmental Benefits

An FCV running on hydrogen does not produce any emissions. It is, however, important to compare the emission for the whole product and energy-carrier life cycle in a cradle-to-grave perspective. The emissions depend, of course, on how the hydrogen or the methanol is produced. In this chapter, it is compared with the methanol that is produced from biomass or natural gas. The methanol is either reformed to hydrogen on board the vehicle or converted into hydrogen at small central units. The list of some FCV models are given in Table 8.6.

The FCVs give lower global emissions than the ICE vehicle, except for particle emission where methanol from biomass has the highest emission level.

**TABLE 8.6**

List of FCVs in Production

| Vehicle | Country |
|---|---|
| Audi A2H2 | Germany |
| Daimler-Mercedes Benz, B-class | United States |
| Fiat—Panda | Spain, Germany, France |
| Ford Focus—FCV | United States |
| GM—Equinox—FCEV | Germany |
| Honda—FCX Clarity | Germany |

*Source:* Adapted from http://www.byd.com; http://www.citroen.com; http://www. drmotor.it/modelli.php; http://www.gm.com; http://www.fiat.com/cgi-bin/ pbrand.dll/FIAT_COM/home.jsp; http://www.ford.com; http://www.honda. com/; http://www.hyundai.com/in/en/main; http://www.renault.com/ pages/index.aspx; http://www.lincoln.com; http://www.lifan.com/English; http://www.marutisuzuki.com; http://www.opel.com; http://www.skoda-auto.co.in/IND/Pages/homepage.aspx; http://www.volvocars.com/in/Pages/ default.aspx.

**TABLE 8.7**

Comparison of Technical Specifications of an FCV with a Conventional Internal Combustion Engine Vehicle

| Vehicle Type | FCV | Internal Combustion Engine Vehicle |
|---|---|---|
| Vehicle model | Honda FCX | Honda Civic |
| Fuel Cell type | Polymer electron membrane fuel cell | Not applicable |
| Power output | 100 kW | 100 kW |
| Battery storage | Lithium ion battery (288 V) | Lead acid battery (12 V) |
| Electric drive | AC synchronous permanent-magnet electric motor | Not applicable |
| Torque (Nm at rpm) | 256 at 0–3056 | 172 at 4300 |
| Mileage | 60 miles/kg | 14.8 km/L |
| Driving range | 240 miles | 460 miles |
| Required fuel | Compressed hydrogen gas | Petrol |
| Fuel capacity/tank pressure | 3.92 kg at 344 bar | 50 L |

*Source:* Adapted from O. Veneri et al., *Journal of Power Sources* 196, 9081–9086, 2011.

In the majority of cases, it is shown that the emission from the FCV running on hydrogen has the lowest emission level. The emission from this vehicle depends to a great extent on how the electricity is produced. The situation would be different if the electricity was produced by fossil fuel [25]. The comparison of technical specifications of an FCV with a conventional ICE vehicle is given in Table 8.7.

## References

1. http://www.byd.com/.
2. http://www.citroen.com/.
3. http://www.drmotor.it/modelli.php.
4. http://www.gm.com/.
5. http://www.fiat.com/cgi-bin/pbrand.dll/FIAT_COM/home.jsp.
6. http://www.ford.com/.
7. http://www.honda.com/.
8. http://www.hyundai.com/in/en/main/.
9. http://www.renault.com/pages/index.aspx.
10. http://www.lincoln.com/.
11. http://www.lifan.com/English/.
12. http://www.marutisuzuki.com/.
13. http://www.opel.com/.
14. http://www.skoda-auto.co.in/IND/Pages/homepage.aspx.
15. http://www.volvocars.com/in/Pages/default.aspx.
16. http://www.bls.gov/green/electric_vehicles/electric_vehicles.pdf.
17. C. C. Chan and K. T. Chau, An overview of power electronics in electric vehicles, *IEEE Transactions of Industrial Electronics*, 44(1), 3–13, 1997.
18. http://www.iea.org/papers/2011/EV_PHEV_Roadmap.pdf.
19. Electric vehicles (EVs), Retrieved January 31, 2010 from http://www.fueleconomy.gov/feg/evtech.shtml, 2009.
20. How hybrids work, Retrieved February 20, 2010 from http://www.fueleconomy.gov/feg/hybridtech.html, 2009.
21. M. Shelef and C. A. Kukkonen, Prospects of hydrogen fuelled vehicles, *Progression of Energy Combustion*, 20, 139–148, 1994.
22. http://www1.eere.energy.gov/hydrogenandfuelcells/fuelcells/pdfs/fc_comparison_chart.pdf.
23. http://automobiles.honda.com/fcx-clarity/specifications.aspx?group=interior.
24. O. Veneri, F. Migliardini, C. Capasso and P. Corbo, Dynamic behaviour of Li batteries in hydrogen fuel cell power trains, *Journal of Power Sources* 196, 9081–9086, 2011.
25. P. Ekdunge and M. Raberg, The fuel cell vehicle analysis of energy use, emissions and cost, *International Journal of Hydrogen Energy*, 23(5), 381–385, 1998.

# 9

## Alternative Fuels for Rail Transportation

### 9.1 Introduction

Rail mode is an important transportation link for medium-to-long distance transportation and freight. The total number of locomotives in India and the world is 5022 and 26,415, respectively, and the approximate power output of each locomotive is in the range of 400–3000 kW [1]. Railways are more energy efficient and environment friendly when compared to other modes of transport. The comparison between rail and road transport is given in Table 9.1.

The average horsepower (hp) of diesel locomotives in India is about 2300–2600 hp [1]. The number of locomotives in India during the years 2005 and 2006 is given in Table 9.2.

### 9.2 Types of Locomotives

There are generally three types of locomotives which are commonly used nowadays:

1. Diesel locomotives
2. Diesel–electric locomotives
3. Electrical locomotives

#### 9.2.1 Diesel Locomotives

A diesel–mechanical locomotive consists of a direct mechanical link between the diesel engine and the wheels through the gear box as shown in Figure 9.1. The typical diesel engine is in the maximum power range of 4000 hp and the transmission is similar to that of an automobile [2]. Table 9.3

**TABLE 9.1**

Comparison between Rail and Road Transportation

| S. No | Rail Transportation | Road Transportation |
|---|---|---|
| 1 | Fast service | Slow service due to traffic congestion and road condition |
| 2 | Load-carrying capacity is very high | Load-carrying capacity is less |
| 3 | Accidents are less due to its structured system | Accidents are high |
| 4 | Fuel economy is very high due to the possibility of operating in the best conditions | Lesser fuel economy due to idle/part load operation due to traffic |
| 5 | It reduces local air and noise pollution | It is a major source of local air pollution |
| 6 | End-to-end transportation is not possible such as from the factory warehouse to the doorstep of the end user | It is possible |

**TABLE 9.2**

Numbers of Locomotives in India

| Types of Locomotives | Number of Locomotives |
|---|---|
| Steam | 44 |
| Diesel | 4793 |
| Electric | 3188 |
| Total | 8025 |

*Source:* Adapted from http://www.irfca.org/docs/.

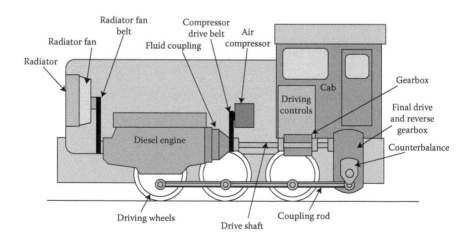

**FIGURE 9.1**

Layout of a diesel locomotive. (Adapted from http://www.railway-technical.com/elec-loco-bloc.shtml.)

**TABLE 9.3**

Technical Specifications of Diesel Locomotives

| Make | Model | Power (hp) | Speed (km/h) | Weight (tonnes) | Starting Torque (kg-force) |
|------|-------|-----------|--------------|-----------------|----------------------------|
| DLW | WDM-3D | 3300 | 160 | 117 | 36,036 |
| GM | YDM-3 | 1390 | 80 | 59 | 14,300 |
| GM + DLW | WDP-4 | 4000 | 160 | 119 | 27,500 |

*Source:* Adapted from http://www.irfca.org/docs/.

shows the specifications of different types of diesel locomotives used in India.

### 9.2.2 Diesel–Electric Locomotive

The generating station consists of a large diesel engine coupled with an alternator producing electricity as shown in Figure 9.2 [2]. Table 9.4 shows the technical specifications of diesel–electric locomotives used in India. Electric power could be used for driving the locomotive and the diesel engine is used for electricity production if the infrastructure is unavailable, for example, in a hill station. An electric locomotive operates with alternating current (AC) power from an overhead line.

## 9.3 Alternative Locomotives

### 9.3.1 Fuel Cell Locomotives

A fuel cell could play a significant role in future rail transportation. As there is a large increase in electric locomotives, the utilisation of diesel engines is reduced. The fuel cell produces electricity from fuel that could be used as supplementary power or as an alternative to the interruption of main grid electrical power. The fuel cell stack module produces power by chemical reaction with both oxygen (cathode) and hydrogen (anode) reactants. The electrical distribution and control systems regulate power output, and control various electrical devices, traction motors and monitor system parameters (Figure 9.3) [3].

### 9.3.2 Biodiesel for Locomotives

Approximately 2 billion litres of diesel fuel are consumed annually by the 4000 freight and passenger locomotives in the Indian railway fleet. The

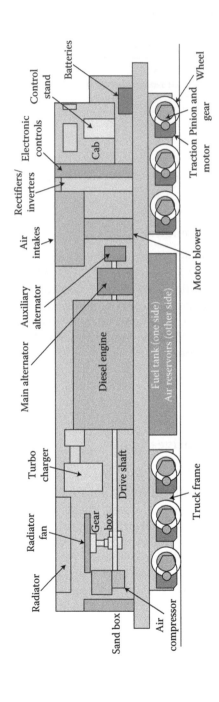

**FIGURE 9.2**
Layout of a diesel electric locomotive. (Adapted from http://www.railway-technical.com/elec-loco-bloc.shtml.)

**TABLE 9.4**

Technical Specifications of Diesel Electric Locomotives Used in India

| Make | Model | Power (hp) | Speed (km/h) | Weight (tonnes) | Starting Torque (kg-force) |
|---|---|---|---|---|---|
| BHEL | WCAG-1 | 2930 DC | 100 | 128 | 29,600 |
| Mitsubishi | YAM-1 | 1740 | 80 | 52 | 19,500 |
| Hitachi | WAG-6C | 6110 | 100 | 123 | 44,950 |

*Source:* Adapted from http://www.irfca.org/docs/.

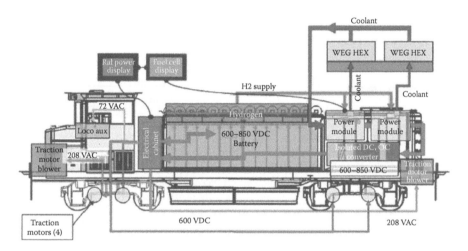

**FIGURE 9.3**

Layout of a fuel cell locomotive. (Adapted from A. R. Miller et al., System design of a large fuel cell hybrid locomotive, *Journal of Power Sources* 173, 935–942, 2007. http://www.sciencedirect.com/science?_ob=MiamiImageURL&_cid=271367&_user=41961&_pii=S0378775307015856&_check=y&_origin=search&_coverDate=15-Nov-2007&view=c&wchp=dGLzVlS-zSkzS&md5=41b741028eedc280ba3750e15f7a2f5d/1-s2.0-S03787753 07015856-main.pdf.)

expenditure of Indian Railways on diesel fuel is 886 million USD. The expenditure on diesel fuel is approximately 18% of the railways' total operating expenses. In view of the significance that diesel fuel has for the railways, alternative fuel sources are being investigated to reduce operating costs, increase economic competitiveness and reduce the environmental impact [4]. Among them, biodiesel is a promising alternative and renewable fuel which reduces emissions such as UHC, PM, CO (carbon monoxide) and $CO_2$ (carbon dioxide) as shown in Table 9.5.

Biodiesel has a promising potential to substitute partially or totally the diesel oil in rail transportation.

**TABLE 9.5**

Emissions of a Locomotive Engine with Different
Biodiesel Blends

| Pollutant | B20 | B100 |
|---|---|---|
| Unburnt hydrocarbon (UHC) (%) | −11.09 | −56.3 |
| Particulate matter (PM) (%) | −18.0 | −55.4 |
| $NO_x$ (%) | +1 | +6 |
| CO (%) | −13 | −43 |
| $CO_2$ (%) | −16 | −78 |

*Source:* Adapted from Bio Diesel—The alternate fuel for Indian
Railways, http://irsme.nic.in/files/Complete_Report_MP_
Misc_150_rev-0-00_Dec-2003.pdf.

# References

1. http://www.irfca.org/docs/.
2. http://www.railway-technical.com/elec-loco-bloc.shtml.
3. A. R. Miller, K. S. Hess, D. L. Barnes and T. L. Erickson, System design of a large fuel cell hybrid locomotive, *Journal of Power Sources* 173 935–942, 2007. http://www.sciencedirect.com/science?_ob=MiamiImageURL&_cid=271367&_user=41961&_pii=S0378775307015856&_check=y&_origin=search&_coverDate=15-Nov-2007&view=c&wchp=dGLzVlS-zSkzS&md5=41b741028eedc280ba3750e15f7a2f5d/1-s2.0-S0378775307015856-main.pdf.
4. Bio Diesel—The alternate fuel for Indian Railways, http://irsme.nic.in/files/Complete_Report_MP_Misc_150_rev-0-00_Dec-2003.pdf.

# 10

## Alternative Fuels for Marine Transportation

### 10.1 Introduction to Marine Fuels

#### 10.1.1 Conventional Fuels

There are three basic types of marine fuels such as *distillate, residual* and *intermediate fuel.*

In the marine industry, distillate fuels, residual fuel and intermediate fuel are commonly called marine gas oil, marine fuel oil and marine diesel fuel (or intermediate fuel oil [IFO]), respectively (Table 10.1). Distillate fuels are used as an auxiliary for small engines, whereas heavy oil is used for a marine diesel engine with preheating of the oil as it has high viscosity.

#### 10.1.1.1 Distillate Marine Fuels

*DMA* is a marine distillate fuel which is the most common compression ignition engine fuel for small- and medium-sized marine engines. *DMB* has some limited amount of contamination. *DMC* is manufactured from either heavier boiling fractions of straight-run distillates or it is blended in marine fuel terminals from DMA and residual fuels [1].

#### 10.1.1.2 Residual Fuels

There are 15 residual fuels according to national and international specifications. *RMA-10* is a residual marine fuel A with a maximum viscosity (at 100°C) of 10 cst. The most common intermediate fuel oil (IFO) grades are called *IFO-180* and *IFO-380*. A marine distillate fuels specification is given in Table 10.2.

*DMA* is the common fuel for tugboats, fishing boats, crew boats, drilling rigs and ferry boats. Ocean-going ships that take residual fuel oil bunkers also take distillate fuels for use in auxiliary engines and sometimes for use in ports. The common fuels are *DMC, IFO-180* and *IFO-380*, depending on the specific engines in service [1].

**TABLE 10.1**

Diesel Fuel Types for Marine Use

| Fuel Type | Fuel Grades | Common Industry Name |
|---|---|---|
| Distillate | DMX, DMA, DMB, DMC | Gas oil or marine gas oil |
| Intermediate | IFO 180 380 | Marine diesel fuel or IFO |
| Residual | RMA-RML | Fuel oil or residual fuel oil |

*Source:* From In-Use Marine Diesel Fuel, United States Environmental Protection Agency, http://www.epa.gov/otaq/regs/nonroad/marine/ci/fr/dfuelrpt.pdf.

**TABLE 10.2**

Marine Distillate Fuels Specification

| Characteristics | Unit | Limit Max | Category ISO-F | | | |
|---|---|---|---|---|---|---|
| | | | DMX | DMA | DMZ | DMB |
| Viscosity at 40°C | mm²/s | Max | 5.5 | 6.0 | 6.0 | 11.0 |
| | mm²/s | Max | 1.4 | 2.0 | 3.0 | 2.0 |
| Microcarbon residue at 10% residue | %m/m | Max | 0.3 | 0.3 | 0.3 | — |
| Density at 15°C | kg/m³ | Max | — | 890 | 890 | 900 |
| Microcarbon residue | %m/m | Max | — | — | — | 0.3 |
| Sulphur | %m/m | Max | 1.0 | 1.5 | 1.5 | 2.0 |
| Water | %v/v | Max | — | — | — | 0.3 |
| Total sediment by hot filtration | %m/m | Max | — | — | — | 0.1 |
| Ash | %m/m | Max | 0.01 | 0.01 | 0.01 | 0.01 |
| Flash point | 0°C | Max | 43.0 | 60.0 | 60.0 | 60.0 |
| Pour point, summer | 0°C | Max | 0 | 0 | 0 | 6 |
| Pour point, winter | °C | Max | –6 | –6 | –6 | 0 |
| Cloud point | °C | Max | –16 | — | — | — |
| Calculated cetane index | | Max | 45 | 40 | 40 | 35 |
| Acid number | mgKOH/g | Max | 0.5 | 0.5 | 0.5 | 0.5 |
| Oxidation stability | g/m³ | Max | 25 | 25 | 25 | 25 |
| Lubricity, corrected wear-scar diameter (wsd 1.4 at 60°C) | μm | Max | 520 | 520 | 520 | 520 |
| Hydrogen sulphide | mg/kg | Max | 2.0 | 2.0 | 2.0 | 2.0 |
| Appearance | | Max | Clear and bright | | — | — |

*Source:* From In-Use Marine Diesel Fuel, United States Environmental Protection Agency, http://www.epa.gov/otaq/regs/nonroad/marine/ci/fr/dfuelrpt.pdf.

## 10.2 Alternative Fuel for Marine Vehicles

Two-thirds of our earth is surrounded by water. The cultivation of micro-algae is conducive in the boundary of an ocean. The following alternative fuels are possible to use in a CI engine as given below:

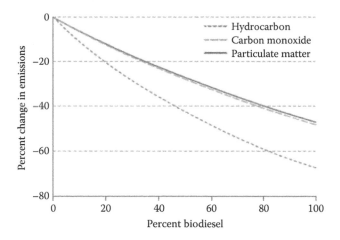

**FIGURE 10.1**
Emission of marine vehicles with biodiesel. (From The Use of Biodiesel Fuels in the U.S. Marine Industry, Maritime Administration, US Department of Transportation, http://www.marad.dot. gov/documents/The_Use_of_Biodiesel_Fuels_in_the_US_Marine_Industry.pdf.)

1. Biodiesel
2. Dimethyl ether
3. Fisher–Tropsch diesel

Biodiesel can be used as an alternative fuel for marine vehicles. Hydrocarbon, carbon monoxide (CO) and particulate matter decreased drastically with biodiesel as shown in Figure 10.1.

## 10.3 Engine/Vehicle Technology

### 10.3.1 Types of Marine Vehicles

Marine vehicles are classified based on the nature of the cargo and the trade routes as given below [3]:

- General cargo ships
- Container ships
- Tankers
- Dry bulk carriers
- Passenger ships
- Tugs

## 10.3.2 Types of Marine Engines

The following engine types are used in ship propulsion as given below [4]:

1. Low-speed diesel
2. Medium-speed diesel
3. High-speed diesel
4. Steam turbine with oil- or coal-fired boiler
5. Steam turbine with a pressurised water nuclear heat source
6. Gas turbine

### 10.3.2.1 Low-Speed Engines

The rated speeds of low-speed engines are normally in the range of 75–150 rpm, making them suitable for direct connection (i.e., without inter- vening speed-reducing gears) to the propulsion shaft of a ship. The engines are massive; a specific mass of 35 kg/kW is typical. Although size and its consequent weight may be a negative characteristic of this engine type, it is not disbarring except in applications where height or exceptionally high powers per unit weight are important. The large cylinder size and low speed are of advantage to the combustion process.

Combustion efficiency and the ability to burn fuel of low quality ('heavy oil', consisting mainly of refinery residuum) are enhanced, and in consequence, the thermal efficiency of the low-speed diesel engine is superior to that of any competitor. The specific fuel consumption of some low-speed engines has now reached 160–170 g/kWh. Given the heating value of the fuel used for attain- ing this figure, the efficiency of the engine is about 50%. As a consequence of high efficiency, coupled with fuel prices much higher than traditional, the low-speed diesel engine captured the major part of the commercial ship pro- pulsion market in the 1970s and 1980s. The technical specifications of some of the commercially used low-speed marine engines are shown in Table 10.3.

**TABLE 10.3**

Technical Specifications of Low-Speed Engines

| Make | Model | Speed (rpm) | Power (kW) | No. of Cylinder | Cylinder Bore (mm) | Piston Stroke (mm) | Fuel Consumption (g/kWh) | Weight (tonnes) |
|------|-------|-------------|------------|-----------------|---------------------|---------------------|---------------------------|-----------------|
| Wartsila | RT-Flex58T | 84–105 | 11,300 at 105 rpm | 5 | 580 | 2416 | 174 | 281 |
| Wartsila | RT-Flex50 | 99–124 | 10,470 at 124 rpm | 6 | 500 | 2040 | 164 | 225 |
| Wartsila | RT-Flex48T | 102–124 | 10,185 at 124 rpm | 7 | 480 | 2000 | 173 | 225 |

*Source:* Adapted from http://www.wartsila.com/en/engines.

**TABLE 10.4**

Technical Specifications for Medium-Speed Engines

| Make | Model | Speed (rpm) | Power (kW) | No. of Cylinder | Cylinder Bore (mm) | Piston Stroke (mm) | Fuel Consumption (g/kWh) | Weight (tonnes) |
|------|-------|-------------|------------|-----------------|--------------------|--------------------|--------------------------|-----------------|
| MAN | L+V48/60Cr | 500 | 7200 | 6 | 480 | 600 | 180 | 106 |
| Wartsila | 50DF | 500 | 5700 | 6 | 500 | 580 | — | 96 |
| MAN | 35/44DF | 720 | 3060 | 6 | 350 | 440 | 187 | — |

*Source:* Adapted from http://www.wartsila.com/en/engines; http://www.getransportation.com/marine/marine-products.html; http://www.mandieselturbo.de.

### 10.3.2.2 Medium-Speed Engines

A competing medium-speed propulsion plant might consist of 2 V engines of 16 cylinders each, turning at 500 rpm, weighing 60–80 × 103 kg, and be 3.5 m in height. An engine speed of 500 rpm is usually much too high for a ship propeller, and hence a reduction gear is nearly always an adjunct of a medium-speed engine. The lesser weight, and especially lesser height, may be important in some applications, and thereby lead to the choice of medium-speed over low-speed engines. The efficiency is likely to be slightly less, especially since the higher speed and smaller cylinders make efficient combustion more difficult, the fuel quality may have to be somewhat higher for the same reason, and the greater number of cylinders may add to the maintenance burden. The first cost of the medium-speed engine is usually less than that of a low-speed engine having the same power rating. The technical specifications of some of the medium-speed marine engines are shown in Table 10.4.

### 10.3.2.3 High-Speed Engines

When the rated speed of a marine engine is above the neighbourhood of 1000 rpm, the engine is often refered as a high-speed engine. A typical rated speed range for such an engine is 1000–1800 rpm, and the engine is likely to be the adaptation of a highway truck engine. Just as the medium-speed engine is smaller than a low-speed engine of equal power, and has more number of cylinders, the high-speed engine is yet smaller and must have more cylinders to produce equal power. For total powers above the vicinity of 1000 kW, and for a typical speed in the 1800- to 2000-rpm range, the number of cylinders may become extreme, even though divided among several (or many) engines, and the use of this class of engine becomes impractical. High-speed engines are therefore confined to vessels of smaller size and power. The technical specifications of some commercially used high-speed marine engines are given in Table 10.5.

**TABLE 10.5**

Technical Specification for High-Speed Engines

| Make | Model | Speed (rpm) | Power (kW) | No. of Cylinder | Cylinder Bore (mm) | Piston Stroke (mm) | Fuel Consumption (g/kWh) | Weight (tonnes) |
|------|-------|-------------|-----------|-----------------|--------------------|--------------------|--------------------------|-----------------|
| GE | 8V228 | 1050 | 1526 | 8 | 228.6 | 266.7 | — | 12.4 |
| CAT | C7 | 2400 | 205 | 6 | 110 | 127 | 217.3 | 0.79 |
| Cummins | QSM11 | 2500 | 526 | 6 | 125 | 147 | 143.8 | 1.18 |

*Source:* Adapted from http://www.getransportation.com/marine/marine-products.html; http://marine.cummins.com/mrn/public_cummins/home.jsp; http://www.cat.com/marine.

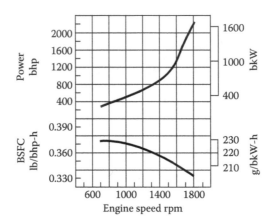

**FIGURE 10.2**
Variation of BSFC and power with respect to engine speed. (Adapted from http://www.cat.com/marine.)

### 10.3.2.4 Two- and Four-Stroke Diesel Engine

Two-stroke diesel engines can also be used for marine engines as there is no problem of scavenging unlike two-stroke spark ignition engines. The stroke length is higher than the bore or diameter of the engine which generates a high torque. The detailed technical specification of a high-speed four-stroke marine diesel engine is given in Figure 10.2 and Tables 10.6 and 10.7.

## 10.4 After-Treatment Technology for an Emission Reduction Selective Catalyst Reduction

Reduction of $NO_x$ from the stationary and mobile combustion sources has become an area of interest. A change in emission norms has forced the

**TABLE 10.6**

Caterpillar Engine Specification

| 16 Cylinders, Four-Stroke-Cycle Diesel | |
|---|---|
| Bore—mm (in) | 170 (6.7) |
| Stroke—mm (in) | 190 (7.5) |
| Displacement—L (cu in) | 69 (4210) |
| Aspiration | Turbocharged-after cooled |
| Rotation (from flywheel end) | Anticlockwise |
| Capacity for liquids—L (U.S. gal) | |
|    Cooling system | 205 (54) |
|    Lube oil system (refill) | 400 (106) |
| Weight, net dry (approximate)–kg (b) including flywheel | 7720 (17,000) |
| Cold start capability | 10°C (50°F) |
| Compression ratio | 14:1 |
| Electronic fuel injection | |
| Meets current EPA/ERRI exhaust emission levels | |

**TABLE 10.7**

Caterpillar Marine Engine Performance Specification

| | |
|---|---|
| Rated kW (bhp) flywheel | 1678 (2250) |
| Full load—rpm | 1800 |
| Low idle—rpm | 600 |
| Operating range—rpm | 1200 |
| Altitude capability—m (ft) max operating altitude (before derate) | 800 (2625) |
| Fuel consumption—L/h (gal/h) idle (600 rpm) | 7.6 (2.0) |
| Full load (1800 rpm) | 406 (107.3) |
| BSFC–g/bkW-h (lb/bhp-h) minimum value (at 1800 rpm) | 203 (0.334) |

scientists and engineers to work in the field of reduction of pollutants such as $NO_x$, CO, unburnt hydrocarbon (UBHC) and many more. Most of the combustion sources now emit $NO_x$, which is a major pollutant. The two methods that are available for the reduction of $NO_x$ are pre-flame and post-flame methods.

Among all the methods shown for $NO_x$ control, the flue gas treatment method is more likely to be used once we have decided on the type of fuel to be used. The first option for $NO_x$ reduction is frozen when the fuel is altered or changed. When we consider other options such as low excess air, exhaust gas recirculation and steam and water injection, there are some problems associated with the performance of the engine such as power loss.

In flue gas treatment technology, there is no effect on the performance of the engine. It involves the designing of a reactor in which flue gas is treated through a reagent to reduce NO (nitrogen oxide) content into $N_2$ (nitrogen). Comparison of various $NO_x$ reduction methods are shown in Table 10.8.

**TABLE 10.8**

Comparison of Various $NO_x$ Reduction Techniques for Stationary

| Parameters | Exhaust Gas Recirculation (EGR) | Water and Steam Injection | Low Excess Air | Selective Catalytic Reduction | Selective Non-Catalytic Reduction |
|---|---|---|---|---|---|
| Process | Exhaust gas from engine exhaust is reinducted into the engine cylinder | Water and steam are inducted into the engine cylinder | Amount of air sent into the engine cylinder is in lesser amounts | Exhaust gas from the cylinder is treated in the presence of a catalyst such as platinum and rhodium | Exhaust gas from the cylinder is treated by ammonia or urea |
| Effect | Less $NO_x$ | Less $NO_x$ | Less $NO_x$ | Less $NO_x$ | Less $NO_x$ |
| Percentage reduction | 50–75% | 45–50% | 30–40% | 25–50% | 25–40% |
| Power loss | Yes | Yes | Yes | No | No |
| Engine modification | Yes | No | No | No | No |
| Retrofitting | Required | Not required | Not required | Required | Required |
| Increase in effectiveness | Limited | Limited | Limited | Unlimited | Unlimited |

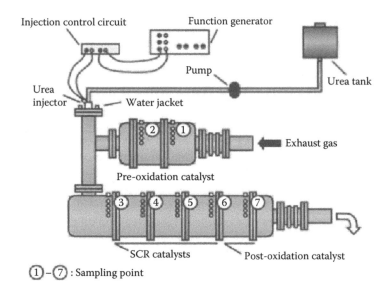

$\textcircled{1}$–$\textcircled{7}$ : Sampling point

**FIGURE 10.3**
Block model for SCR technique. (Adapted from J. Kusaka et al., *Int. J. Engine Res.* 6(1), 11–19, IMechE 2005.)

A block diagram of the SCR (selective catalyst reduction) technique is given in Figure 10.3.

### 10.4.1 Various Reagents That Are Used for Selective Non-Catalytic Reduction

Some of the reagents that are used for selective non-catalytic reduction are

1. Ammonia ($NH_3$)
2. Urea
3. Cynauric acid
4. $NH_3$ blended with sodium carbonate
5. Xenon lamp-based reduction

### 10.4.2 $NH_3$ Process

This was the original SNCR process and has been named as thermal $DeNO_x$. In this process, $NH_3$ is injected into the exhaust gas. The form of $NH_3$ may generally be in the form of water blended ammonia.

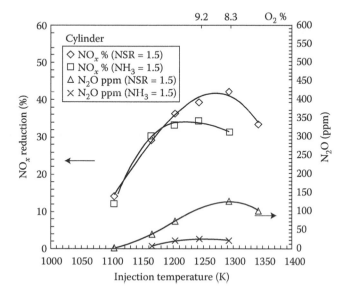

**FIGURE 10.4**

Comparison of $NO_x$ reduction and $N_2O$ emissions between urea and $NH_3$. (Adapted from C. M. Nam and B. M. Gibbs, http://www.sciencedirect.com/science/article/pii/S0082078400803318.)

### 10.4.3 Comparison of NH₃ and Urea Injection Methods

Despite a different decomposition mechanism for $NH_3$ as compared with urea, most $NO_x$ reduction features have been found to be similar as deduced from comparative tests carried out for in-cylinder injections. However, several differences were noticed. Some of them are

1. With urea, the temperature range is much wider
2. At higher oxygen levels, urea performs better than $NH_3$
3. $N_2O$ emissions are significant in the case of urea

Figure 10.4 compares $NO_x$ reduction and $N_2O$ emissions between urea and $NH_3$. Apart from injecting urea and $NH_3$ into the cylinder, we can also inject secondary reagents such as hydrocarbon in the exhaust of the engine to enhance $NO_x$ reduction. Biodiesel, dimethyl ether and Fisher–Tropsch diesel have the potential to replace partially or totally the diesel fuel used in marine engines.

## References

1. In-Use Marine Diesel Fuel, United States Environmental Protection Agency, http://www.epa.gov/otaq/regs/nonroad/marine/ci/fr/dfuelrpt.pdf.
2. The Use of Biodiesel Fuels in the U.S. Marine Industry, Maritime Administration, US Department of Transportation, http://www.marad.dot.gov/documents/The_Use_of_Biodiesel_Fuels_in_the_US_Marine_Industry.pdf.
3. A. F. Molland, *The Maritime Engineering Reference Book, A Guide to Ship Design, Construction and Operation*, http://www.sciencedirect.com/science/book/9780750689878.
4. J. B. Woodward and T. E. Andersen, Marine engines, http://www.sciencedirect.com/science/article/pii/B012227410500404X.
5. http://www.wartsila.com/en/engines.
6. http://www.getransportation.com/marine/marine-products.html.
7. http://www.mandieselturbo.de.
8. http://marine.cummins.com/mrn/public_cummins/home.jsp.
9. http://www.cat.com/marine.
10. J. Kusaka, M. Sueoka, K. Takada, Y. Ohga, T. Nagasaki and Y. Daisho, A basic study on a urea–selective catalytic reduction system for a medium-duty diesel engine, *Int. J. Engine Res.* 6(1), 11–19, IMechE 2005.
11. C. M. Nam and B. M. Gibbs, Selective non-catalytic reduction of $NO_x$ under diesel engine conditions, http://www.sciencedirect.com/science/article/pii/S0082078400803318.

# 11

## Alternative Fuels for Aviation (Airbus and Helicopter)

### 11.1 Introduction

Air mode is the fastest transportation link as compared to other modes such as land and sea. However, it is too expensive.

### 11.2 Importance of Air-Mode Transportation

It is generally used for the following purposes:

- Emergency services: Medicine, medical treatment, rescue operation during natural calamities such as earthquake, flooding and tsunami
- Personal transportation, including tourism
- Spare parts for important lightweight machinery/system, time-bounded validity of certain medicines and chemicals as they need to be used before the expiry date
- Social status

Air-mode transportation can be classified broadly into two categories as short-distance and long-distance travels:

1. Short-distance travel: Internal combustion engine-coupled helicopter.

   The helicopter is used for personal transportation for a single person or a small group of persons for shorter distance travel, especially where there is no availability of road such as forest area or sea. It is widely used for rescue operations in situations such as a natural calamity, tsunami and fire hazardous situations.

2. Long-distance travel: Airbus

The airbus is used for personal transportation for longer distance travels, such as country to country with supplemented important light-weight goods transportation.

The advantages and disadvantages of air-mode transportation using either a helicopter or an airbus are shown in Table 11.1.

$$\text{Transportation cost (TC)} = \text{distance to be travelled} \times \text{cost/km} \quad (11.1)$$

TC is high for air mode, medium for land mode and less for sea mode.

The question arises which mode of transportation would give economical benefits. The desired transportation mode could be found out by using the utility cost (UC), which indicates the value of the person/goods with respect to time. It may even be accounted for the salary of the person/day, by the productivity and the economic loss of service without goods or spares/ho.

If UC is lesser than TC, air mode is the preferable choice as shown in Equation 11.2, whereas land or sea mode is for untimed activities (Equation 11.3).

$$\text{Air mode: UC} > \text{TC} \quad (11.2)$$

$$\text{Land or sea mode: UC} < \text{TC} \quad (11.3)$$

**TABLE 11.1**

Advantages and Disadvantages of Shorter and Longer Distance Air-Mode Transportation

| S. No. | Activity | Advantages of Helicopter Transportation | Disadvantages of Helicopter Transportation | Advantages of Airbus Transportation | Disadvantages of Airbus Transportation |
|---|---|---|---|---|---|
| 1 | Distance travel | — | Only shorter distance | Longer distance | — |
| 2 | Travel speed | — | Lesser | Very higher (ground speed of 700–940 km/h) | — |
| 3 | Seating capacity | — | Lesser | Higher | — |
| 4 | Take-off and landing | Very easy | — | — | Needs an airport |
| 5 | Rescue operation | Very useful | — | — | Limited purpose |
| 6 | Localised emission and noise pollution | — | Higher | Very less as it could fly at a high altitude up to 10 km | — |
| 7 | Light goods carrier | — | Not economical | Economical | — |

## 11.3 Performance, Combustion and Emission Characteristics of Aero-Engines

Gas turbine is normally used for an aircraft engine which consists of a compressor and combustor. A compressor is used for increasing the mass flow rate of air and the pressure and temperature of air. The compressor consists of a series of blades mounted on a horizontal shaft. The compressor enhances the mass flow rate of air, as well as its pressure and temperature. There are three zones in the combustor such as the primary zone, secondary zone and dilution zone. Twenty per cent of air from the compressor is allowed into the combustion chamber and the remaining air is diverted to the outside slot for cooling of the combustor. The swirl plate increases the swirl motion of the compressed air, which enhances the mixing rate of air and fuel. A spark plug initiates ignition of the mixture and combustion (Figure 11.1). The combustion occurs at the stoichiometry air–fuel ratio. The hot combusted air is diluted with the diverted air in the intermediate and dilution zones. The maximum temperature is limited to lesser than 2000°C due to material limitation.

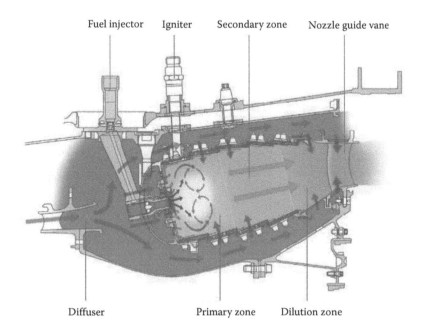

**FIGURE 11.1**
Combustion chamber configuration (Rolls-Royce). (Adapted from *Gas Turbine Technology: Introduction to a Jet Engine*, www.rolls-royce.com.)

## 11.3.1 Emission

Emission parameter of a flight is calculated using the landing and take-off (LTO) cycle as shown in Equation 11.4. A flight which operates at different modes such as idle/taxi, approach, climb and take off needs different power requirement. Subsequently, the emission level from the flight will change based on the mode of the flight operation.

$$\text{Emission parameter} = \sum (\text{EI}_{\text{species}} \times W_{\text{fuel}} \times t_{\text{mode}})/\text{Take-off power} \quad (11.4)$$

Figure 11.2 shows aircraft emissions and climate change. An aircraft engine gives out emissions such as $CO_2$, $NO_x$ (nitrogen oxides), $SO_x$ (sulphur oxides), HC (hydrocarbon), soot and water. These emissions react in the atmospheric medium and some of it transforms into ozone, aerosol, contrails and $CH_4$ (methane). These emissions are known as greenhouse gases that influence climate change and global warming.

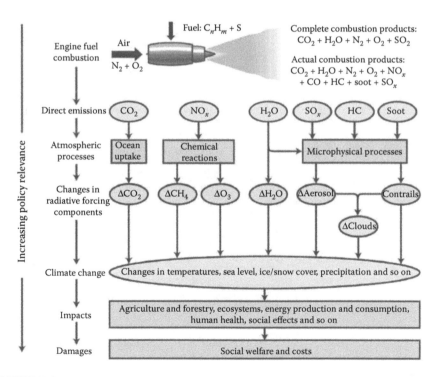

**FIGURE 11.2**
Aircraft emissions and its impact on climate change. (Adapted from D. S. Lee et al., *Journal of Atmospheric Environment*, 43, 3520–3537, 2009.)

## 11.4    Progress in the Use of Alternative Fuels for the Aviation Sector

Some of the world's biggest aviation companies are turning to alternative fuels—made from sources as diverse as hydrogen cells or algae—as soaring oil prices drive the search to build and fly more fuel-efficient planes. The Boeing Company and its European rival Airbus showed off their latest alternative-fuel projects recently at the Berlin Air Show, held against the backdrop of oil prices that hit \$135 a barrel. The Boeing Company showed a one-seater demonstration aeroplane that can fly on batteries and hydrogen fuel cells. Although the plane is still in the early stages of development, the company stated that test flights have shown that 'a manned aeroplane can maintain a straight level flight with fuel cells as the only power source'.

The Boeing Company claimed that the technology could potentially power small manned and unmanned aircrafts. But it said that it 'does not envision fuel cells will provide primary power for large passenger aeroplanes', although it will continue investigating their potential.

The airbus brought along a demonstrator version of its A320 passenger jet that uses fuel cells to power some of the aircraft's steering systems. The company said it sees great potential in fuel-cell applications. Dutch airline

**TABLE 11.2**

Measures of $CO_2$ Emission Reduction for Aviation Industry

| Measures | International Airline Industry Association (IATA) | International Civil Aviation Organization (ICAO) | U.K. Committee on Climate Change (UKCCC) |
|---|---|---|---|
| Fuel economy improvement | 1.5% per year up to 2020 | Average fuel efficiency improvement of 2% for 2020 | Fuel efficiency improvement of 0.8% using current technology |
| Alternative fuels | 10% biofuels for 2017 and 6% mix of second-generation biofuels for 2020 | — | Maximum of 10% by 2050 due to land availability and sustainability issues |

Other measures:
- Low life cycle carbon fuel
- Reducing aircraft weight
- Increasing engine efficiency
- Improving aerodynamic by increasing lift-to-drag ratio
- Increasing average load factor
- Changing fleet mix with fuel-efficient aircraft
- Changing flight distance by modifying network topology
- Optimum cruise speed

KLM meanwhile said that it had signed a contract with Algae Link for fuel made from algae for a pilot project whose first test flight is scheduled in the near future.

Algae Link plans to set up a pair of plants this year—in the Netherlands and Spain—and said that its algae-based kerosene will be mixed with conventional fuel. But KLM's goal is to fuel its entire fleet with kerosene from algae and other plant-based oils. Some analysts predict that a barrel of oil will reach as high as $200 in the next year, leading the industry to look towards lighter, more efficient machinery and alternative fuels. Airports are also trying to improve operations, reduce bottlenecks and improve efficiency. Further, the use of bio-fuels and hydrogen in the aviation sector requires an in-depth study relating to its sustainability and also the impact of emissions.

Table 11.2 shows a sustainable aviation industry by adopting the measures of $CO_2$ emission reduction.

## References

1. *Gas Turbine Technology: Introduction to a Jet Engine*, www.rolls-royce.com.
2. D. S. Lee, D. W. Fahey, P. M. Forster, P. J. Newton, R. C. N. Wite and L. L. Lim, B. Owena, R. Sausen, Aviation and global climate change in the 21st century. *Journal of Atmospheric Environment*, 43, 3520–3537, 2009.

# 12

## Global Warming and Climate Change

## 12.1 Introduction

The world civilisation, human development and technological advancements all move together. The ultimate goal of technological advancement is human comfort and a better lifestyle. Transportation is an important societal need for better human life and economic development. However, the earth's atmosphere warms up due to greenhouse gases (GHGs), which results in serious consequences such as a rise in the sea level due to the melting of ice glaciers, frequent drought and flood, serious effects on the ecosystem. Hence, GHGs need to be addressed for sustainable energy development.

Considerable importance is being given to combating the unregulated emissions of GHG agents such as $CO_2$, $N_2O$, $CH_4$, $O_3$ and CFC (chlorofluorocarbon). As GHG emissions are not under the regulated emissions, much emphasis is not historically given to GHG emissions from the transportation vehicles.

GHG emissions from transport vehicles are mainly $CO_2$, $N_2O$ and $CH_4$. The contribution of GHG through different modes of transportation is given in Figures 12.1 through 12.3. Table 12.1 shows the GHG emission levels of different driving cycles. It is clearly seen from the table that automobiles contribute about 70% of GHG emission among the different transportation sectors. The development of new technology is so far mainly targeted to reduce regulated emissions enforced by the country's legislation. The customer expectation is mainly on fuel economy, comfort and safety. There is an urgent need to reduce GHG emission from all sectors, including the transport sector, in order to avoid global warming consequences. In this direction, the European Union (EU) has taken a number of initiatives to control $CO_2$ emission from transport vehicles. The Kyoto protocol targeted reducing GHG emissions by around 5% between 2008 and 2012 as compared to the 1990 levels.

The GHG emission formation mechanism, control and challenges from transport vehicles are reviewed and discussed in the following sections.

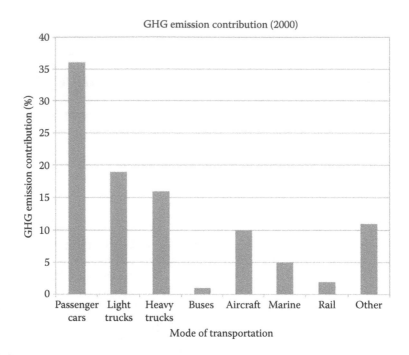

**FIGURE 12.1**
GHG emission contribution by various modes of transportation. (From D. L. Greene and A. Schafer, *Reducing Greenhouse Gas Emissions from U.S. Transportation*, Pew Center on Global Climate Change, www.pewclimate.org.)

### 12.1.1 Effect of Different Modes of Transportation on the Environment

Transport is a key infrastructure sector that acts as a stimulus to economic growth and is an important element of strategies for poverty reduction, regional and national development, and the environmental objective of limiting GHG emissions. Growing energy use by the transport sector is increasingly contributing to climate change and degrading local air quality in developing-country cities. The sector already accounts for a large proportion of public investment, and passenger and freight transport is expected to grow 1.5–2.0 times faster than gross domestic product (GDP) in most developing countries.

### 12.1.2 Impact of Air Transport

Although air transportation accounts for a very small portion (<1%) of worldwide freight, it is growing rapidly. Moreover, with increasing concern about global warming, concern about aircraft emissions has grown. Air transportation can threaten the environment in three important ways. Aircraft emissions at take-off and landing contribute to both conventional air pollution and global warming. Emissions during flight contribute to global warming.

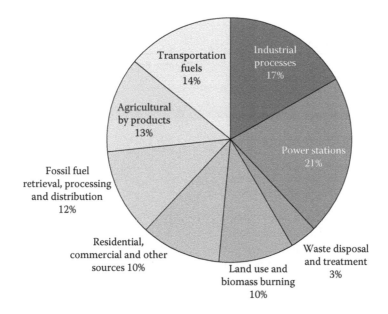

**FIGURE 12.2**
Annual GHG emission by sectors. (From R. Smokers and R. Vermeulen, Review and analysis of the reduction potential and costs of technological and other measures to reduce $CO_2$-emissions from passenger cars, Final report submitted to European Commission, Project number 033.10715/01.01, 2006.)

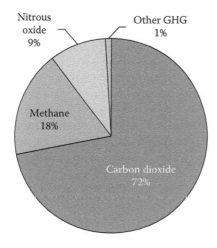

**FIGURE 12.3**
Man-made GHG contribution.

**TABLE 12.1**

Driving Cycle Characteristics

| Driving Cycle | Duration (s) | Average Speed (km/h) | % Time at Idle |
|---|---|---|---|
| U.S. Urban | 1877 | 34.1 | 19.2 |
| U.S. Highway | 765 | 77.6 | 0.7 |
| US06 | 601 | 77.2 | 7.5 |
| European | 1220 | 32.3 | 27.3 |
| Japanese | 660 | 22.7 | 32.4 |

*Source:* Adapted from A. Schafer, J. B. Heywood, M. A. Weiss, *International Journal of Energy*, 31, 2064–2087, 2006.

Low-level pollution is emitted during the aircraft's landing and take-off cycle (LTO). An LTO comprises the descent or approach of the plane from 915 m (3000 ft), its touchdown, landing run, taxi in, idle and shutdown, start-up and idle, checkout, taxi out, take-off and climb out to 915 m. Low-altitude aircraft emissions include nitrogen oxides, carbon monoxide and hydrocarbons. They are converted into ozone and other compounds that comprise smog. While aircraft emissions are minor relative to road traffic, and even relative to other means of transporting goods, they are rising faster than other emission sources, with the growth of air travel and air freight.

Aircraft emissions during high-altitude flight are a significant source of GHGs, although both their quantity and their exact impact are still matters of considerable scientific debate (Vedantham and Oppenheimer 1994, pp. 4–13; John Crayston, personal communication). Global aviation accounts for more than 2% of total anthropogenic carbon dioxide emissions. The altitude at which $CO_2$ is emitted (i.e., during landing and take-off or while in flight) has no bearing on its environmental impact. In contrast, $NO_x$ interacts quickly with other atmospheric chemicals, so its impact depends on the composition of the atmosphere where it is emitted. In the upper troposphere (up to 10 km), where most aircraft miles are logged, $NO_x$ emissions may react with other gases to form ozone, which is a potent GHG at this altitude. Although aviation accounts for only 2% of the global $NO_x$ emissions, their discharge directly into the upper troposphere may greatly increase their impact on ozone formation. Water vapour emitted into the upper troposphere can form ice crystals which may enhance the formation of cirrus clouds that trap heat, acting in much the same manner as greenhouse gases. In the stratosphere (from 17 to 30 or 40 km), where some 20% of aircraft fuel burn may occur, the impact of $NO_x$ emissions on ozone can be positive or negative, depending on the altitude, latitude and season; little is known about the dynamics of these interactions.

The major problem is the noise generated by planes during take-off and landing. Several general strategies are available to reduce the noise generated at landing and take-off. Noise standards categorise aircraft into two 'chapters'. Chapter 2 aircraft are those that were certified based on prototypes built before 1977, while Chapter 3 aircraft were certified based on later prototypes. Chapter 2 aircraft are being phased out gradually, and are expected to be out of use by 2002. However, the U.S. and European efforts to ban Chapter 2 aircraft from their airports in the late 1980s met with resistance from Third World airlines, which are more likely to be flying the older planes. An ICAO-negotiated compromise set the 2002 date and ensured that each individual plane would have a flying life of 25 years. Thus, airport noise should begin to decline as we approach 2002 and the older, noisier planes are no longer in use (John Crayston, personal communication).

### 12.1.3 Impact of Road Transport

The environmental impacts of trucking have received a great deal of attention, particularly in comparison with the impacts of rail. Trucking poses threats to the environment from two major quantifiable sources, air pollution and noise. In addition, the use of trucks contributes to land-use related environmental stresses and to the environmental impacts of accidents.

Vehicle emission factors are based on vehicle tests conducted under protocols established by the European Community, the US EPA and the Japanese government (Cucchi and Bidault 1991). The test cycles vary according to assumptions about the truck idle modes, engine speeds and other driving conditions. The test data are further adjusted to take into account variations in the temperature, grade, speed, weight of load and so on. The calculation of EPA emission factors may offer a typical example (Cambridge Systematic, pp. 2-39 to 2-57). Once emissions from new vehicles within a given truck class have been determined in the laboratory, fixed coefficients are applied to adjust for vehicle the age. This gives a set of emission factors for a vehicle class based on the age of the vehicles. The age distribution of the vehicle fleet in that class is then determined, generally based on the motor vehicle registration data. An assumption is then made of how many miles a vehicle of each age is likely to be driven; the expectation is that older trucks will be driven less than new ones. The products of age distribution and mileage are then used as weights to derive a weighted average emission factor for the vehicle class as a whole. This may be further adjusted for speed or particular driving conditions, as well as to control for cold starts, ambient temperature, and whether local conditions are typically mountainous or flat.

Table 12.2 summarises the emission factors from a number of sources, developed in several different countries. Owing to the variation in the initial test procedures, in the algorithms used to develop the overall truck class emission factors, and possibly the additional modifications made by the authors of these studies, it is not possible to determine whether the differences among these factors reflect actual differences among countries, or variations in the estimation method. However, they do give a rough idea of the amounts of pollution involved.

More disaggregated data on emission factors are available from the U.S. EPA. They are expressed in grams per mile, rather than grams per ton–kilometre.

**TABLE 12.2**

Truck Air Pollution Emission Factors (Grams per Tonne–Kilometre)

| | Kurer (Germany) | | Schoemaker and Bouman (Netherlands) | | | | Whitelegg (Europe) | Befahy (Belgium) | OECD (Europe) |
|---|---|---|---|---|---|---|---|---|---|
| | Local | Long-haul | Trucks | Trucks and Trailers | Trucks and Semi-Trailers | Road Freight Overall | Road Freight Overall | Trucks and Semi-trailers > 10 tonnes | Long-Distance Trucks |
| CO | 1.86 | 0.25 | 2.24 | 0.54 | 0.34 | 0.90 | 2.4 | 2.10 | 0.25 |
| $CO_2$ | 255 | 140 | 451 | 109 | 127 | 211 | 207 | 140 | 140 |
| HC | 1.25 | 0.32 | 1.57 | 0.38 | 0.34 | 0.68 | 0.3 | 0.92 | 0.32 |
| $NO_X$ | 4.1 | 3.0 | 5.65 | 1.37 | 2.30 | 2.97 | 3.6 | 1.85 | 3.0 |
| $SO_2$ | 0.32 | 0.18 | 0.43 | 0.10 | 0.11 | 0.20 | n.a | n.a | 0.18 |
| Particulates | 0.30 | 0.17 | 0.90 | 0.22 | 0.19 | 0.39 | | 0.04 | 0.17 |
| VOC | | | | | | | 1.1 | | |

*Source:* Adapted from Environmental Effects of Freight, Organisation for Economic Co-operation and Development, http://www.oecd.org/dataoecd/14/3/2386636.pdf; R. Kürer 1993, Environment, Global and Local Effects in ECMT, Table 5); T. J. H. Schoemaker and P. A. Bouman. 1991, *Facts and Figures on Environmental Effects of Freight Transport in the Netherlands*, in Kroon et al., eds. (p. 57, Tables 14 and 15); Whitelegg, J. 1993, Transport for a sustainable future—The case for Europe, cited in Commission des Communautés Européennes (July 1995), Le transport maritime à courte distance: Perspectives et defies, Communication de la Commission au Parlement Européen, au Conseil, au Comité économique et social et au Comité des régions, Brussels, 5 July 1995 COM(95)317 final, p. 5. HC data are for methane only; F. Befahy, 1993. Environment, Global and Local Effects in ECMT (Table 4); OECD Environment Directorate, Environmental Policy Committee (June 1993), The Social Costs of Transport: Evaluation and Links with Internalisation Policies, p. 19.

*Note:* n.a. = not available.

Emissions of some pollutants vary with speed of travel, while for others the variation across vehicle types is more important, as shown in Tables 12.2 and 12.3. It should be noted that in the United States, cleaner new truck engines (and rebuilt engines) will be required in the model year 2004. Table 12.4 shows the U.S. EPA particulate and $SO_2$ emission factors.

Trucks are a significant source of road noise, and they may be a more significant source of noise than other modes of freight transport (Table 12.5). Kürer has provided data on the actual level and quantity of noise produced by trucks, using data from the (former) Federal Republic of Germany. Individual vehicles would produce the following volumes of noise when passing at a distance of 25 m.

**TABLE 12.3**

U.S. EPA Truck Emission Factors for Selected Criteria Pollutants

|  | VOC (g/mile) | | CO (g/mile) | | $NO_x$ (g/mile) | |
|---|---|---|---|---|---|---|
| Speed (mph) | HDGV | HDDV | HDGV | HDDV | HDGV | HDDV |
| 15 | 11,726 | 3162 | 163,533 | 16,063 | 5875 | 17,192 |
| 25 | 7061 | 2175 | 95,298 | 9588 | 6416 | 14,132 |
| 30 | 5892 | 1865 | 79,005 | 7931 | 6686 | 13,513 |
| 35 | 5109 | 1634 | 69,201 | 6866 | 6957 | 13,389 |
| 40 | 4579 | 1464 | 64,040 | 6220 | 7227 | 13,745 |

*Source:* Adapted from Mullen, Maureen, E.H. Pechan and Associates, Springfield Virginia, Letter (and accompanying table) to Sharon Nizich, EPA, Research Triangle Park, NC, 26 April, 1996; Environmental Effects of Freight, Organisation for Economic Co-operation and Development, http://www.oecd.org/dataoecd/14/3/2386636.pdf.

*Note:* HDGV = heavy duty gasoline vehicle. HDDV = heavy duty diesel vehicle.

**TABLE 12.4**

U.S. EPA Particulate and $SO_2$ Emission Factors (Grams per Mile)

| Pollutant | HDGV | 2BHDDV | L HDDV | MHDDV | H HDDV |
|---|---|---|---|---|---|
| Exhaust PM10 | 0.163 | 0.322 | 0.972 | 1.107 | 1.433 |
| Break wear PM10 | 0.013 | 0.013 | 0.013 | 0.013 | 0.013 |
| Tyre wear PM10 | 0.012 | 0.008 | 0.012 | 0.012 | 0.036 |
| $SO_2$ | 0.214 | 0.223 | 0.384 | 0.433 | 0.539 |

*Source:* Adapted from Mullen, Maureen, E.H. Pechan and Associates, Springfield Virginia, Letter (and accompanying table) to Sharon Nizich, EPA, Research Triangle Park, NC, 26 April 1996; Environmental Effects of Freight, Organisation for Economic Co-operation and Development, http://www.oecd.org/dataoecd/14/3/2386636.pdf.

*Note:* HDGV = heavy duty gasoline vehicle.
2BHDDV = Class 2 (8501–10,000 Lb) heavy duty diesel vehicle.
L HDDV = light (10,001–19,500 Lbs) heavy duty diesel vehicle.
MHDDV = medium (19,501–33,000) heavy duty diesel vehicle.
H HDDV = heavy (>33,000 Lbs) heavy duty diesel vehicle.

**TABLE 12.5**

Noise Levels of Individual Passing Trucks (Decibel (A))

| Truck Size | Noise Level in Built-Up Areas (Travelling at 30–60 km/h) | | Noise Level on Road (Travelling at 60–100 km/h) | |
| --- | --- | --- | --- | --- |
| | Mean | 95th Percentile | Mean | 95th Percentile |
| Petrol delivery van | 75 | 78 | 77–78 | 82–83 |
| Diesel delivery van | 75 | 80 | 79 | 84–85 |
| Truck under 70 kW | 78–79 | 83–84 | 81–82 | 85–86 |
| Truck 70–105 kW | 79–80 | 84–85 | 82–83 | 88 |
| Truck 105–150 kW | 81–82 | 86–87 | 83–84 | 89 |
| Truck over 105 kW | 84 | 90–91 | 84 | 91–92 |

*Source:* Adapted from R. Kürer, 1993. Environment, Global and Local Effects in ECMT (Figure 6); Environmental Effects of Freight, Organisation for Economic Co-operation and Development, http://www.oecd.org/dataoecd/14/3/2386636.pdf.

### 12.1.4 Impact of Rail Transport

Railway travel is generally held up as a less environmentally damaging mode of land transportation than trucking. Data on air pollution certainly confirm this (Tables 12.6 and 12.7). For noise, it is somewhat less evident, but rail may be less harmful in that respect as well.

Trains in North America tend to be powered by diesel-fired electric generators, which then use electric power to move the locomotive which pulls the rest of the train. In Europe, most trains are electric only; it is therefore the structure of power generation which determines the air pollution characteristics. The U.S. EPA emission factors for locomotives differentiate among engine types for some pollutants, and the average emission factors explicitly reflect the composition of the U.S. fleet. In the United States, however, emission controls on locomotives will be imposed in the future.

The noise nuisance posed by rail transport is generally considered to be less than that posed by trucks.

This is in large measure because railway noise is intermittent, whereas highway noise (including trucks) tends to be fairly constant. However, estimates of the noise created by an individual train passing are higher than those for trucks. The U.S. DOT/BTS (1994) gives the noise level of a diesel train at 100 dB (A), as compared with 90–95 for trucks. Kürer places the average noise level of a train at 90 dB (A) as compared with 71–74 for trucks. On the other hand, his estimate of decibels per tonne is slightly lower for trains than for trucks: 63 for the former as compared with 64 for the latter (Kürer, 1993).

**TABLE 12.6**

Rail Air Pollution Emissions (Grams per Tonne–Kilometre)

| Pollutant | Kurer[a] | Schoemaker and Bouman[b] | Whitelegg[c] | Befahy[d] | OECD[e] | Truck Emissions |
|---|---|---|---|---|---|---|
| CO | 0.15 | 0.02 | 0.05 | 0.06 | 0.15 | 0.25–2.4 |
| $CO_2$ | 48 | 102 | 41 | | 48 | 127–451 |
| HC | 0.07 | 0.01 | 0.06 | 0.02 | 0.07 | 0.3–1.57 |
| $NO_x$ | 0.4 | 1.01 | 0.02 | 0.40 | 0.4 | 1.85–5.65 |
| $SO_2$ | 0.18 | 0.07 | | | 0.18 | 0.10–0.43 |
| Particulates | 0.07 | 0.01 | | 0.08 | 0.07 | 0.04–0.90 |
| VOC | | | 0.08 | | | 1.1 |

*Source:* Adapted from Environmental Effects of Freight, Organisation for Economic Co-operation and Development, http://www.oecd.org/dataoecd/14/3/2386636.pdf.

[a] R. Kürer, 1993. Environment, Global and Local Effects in ECMT (Table 5).

[b] T. J. H. Schoemaker and P. A. Bouman. 1991, *Facts and Figures on Environmental Effects of Freight Transport in the Netherlands*, in Kroon et al., eds (p. 57, Tables 14 and 15).

[c] Whitelegg, J. 1993, Transport for a sustainable future—The case for Europe, cited in Commission des Communautés Européennes (July 1995), Le transport maritime à courte distance: Perspectives et defies, Communication de la Commission au Parlement Européen, au Conseil, au Comité économique et social et au Comité des régions, Brussels, 5 July 1995 COM(95)317 final, p. 5. HC data are for methane only.

[d] F. Befahy, 1993. Environment, Global and Local Effects in ECMT (Table 4).

[e] OECD Environment Directorate, Environmental Policy Committee (June 1993), The Social Costs of Transport: Evaluation and Links with Internalisation Policies.

### 12.1.5 Impact of Marine Transportation

Shipping poses threats to the environment both on inland waterways and on the ocean. These problems come from six major sources: routine discharges of oily bilge and ballast water from marine shipping; dumping of non-biodegradable solid waste into the ocean; accidental spills of oil, toxics or other cargo or fuel at ports and while underway; air emissions from the vessels' power supplies; port and inland channel construction and management; and ecological harm due to the introduction of exotic species transported on vessels.

Ships are designed to move safely through the water when they are filled with cargo. When empty, they fill their tanks with ballast water in order to weigh them down and so stabilise them as they cross the ocean. Before entering the port where they are to load up, they discharge the ballast water, whose weight will be replaced with freight. The water discharged is typically somewhat unclean, being contaminated with oil and possibly other wastes within the ballast tanks. Its discharge is therefore a source of water pollution. It should be noted, however, that segregated ballast tanks, which are required on newer tank vessels, reduce or eliminate the oily ballast problem. A similar source of pollution is bilge water; this is seepage which collects in the hold of a ship and must be discharged regularly. On oil tankers, the bilge water is

**TABLE 12.7**

U.S. EPA Railway Air Pollutant Emission Factors (Kilogram per Litre of Fuel)

| Pollutant | 2-Stroke Supercharged Switch | 4-Stroke Switch | Engine Type 2-Stroke Supercharged Road | 2-Stroke Turbocharged Road | 4-Stroke Road | Average (Based on U.S. Locomotive Fleet) |
|---|---|---|---|---|---|---|
| Particulate | | | | | | 3.0 |
| $SO_x$ | | | | | | 6.8 |
| CO | 10 | 46 | 7.9 | 19 | 22 | 16 |
| HC | 23 | 17 | 18 | 3.4 | 12 | 11 |
| $NO_x$ | 30 | 59 | 42 | 40 | 56 | 44 |
| Aldehydes | | | | | | 0.66 |
| Organic acids | | | | | | 0.84 |

*Source:* Adapted from US Environmental Protection Agency (September 1985), Compilation of Air Pollutant Emission Factors Volume II: Mobile Sources, AP-42, Fourth Edition. Office of Air and Radiation, Office of Mobile Sources, Test and Evaluation Branch, Ann Arbor, MI. P. II-2-1 to II-2-2. Tables II-2-1 and II-2-2; Environmental Effects of Freight, Organisation for Economic Co-operation and Development, http://www.oecd.org/dataoecd/14/3/2386636.pdf.

typically contaminated with oil, which seeps out of the cargo tanks; thus, this is also a source of oil pollution. Such discharges are referred to as 'operational' pollution because they have long been considered a part of the normal operating procedures both of oil tankers and of other ships managing their fuel. Oily discharges, even those hundreds of kilometres from the coast, wash up on beaches and shorelines, killing birds and contaminating tourist facilities. Quantitative data are not available on the magnitude of the ballast water problem, although the MARPOL (short form for marine pollution, international convention for the prevention of pollution from ships) 15 ppm and 60 L/mile standards can be used as a lower bound for tanker discharges per mile or kilometre in predicting the environmental impact of increased oil transport. The requirement for reception facilities in ports, however, has apparently had little effect. Developed country ports typically have the required facilities, but those of less developed countries—including some major oil exporters where ballast water discharges by empty tankers coming in to load up are an important problem—have not made the necessary investments. This means that tanker captains are unable to dispose of oily ballast water properly in some ports.

Spills from waterborne vessels are one of the major sources of water pollution from shipping.

They are of several types. Cargo spills occur frequently while loading or unloading in port, owing to handling errors or equipment problems. Such spills are typically relatively small in volume. They may be of any kind of cargo, though those of petroleum products (primarily cargo rather than fuel) and other chemicals are most common. Spills of non-hazardous cargo are more common than spills of toxics or flammable materials, because the precautions taken in handling dangerous products tend to promote much greater vigilance and far fewer careless spills.

Air pollution is not one of the major environmental consequences of shipping. Nevertheless, almost all commercial freighters are powered by combustion engines, so they do emit air pollutants. These occur under two distinct sets of circumstances: while underway and while docked (for light, heat, ventilation, etc.). For ocean-going vessels, emissions while in port are of greater concern than those while underway, because they are more likely to affect adjacent populations; at sea, of course, there is no adjacent population. Several sets of emission factors are available per tonne-kilometre of freight carried for marine vessels, as shown in the table below. The sources do not specify whether these factors apply to inland or ocean shipping, or a combination of the two. They are, however, comparable to the Dutch data cited in Table 12.8, which are explicitly for inland vessel traffic. It should be noted that in September 1997 a new annex to MARPOL 73/78 was adopted to include air pollutant emission levels for marine vessels in the international convention: Annex 6 on The Prevention of Air Pollution from Ships.

The U.S. Environmental Protection Agency (1985) has estimated more detailed factors for calculating vehicle emissions per unit of fuel consumed. They separate vessels into several categories: commercial steamships,

**TABLE 12.8**

Marine Air Pollutant Emission Factors (Grams per Tonne–Kilometre)

| Pollutant | Befahy | OECD | Whitelegg |
|---|---|---|---|
| CO | 0.20 | 0.018 | 0.12 |
| $CO_2$ | | 40 | 30 |
| HC | 0.08 | 0.08 | 0.04 |
| $NO_x$ | 0.58 | 0.50 | 0.40 |
| $SO_2$ | | 0.05 | |
| Particulates | 0.04 | 0.03 | |
| VOC | | | 0.1 |

*Source:* Adapted from F. Befahy, 1993. Environment, Global and Local Effects in ECMT (Table 4). Hydrocarbon data are for methane; OCED. 1991. Environmental policy: How to apply economic instruments, cited in OECD Environment Directorate, Environmental Policy Committee (June 1993), The Social Costs of Transport: Evaluation and Links with Internalisation Policies; The Social Costs of Transport. p. 19; Whitelegg, J. 1993, Transport for a sustainable future—The case for Europe, cited in Commission des Communautés Européennes (July 1995), Le transport maritime à courte distance: Perspectives et defies, Communication de la Commission au Parlement Européen, au Conseil, au Comité économique et social et au Comité des régions, Brussels, 5 July 1995 COM(95)317 final, p. 5; Environmental Effects of Freight, Organisation for Economic Co-operation and Development, http://www.oecd.org/dataoecd/14/3/2386636.pdf.

diesel-powered vessels and motor ships used on inland waterways, as well as auxiliary generators used to provide energy while in port. The U.S. EPA emission factors are based on several key assumptions. One is that ocean-going vessels consume 80% of the fuel used underway at slow speeds and the other 20% moving at full power. Another is that diesel vessels consume 20% of their fuel running auxiliary generators while at port and the remainder while underway. A third is that the generators used to provide auxiliary power operate at 50% of their rated capacity on average; this is a sensitive assumption, since electric generators are much more polluting when running below the rated capacity than when running at it. This brings them to the following composite emission factors, detailed in Table 12.7.

Several estimates are available of the air pollutants emitted by vessels which travel on inland waterways. One Dutch source provides aggregate data for all inland vessels, while the U.S. EPA data disaggregate data for three types of vessels. Unfortunately, the Dutch and US units are not the same, so comparison is difficult; however, the Dutch data are comparable to those shown above for maritime activity in general. The marine air pollutant emission factors and norms are given in Tables 12.8 and 12.9.

## 12.1.6 Alternative Fuels and GHGs

Among the many contributors to GHG, the transportation vehicles are the major one. To develop sustainable sources of fuels and to reduce the $CO_2$ emissions, alternative fuels are being introduced as transportation fuels.

**TABLE 12.9**

Emissions from Vessels Travelling on Inland Waterways

| Pollutant | Netherlands (in g/tonne–km) | United States (in kg/10³ L of fuel) | | |
|---|---|---|---|---|
| | | Rivers | Great Lakes | Coastal |
| CO | 0.11 | 12 | 13 | 13 |
| $CO_2$ | 33 | | | |
| HC | 0.05 | 6.0 | 7.0 | 6.0 |
| $NO_x$ | 0.26 | 33 | 31 | 32 |
| $SO_2$ | 0.04 | 3.2 | 3.2 | 3.2 |
| Particulates | 0.02 | c.470 g/h | c.470 g/h | c.470 g/h |

*Source:* Adapted from Dutch data from the Central Bureau voor de Statistiek, T. J. H. Schoemaker and P. A. Bouman. 1991, *Facts and Figures on Environmental Effects of Freight Transport in the Netherlands*, in Kroon et al., eds. 1991, (p. 57); US Environmental Protection Agency (September 1985), Compilation of Air Pollutant Emission Factors Volume II: Mobile Sources, AP-42, Fourth Edition. Office of Air and Radiation, Office of Mobile Sources, Test and Evaluation Branch, Ann Arbor, MI (p. II-3-2); Environmental Effects of Freight, Organisation for Economic Co-operation and Development, http://www.oecd.org/dataoecd/14/3/2386636.pdf.

The well-to-wheel (WTW) analysis enable us to find out life cycle emission including GHGs for conventional as well as alternative fueled vehicles [12]. The WTW analysis comprises of well to tank and tank to wheel. $CO_2$ emission from the fuel cell and electric vehicle may be zero. But net $CO_2$ emission is higher with hydrogen derived from natural gases as shown in Table 12.10. It can be observed from the table that the GHG emission from the electric vehicle is about 136 g/km. Localised zero $CO_2$ emission from the electric vehicle may be possible but net $CO_2$ emission may be higher than conventional fuels as the electricity for an electric vehicle battery is generally produced by coal-based power plants. The net $CO_2$ emission reduction from fuel production as well as vehicles poses a big challenge. It may be observed from Table 12.10 that the GHG emission reduction is strongly dependent on fuel production from the nature of resources and vehicle technology.

Biofuels such as biodiesel and ethanol have high potential for $CO_2$ emission reduction. $CO_2$ emission (46.7 Mt/year $CO_2$ equivalent) was achieved in Brazil due to the ethanol and bagasse substitution from fossil fuels [13]. However, the costs of these fuels are still higher than conventional fuels. In general, $CO_2$ emission by the use of biofuels will be recycled by the crop plant resulting in no further new addition into the atmosphere. Hence, biofuels could play a vital role in $CO_2$ emission control in the near future.

Biodiesels are the biofuels that are produced from plants. Hence, net $CO_2$ emission is dropping zero. Many researchers have reported the increase in $NO_x$, with utilisation of biodiesels. But another aspect of the utilisation of

**TABLE 12.10**

WTW Energy Use and Greenhouse Gas Emissions from Vehicles Fuelled by Natural Gas Derivatives

| Fuel | Vehicle | GHG (g/km) |
|---|---|---|
| Compressed natural gas (CNG) | ICE | 148 |
| | Hybrid | 97 |
| | Fuel cell | 96 |
| Fisher–Tropsch diesel derived from natural gas | ICE | 189 |
| | Hybrid | 127 |
| | Fuel cell | 157 |
| Methanol from natural gas | Fuel cell | 117 |
| Dimethyl ether derived from natural gas | ICE | 163 |
| | Hybrid | 109 |
| | Fuel cell | 135 |
| Compressed hydrogen derived from natural gas | ICE | 178 |
| | Hybrid | 120 |
| | Fuel cell | 109 |
| Liquefied hydrogen derived from natural gas | ICE | 226 |
| | Hybrid | 152 |
| | Fuel cell | 139 |
| Hydrogen stored in metal hydrides | ICE | 157 |
| | Hybrid | 105 |
| | Fuel cell | 120 |
| Electricity | EV | 136 |
| Gasoline | ICE | 209.2 |
| | Hybrid | 140 |
| Diesel | ICE | 182.7 |
| | Hybrid | 130 |

*Source:* Adapted from J. Louis, Well-to-wheel energy use and greenhouse gas emissions for various vehicle technologies, Society of Automotive Engineers, SAE paper no. 2001-01-1343, 2001.

**TABLE 12.11**

Direct Greenhouse Gas Emissions from Passenger Cars on Petrol, Diesel, LPG and CNG under Real-World Driving Conditions

| Fuel | $CO_2$ (g/km) | $CH_4$ (g/km) | $N_2O$ (g/km) | GHG Emission ($gCO_2$ eq. g/km) |
|---|---|---|---|---|
| Petrol | 208.1 | 0.009 | 0.003 | 209.2 |
| Diesel | 180.5 | 0.004 | 0.007 | 182.7 |
| LPG | 189.3 | 0.007 | 0.003 | 190.4 |
| CNG | 168.6 | 0.0074 | 0.001 | 170.6 |

*Source:* Adapted from H. Hass, Well-to-wheels analysis of future automotive fuels and power trains in the European context, a joint study by EUCAR/JRC/CONCAWE, European Commission, 2003.

alternative fuels is the production of $N_2O$, which is primarily generated at a lower temperature (<950°C) than $NO_x$ emission in the internal combustion engine (ICE) cylinder. The global warming potential (GWP) of $N_2O$ is about 320 times the $CO_2$ equivalent. The other influencing parameters are the operating conditions, excess air level and catalytic activity.

$CH_4$ emission: $CH_4$ is typically formed due to incomplete combustion. This emission may be due to partial combustion, quenching and unburnt hydrocarbon. This level may be higher in the case of a CNG fuelled vehicle.

The typical value of the $CO_2$, $CH_4$ and $N_2O$ emission from vehicles for different fuels is given in Table 12.11. It can be seen from the table that a CNG fuelled vehicle gives the lowest $CO_2$ emission as compared to petrol, diesel and LPG (liquefied petroleum gas) fuelled vehicles. It is due to the fact that CNG has the lowest carbon-to-hydrogen ratio of all fossil fuels. It may be noted that the GWP of $N_2O$ and $CH_4$ emission is about 320 and 63 times more than its $CO_2$ equivalent, respectively. The typical GHG emissions for petrol, diesel LPG and CNG are given in Table 12.11.

## 12.2 Mechanism of Global Warming

Biodiesels are biofuels that are produced from plants. So the net $CO_2$ emission is nearly zero. Many researchers have reported the increase in $NO_x$, with utilisation of biodiesels. But another aspect of the utilisation of alternative fuels is the production of $N_2O$, which is mainly generated at a relatively lower temperature (<950°C) as compared to $NO_x$ emission in the ICE. The GWP of $N_2O$ is about 320 times higher than its $CO_2$ equivalent (Figure 12.4). The other influencing parameters are (1) operating conditions, (2) excess air level and (3) catalytic activity and so forth.

*$CH_4$ emission*: $CH_4$ is typically formed due to incomplete combustion. This emission may be due to partial combustion, quenching and unburnt hydrocarbon. This level may be higher in case of a CNG fuelled vehicle.

## 12.3 Control Avenues for GHG Emissions

Improving fuel consumption and regulations: $CO_2$ emission reduction is directly related to improving the fuel economy. This means less fuel is used, which would lead to lower $CO_2$ emission. Improvement of fuel efficiency can be achieved in many ways such as

- Use of low-weight materials with high strength.
- Reducing the aerodynamic losses.

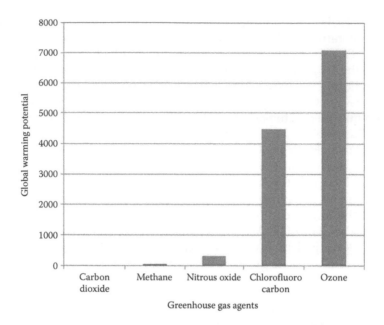

**FIGURE 12.4**
GHG agents and global warming potential. (From K. A. Subramaniam, L. M. Das, M. K. G. Babu, Control of GHG emissions from transport vehicles: Issues and challenges, SAE International 2008-28-0056.)

- Improved energy efficiency of car components (e.g., power steering, air conditioning, alternator).
- Infrastructure-related solutions, ensuring traffic flows and avoiding congestion.
- Traffic management: Eco-driving (shifting gears earlier, anticipating traffic flow and turning off engines at short stops) can achieve fuel savings of 10% per driver. The percent time at idling is higher with the European cycle [16]. It may lead to heavy fuel penalty. It is the highest in the case of India as shown in Table 12.1. In addition, the average speed is also very low as compared to the United States and other countries. It may be noted that a low or partial load operation of any IC engine would lead to low thermal efficiency. Hence, it can be improved by better traffic management system.

Hybrid technology has a high potential of $CO_2$ emission reduction as shown in Table 12.12. The $CO_2$ emission reduction by a port injection spark ignition engine with hybrid is about 47.6%, whereas by direct injection compression with hybrid it is about 31.2%.

**TABLE 12.12**

Percent Reduction of $CO_2$ Emission with Different Engine Technologies

| Technology | $CO_2$ Emission | % $CO_2$ Reduction |
|---|---|---|
| Port injected spark ignition engine (PISI) | 166.2 | 0 |
| PISI + turbo + stop and go | 139.4 | 26.8 |
| Direct injection spark ignition (DISI) | 155.2 | 11 |
| DISI + turbo + stop and go | 137.9 | 28.3 |
| PISI + hybrid electric vehicle (HEV) | 118.6 | 47.6 |
| DISI + HEV | 119.6 | 46.6 |
| Direct injection compression ignition (DICI) | 134.6 | 0 |
| DICI + stop and go | 126.1 | 8.5 |
| DICI + HEV | 103.4 | 31.2 |

*Source:* Adapted from www.concawe.be.

*Battery–electric vehicle*: The main advantage of a battery vehicle is local zero emissions. Depending on the battery type and voltage used, battery recharging can take 4–14 h. The main disadvantages of electric vehicles are the length of time needed for recharging, shorter driving ranges, weight of the battery and its initial cost. Power for a battery is dependent on electricity, which is produced mostly from a coal-based power plant. The local pollution from a battery-operated vehicle may be zero but the net $CO_2$ emission will not be changed unless electricity is produced from non-carbon-based fuel resources.

*Hybrid vehicles*: It is a combination of ICEs and electric motors. The main drawback of conventional ICE is in poor performance at low or part load conditions. In addition, higher fuel consumption due to a higher idling speed and frequent stop and go operation results in poor performance and high emissions. In case of hybrid vehicles, IC engines will always be operated at the optimum conditions, resulting in better performance and lower emission. The motivation behind the hybrid propulsion system is to achieve very low emission, potential improvement of fuel economy up to 25%, energy recovery from braking and a high $CO_2$ emission reduction potential.

*Fuel cell vehicles*: It is basically an electrochemical device that converts the chemical energy into electricity. Zero emission from a fuel cell vehicle may be possible when the use of hydrogen is derived from non-carbon fuel sources. A fuel cell could achieve up to 40–70% efficiency. Fuel cycle efficiency (WTW analysis) of hydrogen derived from fossil fuels is very less as compared to that of diesel or petrol. The cost of a fuel cell was estimated to be around $4000–5000/kW as compared to IC engines with a cost around $10–15/kW [18]. A number of technical issues need to be solved before full-scale commercialisation of fuel cell vehicles can be considered.

European Automobile Manufacturers Association (EAMA) has suggested some technology options to improve the fuel economy of vehicles as given below [19], which could reduce the fuel consumption:

- New generation energy management systems
- New generation of turbocharged petrol engines
- New generation of direct injection systems (high precision injection)
- Introduction of start–stop function in small petrol-engine cars
- Brake energy recuperation
- New generation of variable valve control
- Double clutch (dry) transmissions
- Further improvements in ICEs efficiency—including optimised combustion processes, reduction of friction losses and variable valve actuation combined with internal EGR
- Further improvements and penetration of intelligent engine management systems allowing optimised engine warm-up and cooling as well as better auxiliary energy management
- Further penetration of the gasoline direct injection (GDI) system
- Increased refinement and penetration of advanced gearboxes— including a continuously variable transmission system, robotised gearboxes, 6 speed manuals, dual clutch transmissions, 6/7 speed automatics
- Increased penetration of high-efficiency auxiliaries—including electric and electro-hydraulic, power-assisted steering systems, more efficient HVAC (heating, ventilation and air conditioning) systems and high-efficiency alternators
- Aerodynamic improvements
- Continued greater utilisation of lightweight design and materials
- Further improvements in energy control management systems, including load levelling
- Increased application of low friction tyres
- Development of engines for higher biofuel blending ratios
- Continued technical development and offerings of alternative fuel vehicles (AFVs)

Clean development mechanism (CDM) has been the main market based on international instruments to deal with mitigating climate change emissions in the developing world. Through CDM, industrialised countries can invest in emissions reductions in developing countries, with the explicit dual purpose of providing cost-effective emission reductions to the industrialised world, while contributing towards developing countries sustainable development goals.

Beside these, other control measures are also being taken care of by international forums and summits, like the Kyoto protocol and G-8 Nations.

---

## 12.4 The Proposed GHG Emission Standards

### 12.4.1 GHGs Emission Standards

#### 12.4.1.1 *Roadmap and Strategy for $CO_2$ Emission Reduction by Other Countries*

Some countries are setting the regulation for controlling $CO_2$ emission from vehicles. The country's targeted $CO_2$ emission by year and their strategy are given in Table 12.13.

### PROBLEMS

### Example 12.1

The federal emission standard for CO from new vehicles is 2.11 g/km. Suppose a highway has 10 vehicles per second passing a given spot, each emitting 2.11 g/km of CO. If the wind is perpendicular to the highway and blowing at 2.2 m/s, estimate the ground-level CO concentration 100 m from the freeway (Take $\sigma z = 4.6$ m).

### Solution to Example 12.1

A line source dispersion model can be used for calculating the emissions:

The time-averaged pollution concentration downwind from a continuously emitting infinite line source is given by

$$C(x) = \frac{1}{\sqrt{2\pi}} \frac{2q}{\sigma z \, u}$$

where
   $q$ = emission rate of pollutant per unit distance along the line source, milligram per metre second
   $\sigma z$ = vertical dispersion coefficient, metre
   $u$ = average wind speed.

To estimate the CO emission rate per metre of freeway

$$q = 10 \text{ vehicles/s} \times 2.11 \text{ g/km} \times 1 \text{ km}/1000 \text{ m} = 0.021 \text{ g/s–m}$$

**TABLE 12.13**

Implemented/Targeting $CO_2$ Emission Standards per etc

| Country Name | Target $CO_2$ by Year | Strategy |
|---|---|---|
| European Union (EU) | 130 g $CO_2$/km by 2012 | Infrastructure measures, driver behaviour, alternative fuels, and so on, Test method: EU NEDC (New European Driving Cycle) |
| California | 201 g $CO_2$/km by 2009 | Fuel economy: grams per/ mile, Test method: US CAFE |
| The other states such as New York, New Jersey, Massachusetts, Connecticut, Maine, Rhode Island, Vermont, Oregon and Washington are being adopted under the California's GHG emission regulation | 141 g $CO_2$/km by 2013 127 g $CO_2$/km by 2016 | |
| Japan | 138 g $CO_2$/km by 2015 | Integrated approach: 48% by vehicle technology 52% by infrastructure adjustments such as dynamic traffic lights, lower road rolling resistance, tax incentives for purchase of low-emission vehicles and others, Test method: Japan 10–15 |
| United States | — | Fuel, test method: U.S. CAFE |
| China | — | Fuel, weight-based, test method: E.U. NEDC |
| Australia | — | Fuel, engine size, test method: E.U. NEDC |
| Taiwan, South Korea | — | Fuel, engine size, Test method: U.S. CAFE |
| France | 140 g $CO_2$/km for new vehicle | 20% mandatory purchase of LPG, NGV or electric vehicle |
| Belgium | 145 g $CO_2$/km for diesel and 160 g $CO_2$/km for petrol cars | — |
| United Kingdom | 10% of fleet cars by 2006 | Alternatively fuelled vehicles |

*Source:* Adapted from D. L. Green and A. Schafer, *Reducing Greenhouse Gas Emissions from U.S. Transportation*, Pew Center on Global Climate Change, www.pewclimate.org; R. Smokers and R. Vermeulen, Review and analysis of the reduction potential and costs of technological and other measures to reduce $CO_2$-emissions from passenger cars, Final report submitted to European Commission, Project number 033.10715/01.01, 2006; M. P. Wash, Global trends in diesel emissions control—A 1999 update, Society of Automotive Engineers, SAE Paper No. 1999-01-0107, 1999; Review and analysis of the reduction potential and costs of technological and other measures to reduce $CO_2$-emissions from passenger cars, Final report, TNO Science and Industry (Assignor European Union, 31 October 2006).

Since vertical dispersion coefficient, σz = 4.6 m

We know that concentration of pollution emitted by a line source is given by

$$C(x) = \frac{1}{\sqrt{2\pi}} \frac{2q}{\sigma z \, u}$$

$$C(x) = \frac{2 \times 0.021(\text{g/m}).\text{s} \times 10^3 \, \text{mg/g}}{\sqrt{2\pi} \times 4.6\,\text{m} \times 2.2\,\text{m/s}} = 1.7\,\text{mg/m}^3$$

**Example 12.2**

A PHEV gets 21.25 km/L while running on gasoline that costs Rs. 68/L. If it takes 0.25 kWh to drive 1.609 km on electricity, compare the cost of fuel for gasoline and electricity. Assume that electricity is purchased at a rate of Rs. 4/kWh.

**Solution to Example 12.2**

On per kilometre basis, the cost of gasoline is

$$\text{Gasoline} = \frac{\text{Rs. } 68/\text{L}}{21.25\,\text{km/L}} = \text{Rs. } 3.2/\text{km}$$

The cost to run on electricity is

$$\text{Electricity} = \frac{0.25\,\text{kWh}}{1.609\,\text{km}} \text{Rs. } 4/\text{kWh} = \text{Rs. } 0.62/\text{km}$$

**Example 12.3**

From WTW analysis, it was found that a $CO_2$ emission from gasoline is about 3.17 $\text{kgCO}_2/\text{L}$ and that of average grid electricity is 650 $\text{gCO}_2/\text{kWh}$. Calculate the $\text{gCO}_2/\text{km}$ emissions from

a. A gasoline vehicle which travelled 10 km on 1 L gasoline
b. A hybrid electric vehicle (HEV) which travelled 21 km on 1 L gasoline
c. A plug in HEV which travelled 1/2 km on gasoline at an average of 20 km/L and 1/2 km on electricity at 0.155 kWh/km
d. An electric vehicle which travelled 1 km at 0.155 kWh/km

Also conclude which is the best possible option for mitigating $CO_2$ emission from a vehicle.

**Solution to Example 12.3**

a. Gasoline vehicle

$$CO_2 \text{ emission from gasoline} = 3.17 \text{ kg } CO_2/L$$

$$\text{Distance travelled} = 10 \text{ km/L}$$

$$g \, CO_2 \text{ emission the from vehicle} = \frac{3.17 \times 1000}{10} = 317 \text{ gCO}_2/\text{km}$$

b. HEV

$$CO_2 \text{ emission from gasoline} = 3.17 \text{ kg } CO_2/L$$

Since electricity for the vehicle will be produced in the vehicle itself by gasoline, we will consider only the $CO_2$ emissions by gasoline.

$$\text{Distance travelled} = 21 \text{ km/L}$$

$$g \, CO_2 \text{ emission from the vehicle} = \frac{3.17 \times 1000}{21} = 150.95 \text{ g } CO_2/\text{km}$$

c. Plug-in hybrid electric vehicle (PHEV)

$$CO_2 \text{ emission from gasoline} = 3.17 \text{ kg } CO_2/L$$

$$CO_2 \text{ emission from electricity} = 650 \text{ g } CO_2/\text{kWh}$$

$$\text{Average on gasoline} = 20 \text{ km/L}$$

$$\text{Average on electricity} = 0.155 \text{ kWh/km}$$

$$\text{Distance travelled} = 0.5 \text{ km}$$

$g \, CO_2$ emission by gasoline for 0.5 km travel

$$= \frac{3.17 \times 1000}{20} \times 0.5 = 79.25 \text{ g } CO_2/\text{km}$$

$g \, CO_2$ emission by electricity for 0.5 km travel

$$= 650 \times 0.155 \times 0.5 = 50.375 \text{ g } CO_2/\text{km}$$

$$\text{Total g } CO_2/\text{km emission from PHEV} = (79.25 + 50.375) \text{ g } CO_2/\text{km}$$

$$= 129.625 \text{ g } CO_2/\text{km}$$

d. Electric vehicle (EV)

$$CO_2 \text{ emission from electricity} = 650 \text{ g } CO_2/\text{kWh}$$

$$\text{Average on electricity} = 0.155 \text{ kWh/km}$$

$$\text{Distance travelled} = 1 \text{ km}$$

$$\text{g } CO_2 \text{ emission by electricity for 1 km travel}$$

$$= 650 \times 0.155 \times 1 = 100.75 \text{ g } CO_2/\text{km}$$

Hence, here we can conclude that the use of EV can help in mitigating $CO_2$ emission from vehicles.

### Example 12.4

Find the carbon intensity of methane-based fossil fuel on its higher heating value of 890 kJ/mol and lower heating value of 802 kJ/mol.

### Solution to Example 12.4

Balanced chemical oxidation reaction of methane:

$$CH_4 + 2O_2 = CO_2 + 2H_2O$$

Burning 1 mol of $CH_4$ liberates 890 kJ of energy while producing 1 mol of $CO_2$. Since 1 mol of $CO_2$ has 12 g of carbon, the HHV carbon intensity of methane is

$$\text{HHV carbon intensity} = \frac{12g}{890 \, Kj} = 0.0135 \text{ gC/kj} = 13.5 \text{ gC/MJ}$$

Similarly, the LHV carbon intensity would be

$$\text{LHV carbon intensity} = \frac{12g}{802 \, kJ} 0.015 \text{ gC/kJ} = 15.0 \text{ gC/MJ}$$

### Example 12.5

Estimate the increase in atmospheric $CO_2$ if 125,500 EJ of coal were to be burnt. Assume a constant airborne fraction of 38%. Also calculate the same for 125,500 EJ of natural gas. (Use LHV carbon intensity of coal = 25.8 gC/MJ, LHV carbon intensity of natural gas = 15.3 gC/MJ.)

**Solution to Example 12.5**

For coal

the carbon content = 125,500 EJ × 25.8 gC/MJ × 1012 MJ/EJ
$$\times 10\text{--}15 \text{ GtC/gC}$$
$$= 3238 \text{ GtC}$$

Converting this to $CO_2$ including the airborne fraction of 0.38. Also, since 1 ppm $CO_2$ = 2.12 GtC

$$\Delta \, CO_2 = \frac{3238 \, GtC \times 0.38}{2.12(GtC/ppm)CO_2} = 580 \, ppm \, CO_2$$

For natural gas

the carbon content = 125,500 EJ × 15.3 gC/MJ × 1012 MJ/EJ
$$\times 10\text{--}15 \text{ GtC/gC} = 1920 \text{ GtC}$$

Converting this to $CO_2$ including the airborne fraction of 0.38. Also, since 1 ppm $CO_2$ = 2.12 GtC

$$\Delta \, CO_2 = \frac{1920 \, GtC \times 0.38}{2.12(GtC/ppm)CO_2} = 344 \, ppm \, CO_2$$

**UNSOLVED PROBLEMS**

1. A freeway has 15,000 vehicles per hour passing a house at a distance 400 m away. Each car emits an average of 0.932 g/km of $NO_x$, and winds are blowing at 2 m/s across the freeway towards house. Estimate the $NO_x$ concentration at the house on a clear summer day near noon (take σz 84).

2. The PHEV and EV with an average 0.155 kWh/km get their electricity from 60% efficient natural gas-fired combined cycle power plant delivering electricity through a 96% efficient grid. Natural gas emits 14.4 gC/MJ and 1 kWh = 3.6 MJ.
   a. Find the g $CO_2$/km of EV
   b. Find the g $CO_2$/km of PHEV

3. The total resource bases of coal, petroleum and natural gas are 125,500, 24,600 and 36,100 EJ, respectively, having LHV carbon intensities of 25.8, 20.0 and 15.3 gC/MJ, respectively. Assuming an airborne fraction of 50%, calculate the total increase in atmospheric $CO_2$ that would be caused by burning all of the natural gas, petroleum and coal.

4. For each of the following fuels, find the carbon intensity based on higher heating values (HHV), given:

a. Ethane, $C_2H_6$, HHV = 1542 kJ/mol

b. Propane, $C_3H_8$, HHV = 2220 kJ/mol

c. $n$-Butane, $C_4H_{10}$, HHV = 2878 kJ/mol

## References

1. D. L. Green and A. Schafer, *Reducing Greenhouse Gas Emissions from U.S. Transportation*, Pew Center on Global Climate Change, www.pewclimate.org.

2. R. Smokers and R. Vermeulen, Review and analysis of the reduction potential and costs of technological and other measures to reduce $CO_2$-emissions from passenger cars, Final report submitted to European Commission, Project number 033.10715/01.01, 2006.

3. A. Schafer, J. B. Heywood, M. A. Weiss, Future fuel cell and internal combustion engine automobile technologies: A 25-year life cycle and fleet impact assessment, *International Journal of Energy*, 31, 2064–2087, 2006.

4. Environmental effects of Freight, Organisation for Economic Co-operation and Development, http://www.oecd.org/dataoecd/14/3/2386636.pdf.

5. R. Kürer, Environment, Global and Local Effects in ECMT, 1993.

6. T. J. H. Schoemaker, and P. A. Bouman (1991), Facts and Figures on Environmental Effects of Freight Transport in the Netherlands, in Kroon et al., eds.

7. Commission des Communautés Européennes (July 1995), "Le transport maritime à courte distance: Perspectives et défis" Communication de la Commission au Parlement Européen, au Conseil, au Comité économique et social et au Comité des régions, Brussels, 5 July 1995, COM(95)317 final.

8. F. Befahy, Environment, Global and Local Effects in ECMT, 1993.

9. OECD Environment Directorate, Environmental Policy Committee (June 1993), The Social Costs of Transport: Evaluation and Links with Internalisation Policies.

10. Mullen, Maureen, E. H. Pechan and Associates, Springfield Virginia, Letter (and accompanying table) to Sharon Nizich, EPA, Research Triangle Park NC, 26 April 1996.

11. US Environmental Protection Agency (September 1985), Compilation of Air Pollutant Emission Factors Volume II: Mobile Sources, AP-42, Fourth Edition. Office of Air and Radiation, Office of Mobile Sources, Test and Evaluation Branch, Ann Arbor, MI.

12. J. Louis, Well-to-wheel energy use and greenhouse gas emissions for various vehicle technologies, Society of Automotive Engineers, SAE paper no. 2001-01-1343, 2001.

13. De Carvalho, Macedo I, Greenhouse gas emissions and energy balances in bio-ethanol production and utilization in Brazil, *Journal of Biomass and Bioenergy*, 14, 77–81, 1998.

14. H. Hass, Well-to-wheels analysis of future automotive fuels and power trains in the European context, a joint study by EUCAR/JRC/CONCAWE, European Commission, 2003.
15. K. A. Subramaniam, L. M. Das, M. K. G. Babu, Control of GHG emissions from transport vehicles: Issues and challenges, SAE International 2008-28-0056.
16. M. P. Wash, Global trends in diesel emissions control—A 1999 update, Society of Automotive Engineers, SAE Paper No. 1999-01-0107, 1999.
17. www.concawe.be.
18. Arita, Technical issues of fuel cell systems for automotive application, *Journal of Fuel Cells*, 2, 10–14, 2002.
19. European Automobile Manufacturer Association (ACEA), http://www.acea.be/.
20. Review and analysis of the reduction potential and costs of technological and other measures to reduce $CO_2$-emissions from passenger cars, Final report, TNO Science and Industry (Assignor European Union, 31 October 2006).

# *Index*

Milton Keynes UK
Ingram Content Group UK Ltd.
UKHW031138141024
449569UK00024B/1240